Transgenic Organisms and Biosafety

Springer
Berlin
Heidelberg
New York
Barcelona
Budapest
Hong Kong
London
Milan
Paris
Santa Clara
Singapore
Tokyo

E.R. Schmidt T. Hankeln (Eds.)

Transgenic Organisms and Biosafety

Horizontal Gene Transfer, Stability of DNA, and Expression of Transgenes

With 69 Figures and 47 Tables

Springer

Professor Dr. rev. nat. Erwin R. Schmidt
Johannes Gutenberg-Universität Mainz
Institut für Molekulargenetik,
Gentechnologische Sicherheitsforschung und Beratung
Johannes-J.-Becher-Weg 32
55099 Mainz
Germany

Dr. rev. nat. Thomas Hankeln
Johannes Gutenberg-Universität Mainz
Institut für Molekulargenetik,
Gentechnologische Sicherheitsforschung und Beratung
Johannes-J.-Becher-Weg 32
55099 Mainz
Germany

ISBN 3-540-61077-4 Springer-Verlag Berlin Heidelberg New York

Library of Congress Cataloging-in-Publication Data. Transgenic organisms and biosafety, horizontal gene transfer, stability of DNA, and expression of transgenes/E.R. Schmidt, T. Hankeln (eds.). p. cm. Includes bibliographical references and index. ISBN 3-540-61077-4 (softcover) 1. Transgenic organisms. 2. Transgenic organisms – Safety measures. I. Schmidt, E.R. (Erwin R.) II. Hankeln, Thomas, 1959- .QH442.6.T727 1996 575.1'0724 – dc20 96-14200

Cover design: Design & Production, Heidelberg

Typesetting: Scientific Publishing Services (P) Ltd, Madras

SPIN: 10482733 27/3136/SPS – 5 4 3 2 1 0 – Printed on acid-free paper

Preface

Gene/biotechnology and the broad use of genetically modified organisms is rapidly developing into a commercial enterprise. In spite of the pace of scientific progress in this field, a partly emotional debate is still being carried on by the general public and on the part of the consumer about the potential risks linked to this new technology. Especially when genetically modified organisms (GMOs) are tested in the open field, the question of "safety" is controversially discussed among different groups in society. Environmentalists are worried about unforeseeable dangers inherent in the release of a living organism, assuming that special risks are linked to organisms which have been modified by molecular biologists using tools developed in the laboratory. Frequently quoted potential hazards are the phenomena of "gene escape" and "genetic pollution", meaning the very interesting question of natural (horizontal) gene transfer. This phenomenon of *horizontal gene transfer* is in fact well known to scientists studying prokaryotes. Many prokaryotes are able to exchange DNA readily between individuals of different species or are even able to take up DNA from the extracellular environment by the process of genetic transformation, which has already been known for more than 65 years. For eukaryotes, however, the evidence of natural horizontal gene transfer is very limited. There are some well-known examples, such as the transfer of T-DNA from *Agrobacterium tumefaciens* to plant cells or the transfer and integration of viral DNA into eukaryotic genomes which can carry some non-viral eukaryotic DNA and thus mediate a horizontal gene transfer. Other possible cases of horizontal gene transfer are much less well documented.

Further risks frequently discussed in the context of release and commercialisation of GMOs are the stability of transgenes in the host genome and the continuous, stable expression of these transgenes. The mere loss of a transgene or its instability is not usually considered to be a biosafety issue. However, there can be some scenarios in which the stability and the guaranteed expression of the transgene is of extraordinary importance. One example might be transgenic rice, which has a reduced amount of allergenic rice proteins. In this case, the synthesis of the allergenic proteins is suppressed by the expression of antisense copies of the allergenic protein genes. If the expression of the antisense message is unstable, then the absence of the allergen is not guaranteed and thus poses a risk to the consumer of the product. There are many more conceivable scenarios in which the instability of the transgenic DNA or of its expression might be relevant to the biosafety of the GMO or the product thereof. Furthermore, the stability of transgenes is important for the commercial value of a transgenic organism, because the expression of the transgene might be the basis that makes it attractive for the producer and/or consumer.

Horizontal gene transfer, stability of transgenic DNA within and outside of cells/organisms and the expression of transgenes were the major topics in a workshop held at Schloss Waldthausen in Mainz (Germany) in December 1994. The contributions made at this workshop are summarized in this book. The articles either describe experimental approaches or are a mixture of a review on the state of knowledge and experimental data on the relevant questions. The book is divided into four sections:

1. Horizontal Gene Transfer
2. Stability of DNA
3. Expression of Transgenes
4. Field Releases of Transgenic Plants

We hope that the articles show the state of the art in these controversially discussed topics. Especially in the context of increasing numbers of field trials and the increasing commercialisation of GMOs around the world, the articles in this book should make a scientific contribution to the debate which can be the basis for a more rational discussion. For this reason, we also decided to invite representatives of several biotechnology companies to contribute in order to obtain an impression of their scientific point of view. Some of the articles are clearly written only for the scientific community, but some are also meant to stimulate an open and frank public discussion about the potential or actual risk of gene technology applied to organisms not restricted to safe containment in the laboratory. We would like to thank all the contributors to and participants of the Mainz Biosafety Workshop; it was a pleasure for us, the organisers, to have so many competent scientists gathered together, and we are pleased that we have finally succeeded in collecting all the manuscripts now included in this book.

Acknowledgements. The Workshop on Biosafety of Transgenic Organisms, held in Mainz, December 1994, would not have been possible without the generous financial support from the European Commission DG XII, which we gratefully acknowledge. The publishing of the workshop contributions in this book was generously sponsored by the Stiftung für Innovation Rheinland-Pfalz (Ministry of Education and Science Rheinland-Pfalz). Last but not least, we would like to thank our secretaries Mrs. Beth Ihm and Mrs. Heide Wittke, who did a marvellous job in correcting the English in some of the manuscripts and preparing the final versions for publication.

Mainz, October 1995

E.R. Schmidt
T. Hankeln

Contents

List of Contributors

Ahrenholtz, I.
Genetik, Fachbereich Biologie, Carl von Ossietzky-Universität, Postfach 2503, 26111 Oldenburg, Germany

Ahuja, M.R.
Federal Research Centre for Forestry and Forest Products, Institute of Forest Genetics, Sieker Landstr. 2, 22927 Grosshansdorf, Germany

Anxolabéhère, D.
Laboratoire de Dynamique du Génome et Evolution, Institut J. Monod, Universités Paris VI et Paris VII, 75251 Paris Cedex 05, France

Backhaus, H.
Institut für Biochemie und Pflanzenvirologie, Biologische Bundesanstalt für Land- und Forstwirtschaft, Messeweg 11/12, 38104 Braunschweig, Germany

Bartsch, D.
Department of Biology V (Ecology, Ecotoxicology and Ecochemistry), Technical University of Aachen, Worringerweg 1, 52056 Aachen, Germany

Bazin, C.
Laboratoire de Biologie et Génétique Evolutives, CNRS, 91198 Gif/Yvette Cedex, France

Blum, S.A.E.
Genetik, Fachbereich Biologie, Carl von Ossietzky-Universität, Postfach 2503, 26111 Oldenburg, Germany

Bode, J.
Institut für Biotechnologische Forschung mbH, Genetik von Eukaryonten, Mascheroder Weg 1, 38124 Braunschweig-Stöckheim, Germany

Brinkmann, S.
Free University of Berlin, Department of Biology, Institute of Applied Genetics, Albrecht-Thaer-Weg 6, 14195 Berlin, Germany

Broer, I.
Universität Bielefeld, Fakultät für Biologie, Lehrstuhl für Genetik, Universitätsstraße, Postfach 100131, 33501 Bielefeld, Germany

Burgermeister, W.
Biologische Bundesanstalt für Land- und Forstwirtschaft, Institut für Biochemie und Pflanzenvirologie, Messeweg 11/12, 38104 Braunschweig, Germany

Capy, P.
Laboratoire de Populations, Génétique et Evolution, CNRS, BP1, 91198 Gif/Yvette Cedex, France

Cherdshewasart, W.
Biology Department, Chulalongkorn University, Bangkok 10330, Thailand

Commandeur, U.
Biologische Bundesanstalt für Land- und Forstwirtschaft, Institut für Biochemie und Pflanzenvirologie, Messeweg 11/12, 38104 Braunschweig, Germany

Dale, P.J.
John Innes Center, Colney Lane, Norwich NR4 7UJ, United Kingdom

Dietz-Pfeilstetter, A.
Biologische Bundesanstalt für Land- und Forstwirtschaft, Institut für Biochemie und Pflanzenvirologie, Messeweg 11/12, 38104 Braunschweig, Germany

Doerfler, W.
Institut für Genetik, Universität zu Köln, Weyertal 121, 50931 Köln, Germany

Dröge-Laser, W.
The Salk Institute for Biological Studies, Plant Biology Laboratory, 10010 North Torrey Pines Road, La Jolla, CA 92037, USA

Emonds, A.
Department of Biology V (Ecology, Ecotoxicology and Ecochemistry), Technical University of Aachen, Worringerweg 1, 52056 Aachen, Germany

Engelen, B.
Institut für Biochemie und Pflanzenvirologie, Biologische Bundesanstalt für Land- und Forstwirtschaft, Messeweg 11/12, 38104 Braunschweig, Germany

Fechteler, K
Institut für Genetik, Universität zu Köln, Weyertal 121, 50931 Köln, Germany

Fecker, L.F.
Biologische Bundesanstalt für Land- und Forstwirtschaft, Institut für Biochemie und Pflanzenvirologie, Messeweg 11/12, 38104 Braunschweig, Germany

Felske, A.
Institut für Biochemie und Pflanzenvirologie, Biologische Bundesanstalt für Land- und Forstwirtschaft, Messeweg 11/12, 38104 Braunschweig, Germany

Fladung, M.
Federal Research Centre for Forestry and Forest Products, Institute of Forest Genetics, Sieker Landstr. 2, 22927 Grosshansdorf, Germany

Fritsch, E.
Institut für biologischen Pflanzenschutz, Heinrichstr. 243, 64287 Darmstadt, Germany

Gerke, M.
Universität Bielefeld, Fakultät für Biologie, Lehrstuhl für Genetik, Postfach 100131, 33501 Bielefeld, Germany

Gharti-Chhetri, G.B.
Agronomy Department, IAAS, Rampur, Chitwan, Nepal

Gluth, H.
Department of Biology V (Ecology, Ecotoxicology and Ecochemistry), Technical University of Aachen, Worringerweg 1, 52056 Aachen, Germany

Golz, C.
Federal Environmental Agency, P.O.B. 330022, 14191 Berlin, Germany

Grannemann, R.
Gesellschaft für Biotechnologische Forschung mbH, Mascheroder Weg 1, 38124 Braunschweig, Germany

Graser, E.
Biologische Bundesanstalt für Land- und Forstwirtschaft, Institut für Biochemie und Pflanzenvirologie, Messeweg 11/12, 38104 Braunschweig, Germany

Graupner, S.
Genetik, Fachbereich Biologie, Carl von Ossietzky-Universität, Postfach 2503, 26111 Oldenburg, Germany

Haag, C.
Department of Biology V (Ecology, Ecotoxicology and Ecochemistry), Technical University of Aachen, Worringerweg 1, 52056 Aachen, Germany

Hankeln, T.
Institut für Molekulargenetik, gentechnologische Sicherheitsforschung und Beratung, Johannes Gutenberg-Universität Mainz, 55099 Mainz, Germany

Hauser, H.
Gesellschaft für Biotechnologische Forschung mbH, Mascheroder Weg 1, 38124 Braunschweig, Germany

Heller, H.
Institut für Genetik, Universität zu Köln, Weyertal 121, 50931 Köln, Germany

Himmler, G.
Biologische Bundesanstalt für Land- und Forstwirtschaft, Institut für Biochemie und Pflanzenvirologie, Messeweg 11/12, 38104 Braunschweig, Germany

Hoffmann, T.
Environmental State Office of Brandenburg, Postfach 601061, 14410 Potsdam, Germany

Huber, J.
Institut für biologischen Pflanzenschutz, Heinrichstr. 243, 64287 Darmstadt, Germany

Jehle, J.A.
Department of Virology, Wageningen Agricultural University, Binnenhaven 11, P.O.B. 8045, 6709 PD EM Wageningen, The Netherlands

Jekel, M.
Genetik, Fachbereich Biologie, Carl von Ossietzky-Universität, Postfach 2503, 26111 Oldenburg, Germany

Kämmer, C.
Institut für Genetik, Universität zu Köln, Weyertal 121, 50931 Köln, Germany

Kaufmann, A.
Biologische Bundesanstalt für Land- und Forstwirtschaft, Institut für Biochemie und Pflanzenvirologie Messeweg 11/12, 38104 Braunschweig, Germany

Kay, V.
Institut für Biotechnologische Forschung mbH, Genetik von Eukaryonten, Mascheroder Weg 1, 38124 Braunschweig-Stöckheim, Germany

Klehr-Wirth, D.
Gesellschaft für Biotechnologische Forschung mbH, Mascheroder Weg 1,
38124 Braunschweig, Germany

Knoblauch, M.
Institut für Genetik, Universität zu Köln, Weyertal 121, 50931 Köln, Germany

Koenig, R.
Biologische Bundesanstalt für Land- und Forstwirtschaft, Institut für Biochemie und Pflanzenvirologie, Messeweg 11/12, 38104 Braunschweig, Germany

Köhne, S.
Universität Bielefeld, Fakultät für Biologie, Lehrstuhl für Genetik,
Postfach 100131, 33501 Bielefeld, Germany

Koetsier, P.
Institut für Genetik, Universität zu Köln, Weyertal 121, 50931 Köln, Germany

Landsmann, J.
Institut für Biochemie und Pflanzenvirologie, Biologische Bundesanstalt für Land- und Forstwirtschaft, Messeweg 11/12, 38104 Braunschweig, Germany

Langin, T.
Laboratoire de Cryptogamie, Bât. 400, Université Paris XI, 91405 Orsay Cedex, France

Lindemann, J.
Lindemann Consulting, El Cerrito, CA 94530, USA

Lorenz, M.G.
Genetik, Fachbereich Biologie, Carl von Ossietzky-Universität, Postfach 2503,
26111 Oldenburg, Germany

Meixner, M.
Free University of Berlin, Department of Biology, Institute of Applied Genetics,
Albrecht-Thaer-Weg 6, 14195 Berlin, Germany

Meyer, B.
Genetik, Fachbereich Biologie, Carl von Ossietzky-Universität, Postfach 2503,
26111 Oldenburg, Germany

Mitten, D.
Mitten Consulting, 720 Borchard Court, Woodland, CA 95695, USA

Morak, C.
Department of Biology V (Ecology, Ecotoxicology and Ecochemistry), Technical University of Aachen, Worringerweg 1, 52056 Aachen, Germany

Münk, C.
Gesellschaft für Biotechnologische Forschung mbH, Mascheroder Weg 1, 38124 Braunschweig, Germany

Negrutiu, I.
Ecole Normale Supérieure de Lyon, Reconnaissance Cellulaire et Amélioration des Plantes, 46, Allée d'Italie, 69364 Lyon, France

Neuhausen, F.
Institut für Mikrobiologie und Weinforschung, Johannes Gutenberg-Universität Mainz, 55099 Mainz, Germany

Neumann, K.
Universität Rostock, Fachbereich Biologie, Doberanerstraße 143, 18057 Rostock, Germany

Nickel, A.
Institut für Biochemie und Pflanzenvirologie, Biologische Bundesanstalt für Land- und Forstwirtschaft, Messeweg 11/12, 38104 Braunschweig, Germany

Nübel, U.
Institut für Biochemie und Pflanzenvirologie, Biologische Bundesanstalt für Land- und Forstwirtschaft, Messeweg 11/12, 38104 Braunschweig, Germany

O'Connor, J.V.
Genentech Inc., Department of Process Sciences, South San Francisco, CA 94080, USA

Orend, G.
La Jolla Cancer Research Foundation, Cancer Research Center, 10901 North Torrey Pines Road, La Jolla, CA 92037, USA

Petropoulos, C.J.
Genentech Inc., Department of Process Sciences, South San Francisco, CA 94080, USA

Pickardt, T.
Free University of Berlin, Department of Biology, Institute of Applied Genetics, Albrecht-Thaer-Weg 6, 14195 Berlin, Germany

Pohl-Orf, M.
Department of Biology V (Ecology, Ecotoxicology and Ecochemistry), Technical University of Aachen, Worringerweg 1, 52056 Aachen, Germany

Redenbaugh, K.
Calgene, Inc. 1920 Fifth St., Davis, CA 95616, USA

Riedel-Preuß, A.
Biologische Bundesanstalt für Land- und Forstwirtschaft, Institut für Biochemie und Pflanzenvirologie, Messeweg 11/12, 38104 Braunschweig, Germany

Saat, T.A.W.M.
Vander Have Research, P.O. Box, 4410 AA Rilland, The Netherlands

Scheffler, J.A.
John Innes Center, Colney Lane, Norwich NR4 7UJ, United Kingdom

Schieder, O.
Free University of Berlin, Department of Biology, Institute of Applied Genetics, Albrecht-Thaer-Weg 6, 14195 Berlin, Germany

Schmidt, E.R.
Institut für Molekulargenetik, gentechnologische Sicherheitsforschung und Beratung, Johannes Gutenberg-Universität Mainz, 55101 Mainz, Germany

Schmidt, M.
Department of Biology V (Ecology, Ecotoxicology and Ecochemistry), Technical University of Aachen, Worringerweg 1, 52056 Aachen, Germany

Schmitt, M.J.
Institut für Mikrobiologie und Weinforschung, Johannes Gutenberg-Universität Mainz, 55099 Mainz, Germany

Schröer, J.
Institut für Genetik, Universität zu Köln, Weyertal 121, 50931 Köln, Germany

Schubbert, R.
Institut für Genetik, Universität zu Köln, Weyertal 121, 50931 Köln, Germany

van der Hoeven, C.
Biologische Bundesanstalt für Land- und Forstwirtschaft, Institut für Biochemie und Pflanzenvirologie, Messeweg 11/12, 38104 Braunschweig, Germany

Vlak, J.M.
Department of Virology, Wageningen Agricultural University, Binnenhaven 11, 6709 PD Wageningen, The Netherlands

Wackernagel, W.
Genetik, Fachbereich Biologie, Carl von Ossietzky-Universität, Postfach 2503, 26111 Oldenburg, Germany

Wendt-Potthoff, K.
GKSS Institut für Gewässerforschung, Am Biederitzer Busch 12, 39114 Magdeburg, Germany

Winterpacht, A.
Kinderklinik, Johannes Gutenberg-Universität, 55101 Mainz, Germany

Wirth, M.
Gesellschaft für Biotechnologische Forschung mbH, Mascheroder Weg 1, 38124 Braunschweig, Germany

Wolf, K.
Institut für Biologie IV (Mikrobiologie), Rheinisch-Westfälische Technische Hochschule, Worringer Weg, 52056 Aachen, Germany

Wurm, F.M.
Ecole Polytechnique Fédérale de Lausanne (EPFL), Département de Chimie, Biotech. Chem. IV, 1015 Lausanne, Switzerland

I Horizontal Gene Transfer

1 Gene Transfer Between Organelles and the Nucleus in Lower Eukaryotes

K. Wolf

Institut für Biologie IV (Mikrobiologie), Rheinisch-Westfälische Technische Hochschule, Worringer Weg, 52056 Aachen, Germany

Introduction

The vertical transmission of DNA or RNA sequences from parent to offspring provides the fundament for the genetic component of the evolution of life. There is, however, increasing evidence for lateral or horizontal transfer of sequences across taxonomic boundaries (Kidwell 1992; Smith et al. 1992; Hilario and Gogarten 1993). One aspect of horizontal transfer is the movement of genetic information from the nucleus to the organelles and, vice versa, from the organelles to the nucleus. In this short overview I shall concentrate on these movements during evolution and on experimental approaches to study their mechanisms.

Mitochondria and Chloroplasts as Semiautonomous Organelles

The genomes of mitochondria are variable within a considerable size range (for a review, see Wolf and Del Giudice 1988). There is a basic set of genes comprising the two ribosomal RNAs, a minimal number of tRNAs, three subunits of cytochrome c oxidase, apocytochrome b, and two to three subunits of ATPase. Major polypeptides encoded by plastid DNA include the large subunit of ribulose 1,5-bisphosphate carboxylase (the soluble CO_2-fixing enzyme of photosynthesis), at least three ATP synthase subunits, and several thylakoid membrane proteins. Although there is also evidence for gene transfer between nucleus and plastids, I shall focus the interest of this contribution on mitochondria. The prokaryotic character of organelle genetic systems suggests that organelles evolved from bacteria which were endocytosed more than a billion years ago. This event led to the development of mitochondria about 1.5×10^9 years ago, when oxygen entered the atmosphere in substantial amounts. Since most of the organelle genes are in

E.R. Schmidt/T. Hankeln (Eds.)
Transgenic Organisms and Biosafety

the nucleus, a massive transfer of genes from the endosymbiont to the host has taken place (Gray 1983).

Yeast is a favourable object to study the interaction between nucleus and mitochondria. In most organisms, mutations leading to mitochondrial dysfunction or to inability to assemble a functional organelle, are lethal. However, the fact that yeast cells with defective mitochondria can use fermentation as source of energy, opens the possibility of inducing mitochondrial and nuclear mutants with defects in mitochondrial functions.

Mitochondrial Genomes of Yeasts

Bakers yeast *Saccharomyces cerevisiae* (*S. cerevisiae*) is the best characterized eukaryotic system. In this year, the entire nuclear genome will be sequenced and there is a huge number of well characterized nuclear mutants. The molecular techniques of transformation, gene replacement and gene disruption allow the functional analysis of genes. The fission yeast *Schizosaccharomyces pombe* (*S. pombe*) provides a little less developed genetic background, but is of great interest, since it is evolutionarily far from *S. cerevisiae*. Also in fission yeast, all the powerful molecular techniques are feasible. In both yeasts the mitochondrial genomes have been sequenced entirely. A gene map of both yeasts is presented in Fig. 1a,b. Compilation of all published sequences of *S. cerevisiae* mitochondrial genomes revealed a genome size of 74 000–85 000 BP depending on the strain. The maximal range of coding sequences is 35%, i.e., 65% of the genome are non-coding. The mitochondrial genome of fission yeast contains about 19 kilobase pairs, but strains with shorter or longer genomes can be found, depending on their intron content. The basic set of genes is given in the legend of Fig. 1a,b. There are two extra genes in the *S. cerevisiae* mitochondrial genome: the gene *tsl* (*t*RNA *s*ynthesis *l*ocus) encodes the RNA component of a ribonucleoprotein which is involved in maturation of pre-tRNAs, the gene *var1* encodes a protein of the small ribosomal subunit. These two genes are not found in the mitochondrial genome of fission yeast, which indicates that these functions must be chromosomally encoded (Wolf 1987). In *S. pombe* mitochondrial DNA, an yet unassigned reading frame (*urf a*) is found, which very likely encodes a protein involved in maintenance of mitochondrial DNA (Zimmer et al. 1991). This reading frame has no counterpart in any mitochondrial genome sequenced so far. In both yeasts, only few components of the mitochondrium are mitochondrially encoded and synthesized. Most of the mitochondrial components are encoded by nuclear genes translated on cytoplasmic ribosomes, and imported into the mitochondrium.

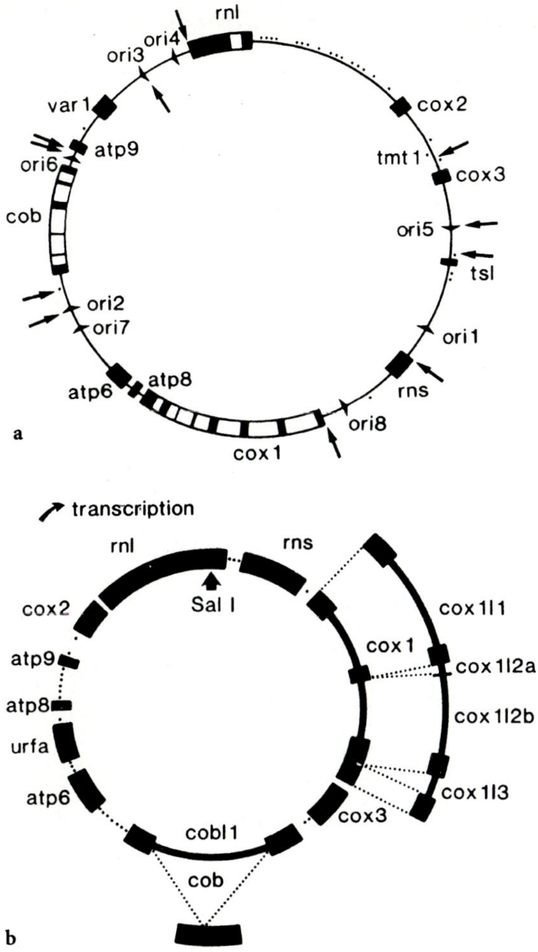

Fig. 1a,b. Gene maps of the mitochondrial genomes of **a** *Saccharomyces cerevisiae* and **b** *Schizosaccharomyces pombe*. **a** Exons are indicated by *black bars*, introns by *open bars*, and intergenic sequences by a *thin line*. The tRNA genes are represented as *dots* (*tmt* is the only tRNA encoded on the opposite DNA strand). *Arrows* mark transcription initiation sites; *one arrow pointing to the outside* indicates a site on the complementary strand. *rnl, rns,* genes encoding the large and the small ribosomal RNA; *atp6, atp8, atp9,* genes encoding the subunits 6, 8, and 9 of the ATPase complex; *cob,* gene encoding apocytochrome b; *cox1, cox2, cox3,* genes encoding the subunits 1, 2, and 3 of cytochrome c oxidase; *var1,* gene encoding a protein of the small ribosomal subunit; *ori,* origins of DNA replication. tRNA genes are abbreviated according to the one-letter code for amino acids (*m,* mitochondrial; *t,* threonine). **b** The *inner circle* represents the genome of strain *50,* the *outer circle* the genome of strain *EF1*. Genes (exons) are drawn as *thick bars,* introns as *thin bars. Dots* indicate tRNA genes. The same symbols are used as in **a**. *Sal*I, conserved restriction site in the 3′ part of *rnl*. The *arrow* indicates the direction of transcription and also the location of the main promotor between *cox2* and *rnl*. *urf a,* unassigned reading frame a

Evolutionary Transfer of Genes from Mitochondria to the Nucleus

It is logical to assume that the endosymbiontic ancestor of mitochondria possessed all the genes necessary for survival, and that there was a continuous horizontal transfer from the endosymbiont to the nucleus of the host. The question arises whether we can identify footprints of the gene transfer during evolution. The gene coding for subunit 9 or dicyclohexylcarbodiimide (DCCD) binding protein of ATPase complex (called *atp9*, for nomenclature see legend to Fig. 1) is present and also translated in *S. cerevisiae* (Hensgens et al. 1979) and *S. pombe* (Lang et al. 1983), but the homologous gene is located in the nucleus of mammals. In *Neurospora crassa*, the *atp9* gene sequences are found in both nuclear and mitochondrial DNA, but only the nuclear sequence is transcribed and translated (Van den Boogaart et al. 1982). The mitochondrial sequence has been denoted the *mal* (*m*itochondrial *A*TPase-proteolipid-*l*ike) gene. A similar situation was found in *Aspergillus nidulans* (Brown et al. 1984); the authors suggested that both versions of this gene might be expressed at different stages of the life cycle. It suffices to say that these cases illustrate the movement of genes from mitochondria to the nucleus, leaving behind mitochondrial pseudogenes.

In yeasts there is no mitochondrial gene encoding subunits of the NADH dehydrogenase, whereas in mammalian mitochondrial genomes 7 polypeptides of the NADH-dehydrogenase complex are mitochondrially encoded (Grivell 1983).

Farrelly and Butow (1983) have also found a contiguous sequence in the nuclear genome with extensive homology to non-contiguous mitochondrial sequences including *var1*, which encodes a protein of the small subunit of mitochondrial ribosomes.

Transport of Nucleic Acids into Mitochondria

From the analysis of mitochondrial genomes it is evident that the vast majority of mitochondrial proteins is involved in energy metabolism and other organellar processes, including nearly all functions concerned with replication and expression of mitochondrial DNA. A tremendous amount of work has been devoted to the mechanism of transport of proteins across the mitochondrial membrane(s) (Schatz 1993; Schwarz and Neupert 1994). For a long time, there have been no serious suggestions that nuclearly encoded mRNA molecules are routinely delivered from the cytosol to the mitochondrium. For a few years, hovever, a body of strong evidence has been accumulating indicating that certain small RNA molecules encoded in the nucleus directly participate in the maintenance and expression of mitochondrial genetic information (reviewed by Nagley 1989).

Definitive evidence for RNA import into mitochondria was provided by Clayton and associates (Xu and Clayton 1992; Schmitt and Clayton 1993; Shadel and Clayton 1993). The mammalian mitochondrial RNA processing endonuclease (MRP) is responsible for the cleavage of some RNA transcripts transcribed from the light strand promotor of mouse mitochondrial DNA. These cleaved transcripts function as primers for DNA-replication on the heavy strand. The precursor RNA molecule of MRP is also involved in nuclear RNA processing. This is a case of an RNA, which is encoded in the nucleus and acts both in the nucleus and in the mitochondria. In this context it is worth mentioning that import of tRNA molecules has been shown in plants (see Nagley 1989) and protozoa (Schneider et al. 1994).

Transport of DNA into mitochondria, i.e., transformation of mitochondria is now possible by high-velocity bombardment of yeast with tungsten or gold microprojectiles (Fox et al. 1991). Strains devoid of mitochondrial DNA *(rho^o)* can now be converted so stable "synthetic" *rho*$^-$ strains by transformation with bacterial plasmids carrying mitochondrial genes. However, the efficiency of mitochondrial transformation varies widely from experiment to experiment and is usually very low (from 1 mitochondrial transformant/500 nuclear transformants to <1/10 000).

Nucleo-mitochondrial Cross-talk

By virtue of the yeast cell surviving without functional mitochondria, it was possible to isolate and characterize a large collection of nuclear mutants impaired in mitochondrial functions. In order to give an idea concerning the number of genes involved, the present status is summarized in Table 1.

There is a considerable body of data on the characterization of nuclear genes involved in mitochondrial functions, which are summarized in the excellent reviews of Costanzo and Fox (1990) and Bolotin-Fukuhara and Grivell (1992).

Table 1. Genes for mitochondrial proteins identified by analysis of nuclear respiratory-deficient (pet) mutants (taken from Bolotin-Fukuhara and Grivell 1992)

Protein/enzyme/complex	Complementation groups
Cytochrome c oxidase	40
QH_2-cytochrome c reductase	23
ATP synthase	9
Translational machinery	17
Other	21
Unidentified	105
Total	215

The other part of the cross-talk, a possible signalling from the mitochondria to the nucleus, is much less elaborated und at present subject of speculation only. Using isochromosomal, respiratory-deficient yeast strains with different lesions in the mitochondrial genome (point mutations, deletions or complete lack of mitochondrial DNA), the laboratory of Butow (Parikh et al. 1987) could demonstrate that the mitochondrial genotype can influence nuclear gene expression. Recently, two genes were identified (Liao and Butow 1993) which are required for this path of communication from mitochondria to the nucleus. In our laboratory we could demonstrate that the resistance of a yeast cell towards the alkaloid lycorine was different between cells with and without mitochondrial DNA, irrespective of the intactness of the mitochondrial genome (Del Giudice et al. 1984; Massardo et al. 1990, 1994).

Horizontal Transfer of Mitochondrial Introns

Mitochondrial genomes of yeast contain two types of introns, which belong either to group I or II. These two types of introns can be distinguished by their secondary structure and their mode of splicing. In this context it is only important to know that several group I and II introns are mobile genetic elements (Wolf 1993; Grivell 1994). In a process called intron homing, they are transferred from intron-containing to intronless alleles of the same gene. They can also be transferred to non-allelic regions in a process called intron transposition. The movement of group I introns is mediated by an intron-encoded endonuclease (for a review see Lambowitz and Belfort 1993). Genetic experiments have shown that intron homing is connected with conversion events on both integration sites. Also for group II introns mobility has been demonstrated (Mueller et al. 1993; Paschke et al. 1994). The intron homing process is connected with conversion of sequences located in the upstream exon (Lazowska et al. 1994). Intron transposition is a process without conversion of flanking sequences (Schmidt et al. 1994). Transposition of group II introns is achieved via an RNA intermediate, followed by integration of the reverse transcribed cDNA in the genome. The reverse transcriptase is encoded by the group II introns. Mörl and Schmelzer (1990) could demonstrate that a group II intron can insert into a foreign nuclear DNA by reversal of the self-splicing reaction in vitro. An interesting case of mitochondrial intron transfer to the nucleus has been observed in *S. pombe* by S. Pietrikovski (pers. communication). A part (probably the whole) intron 2 of the gene encoding subunit I of cytochrome c oxidase was found to be inserted upstream of the gene encoding a *p*-nitrophenylphosphatase.

Horizontal transmission of group I and group II introns between organisms has never been proven, but there are some indications for this process. Closely related introns are present at different locations in different organisms and in other instances introns inserted at the same site are more strongly conserved

Table 2. Homologies of amino acids in the intronic urfs of the highly conserved *cox1* introns of *Schizosaccharomyces pombe* and *Aspergillus nidulans*

	Amino acid positions	Number of amino acids minus insert (216–229)	Homology (%)
'LAGLI' region (first decapeptide)	80–119	40	88
'DADG' region (second decapeptide)	203–318	102	88
Complete reading frame, except region containing secondary structure	80–346	253	70

The homology of the first amino acids was not calculated, since there is a constraint of this region concerning RNA secondary structure. The insert between amino acid 216 and 229 is not considered in the calculation. The LAGLI-DADG region is a conserved region in all intronic reading frames encoding maturases (adapted from Lang 1984)

than exon sequences. The *cob* intron of *Chlamydomonas smithii* mitochondrial DNA has 60% sequence identity to the *Neurospora crassa cob* intron 1, which is located at a different position in the cytochrome b gene (Colleaux et al. 1990). Other suggested cases involve closely related group I introns located at the same position in the otherwise less conserved *cox1* genes of *S. pombe* and *Aspergillus nidulans* (Lang 1984) and nearly identical group II introns at the same position in the *cox1* genes of *Kluyveromyces lactis* and *S. cerevisiae* (Hardy and Clark-Walker 1991). The postulated transfer of the *cox1* group I intron from *Aspergillus nidulans* to *S. pombe* was supported by the fact that the intronic open reading frame contains TGA tryptophan codons, which are used in *Aspergillus nidulans* but not in *S. pombe* mitochondrial genes. Table 2 illustrates the highly homologous introns in *S. pombe* and *Aspergillus nidulans* and Table 3 shows the percentage of amino acid homology of the *cox1* exons in various organisms. Horizontal gene transfer possibly explains the presence of related intronic

Table 3. Percentage of amino acid homology of *cox1* exons

	A. nidulans	*N. crassa*	*S. cerevisia*	*HeLa*
S. pombe	58	57	60	53
A. nidulans	–	88	67	58
N. crassa	–	–	64	57
S. cerevisiae	–	–	–	54

Note the evoutionary distance between *S. pombe* and *S. cerevisiae* (adapted from Lang 1984).
S. pombe, Schizosaccharomyces pombe; *A. nidulans, Aspergillus nidulans*; *N. crassa, Neurospora crassa*; *S. cerevisiae, Saccharomyces cervisiae*; *HeLa,* human cell line

reading frames in the group I intron in the *td* gene of bacteriophage *T4* and the NADH dehydrogenase genes of the filamentous fungi *Podospora anserina* and *Neurospora crassa* (Michel and Dujon 1986). In these cases, the horizontal gene transfer is restricted to the open reading frame and does not involve the entire intron. Concerning the natural habitats of yeasts, fungi and bacteria, the transfer of DNA from one organism to the other is possible.

Escape of Genes from the Mitochondria to the Nucleus – An Experimental Approach

This paragraph is devoted to experiments to identify the mechanisms and the genes involved in the horizontal transfer of genes from mitochondrial to the nucleus. Again, yeast is a valuable organism for this kind of study, since it is easy to eliminate DNA from mitochondria and repopulate mitochondria with genetically engineered genomes. Thorsness and Fox (1990) induced a nuclear *URA3* gene (a gene involved in uracil biogenesis) into the mitochondria of an *ura3*– mutant of *S. cerevisiae.* The *URA3* gene was not expressed in the mitochondria. The gene, however, escaped with a surprisingly high frequency (about 2×10^{-5} per cell per generation) from the mitochondria to the nucleus, where it conferred uracil independence on the cell. The opposite transfer – nucleus to mitochondria – is apparently at least 100 000 times less. One possibility of DNA escape from mitochondria is breakage of the organelle and release of DNA into the cytoplasm. Higher or lower temperatures increase the escape rate, which is in line with the assumption that the integrity of membranes is influenced by the temperature. Recently, Thorsness and Fox (1993) identified a series of mutants with an increased frequency of escape. One of these genes has a considerable similarity with an *Escherichia coli* gene involved in septum formation, again indicating the role of membrane integrity in this process.

Relocation of Mitochondrial Genes to the Nucleus

Since mitochondrial transformation is still cumbersome, some researchers have developed the technique of allotopic expression of organellar genes (reviewed by Nagley and Devenish 1989). This technique is somehow the mimicking of the evolutionary gene transfer process. I shall consider only two examples to illustrate the topic (see Nagley and Devenish 1989, and references therein). The mitochondrial ATP synthase is an enzyme complex which is encoded by nuclear and mitochondrial genes. Subunit 8 of this complex is encoded by the mitochondrial gene *atp8*. In a first step, the *atp8* gene was completely redesigned to adapt it to the standard genetic code. Then the gene was fused to the DNA sequence encoding the transit peptide of the nuclearly encoded subunit 9 of mitochondrial ATPase from *Neurospora crassa*. This chimaeric gene was successfully expressed in the cytoplasm and the gene product was imported into mitochondria. The fact that a mutation in the mitochondrial *atp8* gene was complemented, demonstrated that the protein was functioning inside the mitochondria. Furthermore, the presence of the *atp8* gene product was shown immunochemically. A second protein for which allotopic expression was achieved, is an intron encoded protein, the so-called maturase encoded by the fourth intron of the cytochrome b gene. This maturase protein assists in splicing its own intron from an mRNA precursor. So far, attempts to produce nuclear versions for other mitochondrial proteins have failed and this failure is the subject of a lot of speculation. The ultimate goal to "make functional mitochondria without mitochondrial DNA", as proposed by Piet Borst in 1984 has not been achieved in the last decade.

Consequences of DNA Escape from Human Mitochondria – Cause of Aging and Cancer?

Shay and Werbin (1992) observed that over 100 mitochondrial DNA-like copies were integrated in the nuclear genome of HeLa cells. This could mean that about 10% of them may be in structural or regulatory genes, most of them, however, in non-coding sequences. At present, we have detailed knowledge on several mitochondrial human diseases (Lestienne 1992). Insertion of foreign DNA in regions upstream of genes may disturb their regulation and lead to cancer. Any perturbation of mitochondrial syntheses could lead to a destabilization of mitochondrial membranes. It is evident from studies on yeast (Thorsness and Fox 1993) that the escape rate of mitochondrial DNA from *rho*$^-$ mitochondria is much higher than from *rho*$^+$ mitochondria. As a consequence, mitochondrial membranes from patients with mitochondrial diseases, will be more prone to damage. Older persons accumulate deletions in the mitochondrial DNA, which also could lead to a destabilization of mitochondrial membranes. It is clear that this hypothesis of a mitochondrial contribution to cancer and aging needs a rigorous experimental proof. If this hypothesis is true, it is evident that this might be one of the many causes of aging and cancer.

Conclusions

In this contribution I wanted to demonstrate that there is a continuous flow of genetic information from the formerly endosymbiontic mitochondria to the nucleus. It is evident that there must be a cross-talk between the two systems in order to regulate the concerted biosynthesis of mitochondria. Furthermore, experimental procedures are outlined for the import of nucleic acids into mitochondria and the allotopic expression of mitochondrial genes in the nucleus. The sum of information should lead to the insight that nucleic acids can change their compartments both during evolution and also in an experimental situation. Finally, the transfer of genetic information from mitochondria to a wrong place in the nuclear genome could cause aging and cancer.

References

Bolotin-Fukuhara M, Grivell LA (1992) Genetic approaches to the study of mitochondrial biogenesis in yeast. Antonie van Leeuwenhoek 62: 131–153

Brown TA, Ray AJ, Waring RB, Scazzocchio C, Davies RW (1984) A mitochondrial reading frame which may code for a second form of ATPase subunit 9 in *Aspergillus nidulans.* Curr Genet 8: 489–492

Colleaux L, Michel-Wolwertz MR, Matagne RF, Dujon B (1990) The apocytochrome b gene of *Chlamydomonas smithii* contains a mobile intron related to both *Saccharomyces* and *Neurospora* introns. Mol Gen Genet 223: 288–296

Costanzo MC, Fox TD (1990) Control of mitochondrial gene expression in *Saccharomyces cerevisiae.* Ann Rev Genet 24: 91–113

Del Giudice L, Massardo DR, Manna F, Evidente A, Randazzo G, Wolf K (1984) Differential effect of the alkaloid lycorine of *rho*$^+$, *mit*$^-$, *rho*$^-$, and *rho*^{o} strains of *Saccharomyces cerevisiae.* Curr Genet 8: 493–497

Farelly F, Butow RA (1983) Rearranged mitochondrial genes in the yeast nuclear genome. Nature (Lond) 301: 296–301

Fox TD, Folley LS, Mulero JJ, McMullin TW, Thorsness PE, Hedin LO, Costanzo MC (1991) In: Guthrie C, Fink GR (eds) Guide to yeast genetics and molecular biology. Methods Enzymol 194: 149–165

Gray MW (1983) The bacterial ancestry of plastids and mitochondria. BioScience 33: 693–699

Grivell L (1983) Mitochondrial DNA. Sci Am 248: 73–78

Grivell L (1994) Invasive introns. Curr Biol 4: 161–164

Hardy CM, Clark-Walker GD (1991) Nucleotide sequence of the *cox1* gene in *Kluyveromyces lactis* mitochondrial DNA – evidence for recent horizontal transfer of a group II intron. Curr Genet 20: 99–114

Hensgens LAM, Grivell LA, Borst P, Bos JL (1979) Nucleotide sequence of the mitochondrial structural gene for subunit 9 of the yeast ATPase complex. Proc Natl Acad Sci USA 76: 1663–1667

Hilario E, Gogarten JP (1993) Horizontal transfer of ATPase genes – the tree of life becomes a net of life. BioSystems 31: 111–119

Kidwell MG (1992) Horizontal transfer. Curr Opin Genet Dev 2: 868–873

Lambowitz AM, Belfort M (1993) Introns as mobile genetic elements. Annu Rev Biochem 62: 587–622

Lang BF (1984) The mitochondrial genome of the fission yeast *Schizosaccharomyces pombe*: highly homologous introns are inserted at the same position of the otherwise less conserved *cox1* genes in *Schizosaccharomyces pombe* and *Aspergillus nidulans.* EMBO J 3: 2129–2136

Lang BF, Ahne F, Distler S, Trinkl H, Kaudewitz F, Wolf K (1983). Sequence of the mitochondrial DNA, arrangement of genes and processing of their transcripts in

Schizosaccharomyces pombe. In: Schweyen RJ, Wolf K, Kaudewitz F (eds) Mitochondria. De Gruyter, Berlin, pp 313–329

Lazowska J, Meunier B, Macadre C (1994) Homing of a group II intron in yeast mitochondrial DNA is accompanied by unidirectional coconversion of upstream-located markers. EMBO J 13: 4963–4972

Lestienne P (1992) Mitochondrial DNA mutations in human diseases: a review. Biochimie 74: 123–130

Liao XS, Butow RA (1993) *RTG1* and *RTG2* – 2 yeast genes required for a novel path of communication from mitochondria to the nucleus. Cell 72: 61–71

Massardo DR, Manna F, Del Giudice L, Wolf K (1990) Interactions between the yeast mitochondrial and nuclear genomes: isogenic suppressive and hypersuppressive petites differ in their resistance to the alcaloid lycorine. Curr Genet 17: 455–457

Massardo DR, Manna F, Schäfer B, Wolf K, Del Giudice L (1994) Complete absence of mitochondrial DNA in the petite-negative yeast *Schizosaccharomyces pombe* leads to resistance towards the alkaloid lycorine. Curr Genet 25: 80–83

Michel F, Dujon B (1986) Genetic exchanges between bacteriophage *T4* and filamentous fungi. Cell 46: 323

Mörl, Schmelzer (1990) Integration of group II intron bI1 into a foreign RNA by reversal of the self-splicing reaction in vitro. Cell 60: 629–636

Mueller MW, Allmaier M, Eskes R, Schweyen RJ (1993) Transposition of group II intron aI1 in yeast and invasion of mitochondrial genes at new locations. Nature (Lond) 366: 174–176

Nagley P (1989) Trafficking in small mitochondrial RNA molecules. TIG 5: 67–69

Nagley P, Devenish RJ (1989) Leading organellar proteins along new pathways: the relocation of mitochondrial and chloroplast genes to the nucleus. TIBS 14: 31–35

Parikh VS, Morgan MM, Scott R, Clements LS, Butow RA (1987) The mitochondrial genotype can influence nuclear gene expression in yeast. Science 235: 576–580

Paschke T, Meurer P, Schäfer B, Wolf K (1994) Homing of mitochondrial introns in the fission yeast *Schizosaccharomyces pombe*. Endocytobiosis Cell Res 10: 205–213

Schatz G (1993) The protein import machinery of mitochondria. Protein Sci 2: 141–146

Schmidt WM, Schweyen RJ, Wolf K, Mueller MW (1994) Transposable group II introns in fission and budding yeast: site-specific genomic instabilities and formation of group II IVS plDNAs. J Mol Biol 243: 157–166

Schmitt ME, Clayton DA (1993) Nuclear RNase MRP is required for correct processing of pre-5.8S rRNA in *Saccharomyces cerevisiae*. Mol Cell Biol 13: 7935–7941

Schneider A, Martin J, Agabian N (1994) A nuclear encoded tRNA of *Trypanosoma brucei* is imported into mitochondria. Mol Cell Biol 14: 2317–2322

Schwarz E, Neupert W (1994) Mitochondrial protein import: mechanisms, components and energetics. Biochim Biophys Acta 1187: 270–274

Shadel GS, Clayton DA (1993) Mitochondrial transcription initiation – variation and conservation. J Biol Chem 268: 16083–16086

Shay JW, Werbin H (1992) New evidence for the insertion of mitochondrial DNA into the human genome: significance for cancer and aging. Mutat Res 275: 227–235

Smith MW, Feng DF, Doolittle RF (1992) Evolution by acquisition: the case for horizontal gene transfers. TIBS 17: 489–493

Van den Boogaart P, Samallo J, Agsteribbe E (1982) Similar genes for a mitochondrial ATPase subunit in the nuclear and mitochondrial genome of *Neurospora crassa*. Nature (Lond) 298: 187–189

Thorsness PE, Fox TD (1990) Escape of DNA from mitochondria to the nucleus in *Saccharomyces cerevisiae*. Nature (Lond) 346: 376–381

Thorsness PE, Fox TD (1993) Nuclear mutations in *Saccharomyces cerevisiae* that effect the escape of DNA from mitochondria to the nucleus. Genetics 134: 21–28

Wolf K (1987) Mitochondrial genes of the fission yeast *Schizosaccharomyces pombe*. In: Kinghorn JR (ed) Gene structure in eukaryotic microbes. IRL Press, Oxford, pp 69–91

Wolf K (1993) Mitochondrial introns in yeast – mobile genetic elements. Endocytobiosis Cell Res 10: 55–63

Wolf K, Del Giudice L (1988) The variable mitochondrial genome of ascomycetes: Organization, mutational alterations, and expression. Adv Genet 25: 185–308

Xu BJ, Clayton DA (1992) Assignment of a yeast protein necessary for mitochondrial transcription initiation. Nucleic Acids Res 20: 1053–1059

Zimmer M, Krabusch M, Wolf K (1991) Characterization of a novel open reading frame, *urf a*, in the mitochondrial genome of fission yeast: correlation of *urf a* mutations with a mitochondrial mutator phenotype and possible role of frameshifting in *urf a* expression. Curr Genet 19: 95–102 (1991)

2 Horizontal Transfers and the Evolution of Transposable Elements

P. Capy[1], C. Bazin[1], D. Anxolabéhère[2], and T. Langin[3]

[1]Laboratoire de Biologie et Génétique Evolutives, CNRS, 91198 Gif/Yvette, France
[2]Laboratoire de Dynamique du Génome et Evolution, Institution J. Monod, Universités Paris VI et Paris VII, 75251 Paris Cédex 05, France
[3]Laboratoire de Cryptogamie, Bâtiment 400, Université Paris XI, 91405 Orsay, France

Introduction

Since their discovery by Barbara McClintock in the 1950s (see McClintock 1984 for a review), transposable elements have been detected in all organisms in which they have been studied. Moreover, they may represent from 5% to more than 10% of the genome. Due to their intrinsic capacity to be mobile or mobilisable, these elements can have an impact on the evolution of the genomes as discussed in a special issue of Genetica (vol 86, 1992; see also McDonald 1993). The genomes can be modified in different ways such as changes in the expression of the gene closed to their insertion point (Robins and Samuleson 1992; Smith and Corces 1992), chromosomal rearrangements such as inversions (Lim 1988; Lyttle and Haymer 1992;) or deletion (Sheen et al. 1993), evolution of intron (Wessler 1989; Purugganan and Wessler 1992), and the telomeric structure (Levis et al. 1993; Biessmann et al. 1992).

At the individual level, the mobility of these elements can be deleterious to the organism which contains them. For instance, insertions of a LINE element has been involved in canine tumor (Katzir et al. 1985), or in man in haemophilia (Kazazian et al. 1988; Dombroski et al. 1993), breast, colon and testicular cancer (Morse et al. 1988; Bratthauer and Fanning 1992; Bratthauer et al. 1994; Miki et al. 1992) or Duchene muscular dystrophy (Narita et al. 1993). Finally, the retrotransposons, which transpose via an RNA intermediate using a reverse transcriptase activity, are closely related to the structure of retroviruses. In this respect, from a comparison of their reverse transcriptase domain with those of the retroviruses, Xiong and Eickbush (1990) suggested that retroviruses derived from retrotransposons by the acquisition of an envelope gene.

The evolution of such elements remains puzzling. Indeed, in each class of elements defined by Finnegan (1989; i.e., class I transposing by a RNA and class II by a DNA intermediates), it is possible to observe some relationships between elements described in species which can be phylogenetically distant while ele-

E.R. Schmidt/T. Hankeln (Eds.)
Transgenic Organisms and Biosafety

ments of closely related species can be highly divergent. Such observations are frequent as soon as the evolution of transposable elements is investigated and the interpretation of such results remains the central point of many discussions in terms of horizontal *versus* vertical transfers.

The transmission of genetic material from one species to another without sexuality raises many questions about the genome stability and its evolution. Such a phenomenon, also called horizontal transfer, in contrast to vertical transfer in which the similarity of genes in different species refers to a common ancestor, concerns both structural genes and transposable elements. However, this interpretation of many phylogenetic oddities remains a hypothesis since no clear mechanism of transmission has been yet described.

One of the first cases suggesting this phenomenon, was the description of a close similarity of C-type viral genes between *Cercopithecinae* and *Felidae* (Benveniste and Todaro 1974). Another example is the transmission of *SOD* (superoxide dismutase) between *Leiognathus splendens* (ponyfish) and its symbiotic bacterium *Photobacterium leiognathi* (Martin and Fridovich 1981; Bannister and Parker 1985). Several other cases suggesting a horizontal transmission between eukaryotes to bacterium or from bacterium to plants or yeast are summarized by Smith et al. (1992) and by Avise (1994).

In this review we would like to present the arguments in favor and against the horizontal transfer hypothesis and the alternative hypotheses which could explain some of the strange phylogenies of transposable elements.

Life, Death, and Impact of Transposable Elements

The analyses of the recent invasions of trnsposable elements in species initially deprived of them provide many informations about their “life cycle”. The best known examples are given by the elements *P*, *hobo* and *I* for which recent invasions of *Drosophila melanogaster* are suggested.

Among these three elements, only the data concerning the *P* element are convincing. Indeed, the invasion of this element in natural populations of *D. melanogaster* has been followed from the American continent to Europe and to Asia (Kidwell 1983; Anxolabéhère et al. 1984; Anxolabéhère et al. 1988). Moreover, thanks to old laboratory strains, it was possible to date the beginning of this invasion to the 50s. A molecular comparison of different copies of the *P* element sequenced in several species of the *Drosophilidae* family showed that elements of *D. willistoni* and *D. melanogaster* were similar except for a single base pair in position 32 over 2.9 kb (Daniels et al. 1990; Clark et al. 1994) while these two species shared a common ancestor about 50 MYR ago. On the other hand, in the same family, and for the same element, two other species separated at the same time, exhibit a similarity of about 70% (Hagemann et al. 1990; Miller et al. 1992;

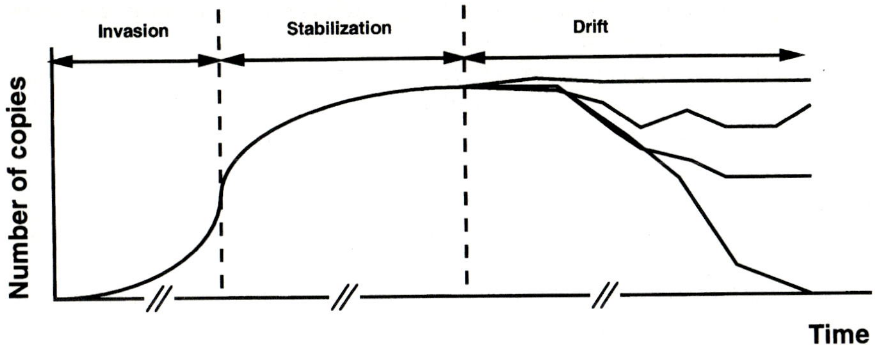

Fig. 1. Summary of the life and death of a transposable element in a natural population or in a species. During the invasion phase and in relation to high transposition rates, the average number of copies per diploid genome increases to a plateau value. This value is reached during the stabilization phase, in which regulation systems of transposition take place. During the last phase (drift), the ratio between autonomous and nonautonomous elements decreases and, due to the low number of rare transposition events, the copies can be lost by random sampling, by ectopic recombination, or by degeneration after a long period of time

Paricio et al. 1992). Therefore, a horizontal transmission between *D. willistoni* and *D. melanogaster* is likely.

From the data available on this element, it is possible to get some informations about its dynamic and regulation in a newly colonized species (see Fig. 1). After a period of diffusion in the species and a stabilisation to an average of about 30 copies per diploid genome, some deleted elements may appear. In the *P-M* system of *Drosophila*, the *KP* deleted element, deriving from an active copy, harbors a deletion which does not allow the synthesis of a functional transposase required for the mobility of the element (Black et al. 1987). Moreover, it seemed that this element may have a regulatory effect on the activity of the autonomous elements as well as different in *vitro*-modified elements (Rio 1990). On the other hand, the genomic position of the autonomous element may also have a regulatory effect such as the element localized in position 1A on the X chromosome (Ronsseray et al. 1991).

The natural occurence of deleted *P* elements is also observed after an artificial introduction in other species. Indeed, the injection of active *P* element in *D. simulans*, a sibling species of *D. melanogaster* initially deprived of this element, is followed by an increase in the number of copies, which is then stabilized before the appearance of deleted elements (Daniels et al. 1989; Montchamp-Moreau 1989). Such data may be in agreement with the results obtained for other elements like *hobo* elements (Periquet et al. 1989) or *mariner* showing that most of the *mariner* elements found in many *Drosophila* species are deleted (Brunet et al., unpublished results). Thus, the appearance of deleted elements could be a normal step in the dynamics of TEs after the invasion phase. Finally, during the last phase (drift) the elements could be lost by genetic drift in species with small effective population size or maintained to a relatively small number of copies (Kaplan et al. 1985). Therefore, the number of copies of an element can be highly variable from one species to another. This may reflect that this element is in different phases of its "life cycle" in the different species, and/or that the species are more or less sensitive to its activity. For instance, the number of copies of *mariner* is very high *Hyalophora cecropia* (Lidholm et al. 1991) and low in *D. simulans* or *D. sechellia* (Capy et al. 1994b).

TEs of several families seem also to be apt to transpose simultaneously at a high frequency during a phenomenon called by Gerasimova et al. (1984, 1985) the "transposition bursts". As suggested by several authors, such a phenomenon does not occur at random, and could be the result of the modifications of environmental conditions and/or populational parameters (Pasyukova et al. 1986; Biémont et al. 1987). In other words, these bursts could generate *de novo* mutations which could be selected and then contribute to a better adaptation of a population to its environment. Such a phenomenon can be illustrated by the work of Mackay showing that the mutations generated by the *P* element, in dysgenic crosses, can be selected in less than ten generations for a quantitative trait (Mackay 1984, 1989; see also Frankham et al. 1991). Similarly mutations due to, *Ty* transposition can also be selectively favored in particular environments (Wilke et al. 1992).

Therefore, the "selfish gene" hypotheses proposed by Doolittle and Sapienza (1980) and Orgel and Crick (1980), are not always well adapted to describe transposable element evolution within and between species. Indeed, even if an element may be responsible for many genomic rearrangements leading to a decrease in the fitness of some individuals, the same element could be involved in the adaptation of natural populations and species to new environmental conditions (for a review, see the special issue of Genetica 1992, vol 86).

Phylogeny of Elements and Phylogeny of Species: The Horizontal Transfer Hypothesis

As mentioned in the introduction, TEs have been described in many organisms, but when the phylogenies of the host species based on morphological traits or DNA polymorphism, are compared to those based on their TEs, some inconsistencies may be observed. The main situations from which the horizontal transfer is proposed are summarized in Fig. 2.

Several examples are provided by the analysis of TE in *Drosophila*. For instance, the analysis of the *mariner* element in the *Drosophilidae* family shows that copies detected in two species, *Drosophila mauritiana* and *Zaprionus tuberculatus*, belonging to 2 different genera (*Drosophila* and *Zaprionus* genera) can be less divergent than copies extracted from species classified in the same genus, i.e., *D. mauritiana* and *D. tsacasi* (Maruyama and Hartl 1991; Brunet et al.,

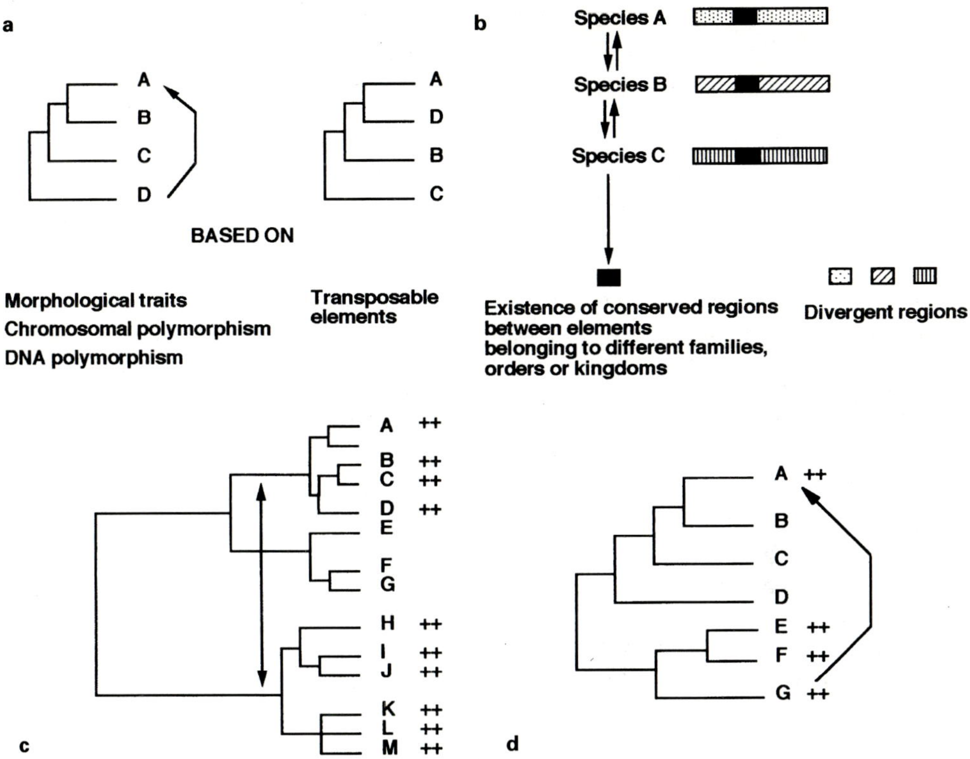

Fig. 2a–d. Situations which suggest the existence of horizontal transfers. Arrows represent the potential horizontal transfers. **a** Inconsistencies between phylognies of species based on conventional traits and based on transposable elements (TEs). **b** Similarities observed between elements extracted from highly divergent species. **c** Patchy specific distribution of the TEs. **d** Recent invasion of a TE

in preparation). Several examples have also been reported by Robertson (1993, 1995) from the analysis of the *mariner* element in *Diptera* and *Hymenoptera*. Another case study is provided by the high similarity (99 % i.e. a single substitution over 2.9 kb) observed between the *P* elements of *D. willistoni* and those of *D. melanogaster*, two species belonging to different groups and separated since 50 MYR; while a similarity of about 70% is observed between *D. melanogaster* and *D. guanche* also diverging since the same period of time (Daniels et al. 1990; Miller et al. 1992). Again, Hagemann et al. (1994) described two distinct subfamilies of *P* elements in *D. bifasciata* (*obscura* group). The comparison of these two types of elements (M and O) with those of *Scaptomyza pallida* (species belonging to another genus) shows that the M-type elements are closely related to the elements described in *S. pallida* suggesting a recent invasion of these elements in *D. bifasciata*.

In plants, similar phenomena have been described. For instance, Flavell (1992) shows that the phylogeny of the retrotransposons belonging to the *Ty1*-

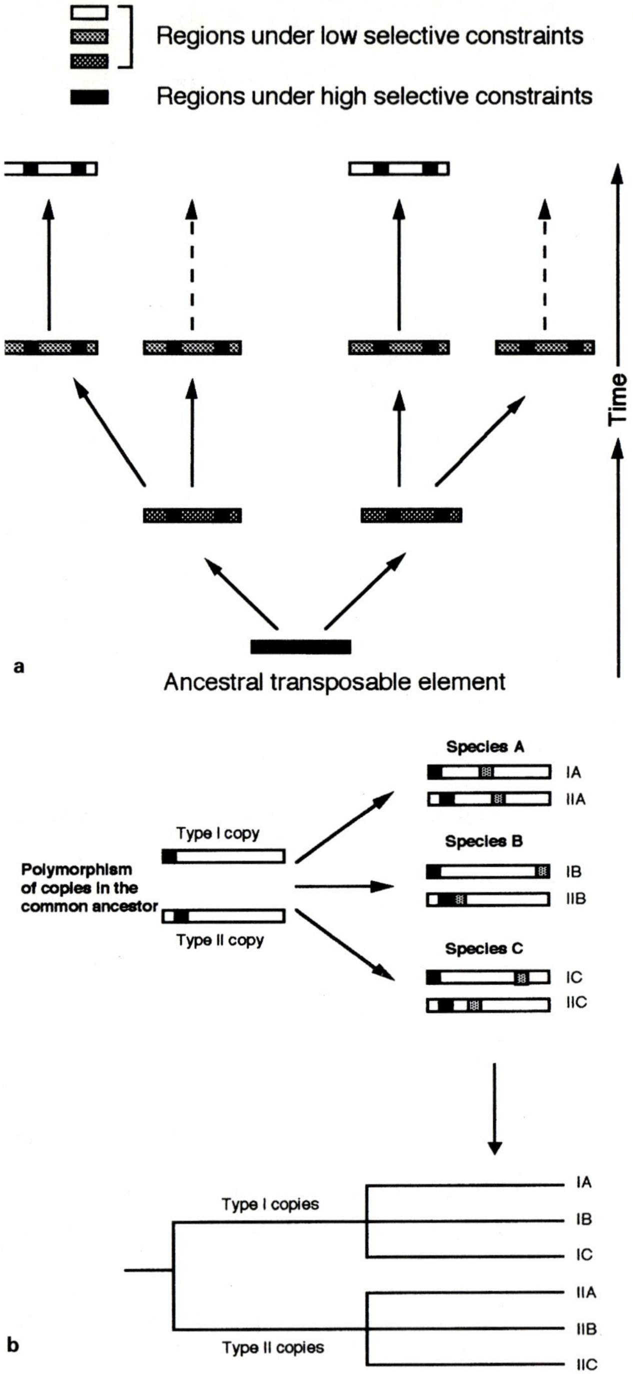
Regions under low selective constraints
Regions under high selective constraints
Time
a
Ancestral transposable element
Species A
IA
IIA
Type I copy
Species B
Polymorphism
of copies in the
common ancestor
IB
IIB
Type II copy
Species C
IC
IIC
Type I copies
IA
IB
IC
IIA
IIB
IIC
b
Type II copies

Fig. 3a,b

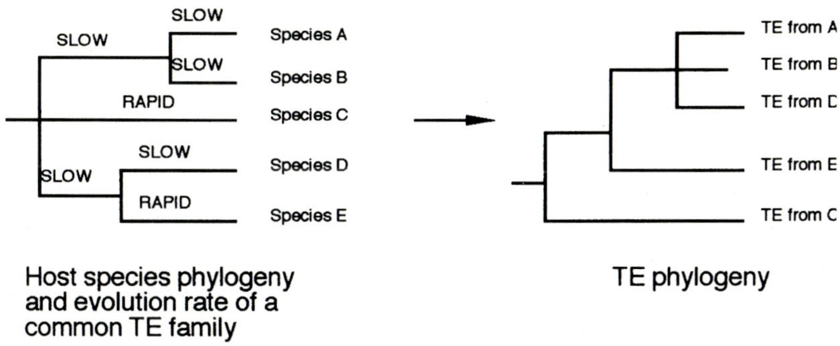

c

Fig. 3a–c. Alternative hypotheses redrawn from Capy et al. (1994a). **a** Existence of an ancestral element distributed in several lineages. This element can be lost along some lineages and maintained in others. Moreover, probably due to the roles played in its autonomy, the different parts of an element, may then evolve at different rates. Therefore, the similarity observed for small regions between elements could be due to the important role played by these regions in comparison with the other parts of the elements. **b** The existence of an ancestral polymorphism could explain the inconsistencies between phylogenies. The elements can be clustered according to their copy of origin rather than their present host species. **c** The variability of the evolution rate of an element in different species may also lead to inconsistencies between the conventional phylogeny of species and a phylogeny based on their transposable elements

copia group in eukaryotes, suggests both vertical transfer between recently separated species and horizontal transmission between distantly related species like potatoes and barley which are separated by about 200 MYR. However, VanderWiel et al. (1993) also proposed a model for the retrotransposon evolution in the absence of horizontal transfer based on loss/amplification of the element in different lineages.

Alternative Hypotheses

For the transposable elements, if the horizontal transmission is probably the right interpretation in a few cases like for the invasion of the *P* element in *D. melanogaster* from *D. willistoni*, several alternative hypotheses may also explain most of the inconsistencies observed between conventional phylogenies of species and those based on their TEs (Capy et al. 1994a; Cummings 1994).

All the alternative hypotheses suggest an ancestral origin of most TEs. But this assumption requires additional hypotheses as shown in Fig. 3. For instance, the loss of TEs by genetic drift may explain the discontinuous distributions of almost all elements among species. Moreover, no argument supports the existence of a constant rate of evolution of a TE in different species. On the contrary, several results show that a given element may have different behaviour (number of copies and transposition rate) according to its host strain, population or species (Biémont et al. 1987; Csink and McDonald 1989; Capy et al. 1994b).

In fungi some phenomena called Methylation Induced Premeiotically (MIP) in *Ascobolus immersus* and Repeat-Induced Point mutation (RIP) in *Neurospora crassa* have been described (Selker et al. 1987; Goyon and Faugeron 1989). These two mechanisms correspond to an alteration of the duplicated sequences and may lead, in *N. crassa*, to a rapid sequence inactivation due to an increase in the C-G to T-A transition rate (Cambareri et al. 1989; for reviews, see also Selker 1990; Rossignol and Faugeron 1994). For instance, the *Tad* element, a LINE-like retrotransposon of *N. crassa* shows numerous C-G to T-A mutations probably due to the RIP phenomenon (Kinsey et al. 1994). Similarly, inactivations of repeated sequences have also been described in plants (Jorgensen 1990) and in eukaryotes (Kricker et al. 1992). All these mechanisms may play an important role in the evolution of repeated sequences such as TEs since a large number of mutations could be accumulated in a short time. Moreover, if the evolution rate of the elements depends on their host species, this could explain some inconsistencies between conventional and TE phylogenies.

Again, assuming an ancestral origin of an element does not mean that this element derives from a single copy. In other words, the term "ancestral origin" may correspond to a set of polymorphic copies. Therefore, as summarized in Fig. 3b, the existence of an ancestral polymorphism may explain many phylogenetic trees of TEs in which copies of two distinct species may be more closely related than copies of the same species (Capy et al. 1994a).

Otherwise, the observation of conserved sequences between elements described in distantly related species can be interpreted as a result of selective pressures on particular regions involved in the activity of the element. In other words, conserved regions should be flanked by less conserved regions which should be under less or no selective pressures. This is probably the case for the similarities observed between retroviruses and retrotransposons (Argos et al. 1986; Xiong and Eickbush 1990; Henikoff 1992; Kulkosky et al. 1992; Skalka 1993; Burke et al. 1993), among the members of the *mariner-Tc1* superfamily (Doak et al. 1994; Robertson 1995; Langin et al. 1995) or for the *Ac*, *Tam3* and *hobo* elements (Calvi et al. 1991).

Finally, a bias could be introduced by the technical procedure used. For instance, the extensive use of PCR amplifications on total genomic DNA may raise some problems. Indeed, this DNA may include the DNA of the species analysed and the DNA of its symbionts. Thus, in this case, the similarity observed between distantly related species could be due to a similarity between their symbionts.

All these alternative hypotheses, based on an ancestral origin of the elements, do not preclude occasional horizontal transmission. For instance, in spite of the recent introduction of *P* element in *D. melanogaster*, the origin of *P* element described in other species of *Drosophila* is probably ancestral (Simonelig and Anxolabéhère 1991; Miller et al. 1992; Clark et al. 1994). The same conclusion can be drawn from the comparison of the *copia*-like elements in plants (Flavell 1992).

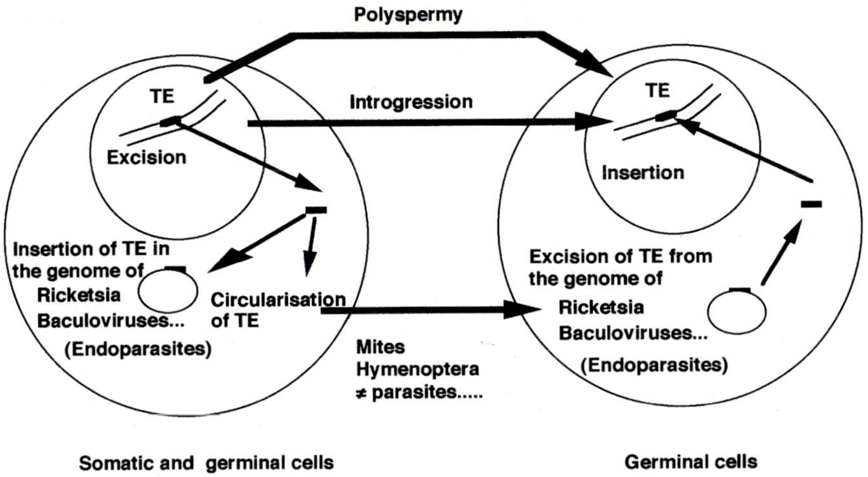

Fig. 4. Hypothetical mechanisms of horizontal transfer between two species A (*left*) and B (*right*). *TE*, transposable element. See text for details

Possible Mechanisms of Horizontal Transfer

While extensively evoked, no mechanism of horizontal transfer has been clearly evidenced. This is probably due to the rarity of this phenomenon. However, several mechanisms summarized in Fig. 4 have been proposed.

Firstly, between closely related species which can mate in natural conditions, an exchange of genetic material may occur in two ways: by introgression, if an interspecific hybrid is fertile, or by polyspermy, i.e., the fecundation of the ovule by two spermatozoids of two different species. The former scenario is well documented, but for the latter very few and contradictory data are available. For instance, in *Drosophila*, an introgression has been demonstrated between *D. mauritiana* and *D. simulans* from an analysis of their mitochondrial DNA (Solignac and Monnerot 1986) while some phylogenetic oddities have been observed between the two species from the analysis of their *mariner* elements (Capy et al. 1992). Such a phenomenon has also been used to transfer some *P* elements from a transgenic strain of *D. simulans* to *D. mauritiana* in laboratory conditions (Montchamp-Moreau et al. 1991). On the other hand, no convincing experiment demonstrates the existence of polyspermy in *Drosophila* (Brandt-Rosquist and Lüning 1984; Palmgren and Lake 1986; see Ashburner 1989 for comments). In other species, only one case of chimeric males, obtained in laboratory conditions, has been clearly described by Sawa and Oishi (1989) in *Athalia rosae*.

The second type of mechanism refers to the natural parasites of a group of species. For instance, in *Drosophila*, the parasitic wasp, *Leptopilina boulardi*, has several hosts (Carton et al. 1986). Therefore, such parasites may act as a natural syringe and may transfer some genetic material from the cytoplasm or the nucleus of one individual to another. Recently, Houck et al. (1991) suggested that the semiparasitic mite *Proctolaelaps regalis* DeLeon could be the potential vector for *P* element between *D. willistoni* and *D. melanogaster*. Such a transfer does not necessarily require the integration of a TE in the genome of vector. It could be transmitted directly as excision product. In this respect, several transposable elements have been identified as circular DNA in the cytoplasm (Pont et al. 1987; Sundaresan and Freeling 1987; Rose and Snutch 1984; Ruan and Emmons 1984).

Among the endosymbionts which could be responsible for such a transfer, the baculoviruses, parasites of insects, could be serious candidates since some part of their genome corresponds to sequences coming from their hosts (see Blissard and Rohrmann 1990). In this respect, some *copia*-like elements have been described in the nuclear polyhedrosis virus (AcNVP) of *Autographa californica* (Miller and Miller 1982). Other bacteria such as *Rickettsia*-like microorganism are also suspected to be potential vectors since many horizontal transfers between their hosts are suggested (Rousset et al. 1992). In fungi a transfer of genetic material from the mycoparasite *Parasitella parasitica* to the host species *Absida glauca* via plasmids was described by Kellner et al. (1993).

Is Horizontal Transmission Frequent?

As stated by Kidwell (1992), the estimation of the frequency of horizontal transmission raises the question of the identification of the vector(s). However, the minimum number of steps from one host to another would be: the excision from the initial host, the transfer to a new host, with or without integration in the genome of a shuttle species, the reintegration in a new host, and especially in its germline cells. With the exception of the excision, for which the rates are probably highly variable in natural populations and species, the probability of each of the other events is very low. Therefore, the succession of all the steps suggests that horizontal transfers are likely rare.

Discussion

From these arguments, for and against the horizontal transmission of TEs between species, it is important to stress that this phenomenon probably exists in few cases, but that in most situations described, this hypothesis remains challenged by alternative interpretations and no strong arguments allow us to choose between them.

On the basis of the existence of horizontal transfers, many authors try to use TEs as tools for gene tagging or as vectors to insert foreign DNA in host species. The first point allows the detection and the analysis of some genes while the latter point is of particular interest especially to modify some genetic characteristics of species having an economical impact.

In *Drosophila*, and almost exclusively in *D. melanogaster*, the *P* element is extensively used as vector for the germline transformation or as mutagenic agent to detect particular genes (Rubin and Spradling 1982), but other elements like *hobo* (Blackman and Gelbart 1989) and *mariner* (Lidholm et al. 1993) are also efficient. However, as stressed by Lidholm et al. (1993) "*the holy grail of insect transposable elements is an element with a broad host range able to serve as a vector for germ line transformation in a wide variety of insect species*" and especially in insects with an economical impact. Indeed, the *P* transformation system does not work outside the *Drosophilidae* family (Engels 1989). For the other systems, no results have been published, but the *mariner* system, which appears widely dispersed, should be tested in medically important insects such as mosquitos (Warren and Crampton 1994).

In maize, several elements have been used to isolate genes by transposon tagging like the *Ac* element (Fedoroff et al. 1984) or the *En* and *Mu1* elements (O'Reilly et al. 1985). Moreover, the maize transposable elements seem to be mobile in other plants like tobacco, carrots, *Arabidopsis thaliana*, (see Fedoroff 1989 for a review). Other systems based on retrotransposons like *Tnt1* could be potentially useful (Grandbastien 1992; Pauls et al. 1994).

In conclusion, thanks to their ability to move naturally in a genome, and to integrate artificially into different genomes, transposable elements are good candidates for horizontal transmission. However, the greatest care must be taken to interpret their phylogenetic oddities since several alternative hypothesis have to be taken into account.

Acknowledgements. The authors thank M. Eden for comments on the early version of the manuscript and the GREG (Groupement de Recherche et d'Etude sur les Génomes, décision no. 48) for its financial support.

References

Anxolabéhère D, Kai H, Nouaud D et al. (1984) The geographical distribution of *P-M* hybrid dysgenesis in *Drosophila melanogaster*. Genet Sel Evol 16: 15–26

Anxolabéhère D, Kidwell MG, Periquet G (1988) Molecular characteristics of diverse populations are consistent with the hypothesis of a recent invasion of *Drosophila melanogaster* by mobile *P* elements. Mol Biol Evol 5: 252–269

Argos P, Landy A, Abremski K et al. (1986) The integrase family of site-specific recombinases: regional similarities and global diversity. EMBO J 5: 433–440

Ashburner M (1989) *Drosophila*: a laboratory handbook. Cold Spring Harbor Laboratory Press, NewYork

Avise JC (1994) Molecular markers, natural history and evolution. Chapman and Hall, New York

Bannister JV, Parker MW (1985) The presence of a copper/zinc superoxide dismutase in the bacterium *Photobacterium leiognathi*: a likely case of gene transfer from eukaryotes to prokaryotes. Proc Natl Acad Sci USA 82: 149–152

Benveniste RE, Todaro GJ (1974) Evolution of C-type viral genes: inheritance of exogenously acquired viral genes. Nature 252: 456–459

Biémont C, Aouar A, Arnault C (1987) Genome reshuffling of the *copia* element in a *Drosophila* inbred line. Nature 329: 742–744

Biessmann H, Valgeirsdottir V, Lofsky A et al. (1992) *HeT-A*, a transposable element specifically involved in "Healing" broken chromosome ends in *Drosophila*. Mol Cell Biol 12: 3910–3918

Black DM, Jackson MS, Kidwell MG, Dover GA (1987) *KP* elements repress hybrid dysgenesis in *Drosophila melanogaster*. EMBO J 6: 4125

Blackman RK, Gelbart WM (1989) The transposable element *hobo* of *Drosophila melanogaster*. In: Berg DE, Howe M (eds) Mobile DNA. American Society for Microbiology, Washington, DC, pp 523–529

Blissard GW, Rohrmann GF (1990) Baculovirus diversity and molecular biology. Annu Rev Entomol 35: 127–155

Brandt-Rosquist K, Lüning KG (1984) Double fertilization in *Drosophila melanogaster*. Hereditas 101: 69–73

Bratthauer GL, Cardiff RD, Fanning TG (1994) Expression of *LINE*–1 retrotransposons in human breast cancer. Cancer 73: 2333–2336

Bratthauer GL, Fanning TG (1992) Active *LINE*–1 retrotransposons in human testicular cancer. Oncogene 7: 507–510

Burke WD, Eickbush DG, Xiong Y et al. (1993) Sequence relationship of retrotransposable elements *R1* and *R2* and between divergent insect species. Mol Biol Evol 10: 163–185

Calvi BR, Hong TJ, Findley SD, Gelbart WM (1991) Evidence for a common evolutionary origin of inverted repeat transposons in *Drosophila* and plants: *hobo*, *Activator*, and *Tam3*. Cell 66: 465–471

Cambareri EB, Jensen BC, Schabtach E, Selker EU (1989) Repeat-induced G-C to A-T mutations in *Neurospora*. Science 244: 1571–1575

Capy P, Koga A, David JR, Hartl DL (1992) Sequence variation among active *mariner* elements in natural populations of *Drosophila simulans*. Genetics 130: 499–506

Capy P, Anxolabéhère D, Langin T (1994a) The strange phylogenies of transposable elements: are the horizontal transfer the only explanation? TIG 10: 7–12

Capy P, Langin T, Bigot Y et al. (1994b) Horizontal transmission versus ancient origin: *mariner* in the witness box. Genetica 93: 161–170

Carton Y, Boulétreau M, vanAlphen JJM, vanLenteren JC (1986) The *Drosophila* parasitic wasp. In: Ashburner M, Carson HL, Thompson JN (eds) The genetics and biology of Drosophila, vol 3. Academic, London, pp 347–394

Clark JB, Maddison WP, Kidwell MG (1994) Phylogenetic analysis supports horizontal transfer of *P* transposable elements. Mol Biol Evol 11: 40–50

Csink AK, McDonald JF (1989) *Copia* expression is variable among natural populations of *Drosophila*. Genetics 126: 375–385

Cummings MP (1994) Transmission patterns of eukaryotic transposable elements: arguments for and against horizontal transfer. TREE 9: 141–145

Daniels SB, Chovnik A, Kidwell MG (1989) Hybrid dysgenesis in *Drosophila simulans* lines transformed with autonomous *P* elements. Genetics 121: 281–291

Daniels SB, Peterson KR, Strausbaugh LD et al. (1990) Evidence for horizontal transmission of the *P* element between *Drosophila* species. Genetics 124: 339–355

Doak TG, Doerder FP, Jahn CL, Herrick G (1994) A proposed superfamily of transposase-related genes: new members in transposon-like elements of ciliated protozoa and a common "D35E" motif. Proc Natl Acad Sci USA 91: 942–946

Dombroski BA, Scott AF, Kazazian HH (1993) Two additional potential retrotransposons isolated from a human *L1* subfamily that contains an active retrotransposable element. Proc Natl Acad Sci USA 90: 6513–6517

Doolittle WF, Sapienza C (1980) Selfish genes, the phenotype paradigm and genome evolution. Nature 284: 601–603

Engels WR (1989) *P* elements in *Drosophila melanogaster*. In: Berg DE, Howe M (eds) Mobile DNA. American Society for Microbiology, Washington, DC, pp 503–517

Fedoroff N, Furtek D, Nelson O (1984) Cloning of the *bronze* locus in maize by a simple and generalizable procedure using transposable controlling element *Ac*. Proc Natl Acad Sci USA 81: 3825–3829

Fedoroff NV (1989) Maize transposable elements. In: Berg DE, Howe M (eds) Mobile DNA. American Society for Microbology, Washington, DC, pp 375–411

Finnegan DJ (1989) Eukaryotic transposable elements and genome evolution. TIG 5: 103–107

Flavell AJ (1992) *Ty1-copia* group retrotransposons and the evolution of retroelements in eukaryotes. Genetica 86: 203–214

Frankham R, Torkamanzehi A, Moran C (1991) *P* element transposon-induced quantitative genetic variation for inebriation time in *Drosophila melanogaster*. Theor Appl Genet 81: 317–320

Gerasimova TI, Matjunina LV, Mizrokhi LJ, Georgiev GP (1985) Successive transposition explosions in *Drosophila melanogaster* and reverse transpositions of mobile dispersed genetic elements. EMBO J 4: 3773–3779

Gerasimova TI, Mizrokhi LJ, Georgiev GP (1984) Transposition bursts in genetically unstable *Drosophila melanogaster*. Nature 309: 714–716

Goyon C, Faugeron G (1989) Targeted transformation of *Ascobolus immersus* and *de novo* methylation of the resulting duplicated DNA sequences. Mol Cell Biol 9: 2818–2827

Grandbastien MA (1992) Retroelements in higher plants. TIG 8: 103–108

Hagemann S, Miller WJ, Pinsker W (1990) *P*-related sequences in *Drosophila bifasciata*: a molecular clue to the understanding of *P* element evolution in the genus *Drosophila*. J Mol Evol 31: 478–484

Hagemann S, Miller WJ, Pinsker W (1994) Two distinct *P* element subfamilies in the genome of *Drosophila bifasciata*. Mol Gen Genet 244: 168–175

Henikoff S (1992) Detection of *Caenorhabditis* transposon homologs in diverse organisms. New Biologist 4: 382–388

Houck MA, Clark JB, Peterson KR, Kidwell MG (1991) Possible horizontal transfer of *Drosophila* genes by the mite *Proctolaelaps regalis*. Science 253: 1125–1129

Jorgensen R (1990) Altered gene expression in plants due to *trans* interactions between homologous genes. TIBTECH 8: 340–344

Kaplan N, Darden T, Langley CH (1985) Evolution and extinction of transposable elements in mendelian populations. Genetics 109: 459–480

Katzir N, Rechavi G, Cohen JB, Unger T, Simoni F, Segal S, Cohen D, Givol D (1985) "Retroposon" insertion into the cellular oncogene *c-myc* in canine transmissible venereal tumor. Proc Natl Acad Sci USA 82: 1054–1058

Kazazian HHJ, Wong C, Youssoufian H, Scott AF, Phillips DG, Antonarakis SE (1988) Haemophila A resulting from *de novo* insertion of *L1* sequences represents a novel mechanism for mutation in man. Nature 332: 164–166

Kellner M, Burmester A, Wöstemeyer A, Wöstemeyer J (1993) Transfer of genetic information from the mycoparasite *Parasitella parasitica* to its host *Absidia glauca*. Curr Genet 23: 334–337

Kidwell MG (1992) Horizontal transfer of *P* elements and other short inverted repeat transposons. Genetica 86: 275–286

Kidwell MG (1983) Evolution of hybrid dysgenesis determinants in *Drosophila melanogaster*. Proc Natl Acad Sci USA 80: 1655–1659

Kinsey JA, Garrett-Engele PW, Cambareri EB, Selker EU (1994) The *Neurospora* transposon *Tad* is sensitive to repeat-induced point mutation (RIP). Genetics (in press)

Kricker MC, Drake JW, Radman M (1992) Duplication-targeted DNA methylation and mutagenesis in the evolution of eukaryotic chromosomes. Proc Natl Acad Sci USA 89: 1075–1079

Kulkosky J, Jones KS, Katz RA, Mack JPG, Skalka AM (1992) Residues critical for retroviral integrative recombination in a region that is highly conserved among retroviral/retrotransposon integrases and bacterial insertion sequence transposases. Mol Cell Biol 12: 2331–2338

Langin T, Capy P, Daboussi MJ (1995) The transposable element, *impala*, a fungal member of the *Tcl-mariner* superfamily. Mol Gen Genet 246: 19–28

Levis RW, Ganesan R, Houtchens K, Tolar LA, Sheen FM (1993) Transposons in place of telomeric repeats at a *Drosophila* telomere. Cell 75: 1083–1093

Lidholm DA, Gudmundsson GH, Boman HG (1991) A highly repetitive *mariner*-like element in the genome of *Hyalophora cecropia.* J Biol Chem 266: 11518–11521

Lidholm DA, Lohe AR, Hartl DL (1993) The transposable element *mariner* mediates germline transformation in *Drosophila melanogaster.* Genetics 134: 859–868

Lim J (1988) Intrachromosomal rearrangements mediated by *hobo* transposons in *Drosophila melanogaster.* Proc Natl Acad Sci USA 85: 9153–9157

Lyttle TW, Haymer DS (1992) The role of transposable element *hobo* in the origin of the endemic inversions in wild populations of *Drosophila melanogaster.* Genetica 86: 113–126

Mackay TFC (1984) Jumping gene meet abdominal bristles: hybrid dysgenesis-induced quantitative variation in *Drosophila melanogaster.* Genet Res 44: 231–237

Mackay TFC (1989) Transposable elements and fitness in *Drosophila melanogaster.* Genome 31: 113–126

Martin JP, Fridovich I (1981) Evidence of natural gene transfer from the ponyfish to its bioluminescent bacterial symbiont *Photobacter leiognathi.* The close relationship between bacteriocuprein and the copper-zinc superoxide dismutase of teleost fishes. J Biol Chem 256: 6080–6089

Maruyama K, Hartl DL (1991) Evidence for interspecific transfer of the transposable element *mariner* between *Drosophila* and *Zaprionus.* J Mol Evol 33: 514–524

McClintock B (1984) The significance of responses of the genome to challenge Science 226: 792–801

McDonald JF (1993) Transposable elements and evolution. Kluwer, Dordrecht

Miki Y, Nishisho I, Horii A et al. (1992) Disruption of the APC gene by a retrotransposal insertion of *L1* sequence in a colon cancer. Cancer Res 52: 643–645

Miller DW, Miller LK (1982) A virus mutant with an insertion of a *copia*-like transposable element. Nature 299: 562–564

Miller WJ, Hagemann S, Reiter E, Pinsker W (1992) *P* element homologous sequences are tandemly repeated in the genome of *Drosophila guanche.* Proc Natl Acad Sci USA 89: 4018–4022

Montchamp-Moreau C (1989) Dynamics of P-M hybrid dysgenesis in *P*-transformed lines of *Drosophila simulans.* Evolution 41: 194–203

Montchamp-Moreau C, Periquet G, Anxolabéhère D (1991) Interspecific transfer of *P* elements by crosses between *Drosophila simulans* and *Drosophila mauritiana.* J Evol Biol 4: 131–140

Morse B, Rotherg PG, South VJ et al. (1988) Insertional mutagenesis of the *myc* locus by a *LINE*-1 sequence in a human breast carcinoma. Nature 333: 87–90

Narita N, Nishio H, Kitoh Y et al. (1993) Insertion of a 5 truncated *L1* element into the 3′ end of exon 44 of the dystrophin gene resulted in skipping of the exon during splicing in a case of Duchenne muscular dystrophy. J Clin Invest 91: 1862–1867

O'Reilly C, Shepperd NS, Pereira A et al. (1985) Molecular cloning of the *a1* locus of *Zea mays* using the transposable element *En* and *Mu1*. EMBO J 4: 877–882

Orgel LE, Crick FHC (1980) Selfish DNA: the ultimate parasite. Nature 284: 604–607

Palmgren B, Lake S (1986) Evidence for the origin of bi-nucleated eggs in *Drosophila*. Hereditas 105: 155–156

Paricio N, Pérez-Alonso M, Martinez-Sebastian MJ, de Frutos R (1992) *P* sequence of *Drosophila subobscura* lack of exon 3 and may encode a 66 kd repressor-like protein. Nucleic Acid Res 19: 6713–6718

Pasyukova EG, Belyaeva ES, Kogan GL, Kaidanov LZ, Gvozdev VA (1986) Concerted tranpositions of mobile genetic elements coupled with fitness changes in *Drosophila melanogaster*. Mol Biol Evol 3: 299–312

Pauls PK, Kunert K, Huttner E, Grandbastien MA (1994) Expression of the tobacco *Tnt1* retrotransposon in heterologous species (submitted)

Periquet G, Hamelin MH, Bigot Y, Hu K (1989) Presence of the deleted *hobo* element *Th* in Eurasian populations of *Drosophila melanogaster*. Genet Sel Evol 21: 107–111

Pont G, Degroote F, Picard G (1987) Some extrachromosomal circular DNAs from *Drosophila* embryos are homologous to tandemly repeated genes. J Mol Biol 195: 447–451

Purugganan M, Wessler S (1992) The splicing of transposable elements and its role in intron evolution. Genetica 86: 295–303

Rio DC (1990) Molecular mechanisms regulating Drosophila *P* element transposition. Annu Rev Genet 24: 543–578

Robertson HM (1993) The *mariner* transposable element is widespread in insects. Nature 362: 241–245

Robertson HM (1995) The *mariner-Tc1* superfamily of transposons in animals. J Insect Phyiol 41: 99–105

Robins DM, Samuleson LC (1992) Retrotransposons the evolution of mammalian gene expression. Genetica 86: 191–203

Ronsseray S, Lehmann M, Anxolabéhère D (1991) The maternally inherited regulation of *P* elements in *Drosophila melanogaster* can be elicited by two *P* copies at cytological site 1A on the X chromosome. Genetics 129: 501–512

Rose AM, Snutch TP (1984) Isolation of the closed circular form of the transposable element *Tc1* in *Caenorhabditis elegans*. Nature 311: 485–486

Rossignol JL, Faugeron G (1994) Gene inactivation triggered by recognition betwen DNA repeats. Experientia 50: 307–317

Rousset F, Vautrin D, Solignac M (1992) Molecular identification of *Wolbachia*, the agent of cytoplasmic incompatibility in *Drosophila simulans*, and variability in relation with host mitochrondrial types. Proc R Soc Lond B 247: 163–168

Ruan KS, Emmons SW (1984) Extrachromosomal copies of transposon *Tc1* in the nematode *Caenorhabditis elegans*. Proc Natl Acad Sci USA 81: 4018–4022

Rubin GM, Spradling AC (1982) Genetic transformation of *Drosophila* with transposable element vectors. Science 218: 348–353

Sawa M, Oishi K (1989) Studies on the sawfly, *Athalia rosae* (Insecta, Hymenoptera, Tenthredinidae) III Fertilization by sperm injection. Zool Sci 6: 557–563

Selker EU (1990) Premeiotic instability of repeated sequences in *Neurospora crassa*. Annu Rev Genet 24: 579–613

Selker EU, Cambareri EB, Jensen BC, Haack KR (1987) Rearrangement of duplicated DNA in specialized cells of *Neurospora*. Cell 51: 741–752

Sheen F, Lim JK, Simmons MJ (1993) Genetic instability in *Drosophila melanogaster* mediated by *hobo* transposable elements. Genetics 133: 315–334

Simonelig M, Anxolabéhère D (1991) A *P* element of *Scaptomyza pallida* is active in *Drosophila melanogaster*. Proc Natl Acad Sci USA 88: 6102–6106

Skalka AM (1993) Retroviral DNA integration: lessons for transposon shuffling. Gene 135: 175–182

Smith MW, Feng DF, Doolittle RF (1992) Evolution by acquisition: the case for horizontal transfers. TIBS 17: 489–493

Smith PA, Corces VG (1992) The suppressor of *Hairy-wing* binding region is required for *gypsy* mutagenesis. Mol Gen Genet 233: 65–70

Solignac M, Monnerot M (1986) Race formation, speciation and introgression within *Drosophila simulans, D mauritiana* and *D sechellia* inferred from mitochondrial DNA analysis. Evolution 40: 531–539

Sundaresan V, Freeling M (1987) An extrachromosomal form of the *Mu* transposons of maize. Proc Natl Acad Sci USA 84: 4924–4928

VanderWiel PL, Voytas DF, Wendel JF (1993) *Copia*-like retrotransposable element evolution in diploid and polyploid Cotton (*Gossypium* L). J Mol Evol 36: 429–447

Warren AM, Crampton JM (1994) *Mariner*: its propects as a DNA vector for the genetic manipulation of medically important insects. Parasitol Today 10: 58–63

Wessler S (1989) The splicing of maize transposable elements from pre-mRNA – a minireview. Gene 82: 127–133

Wilke CM, Maimer E, Adams J (1992) The population biology and evolutionary significance of *Ty* elements in *Saccharomyces cerevisiae.* Genetica 86: 155–176

Xiong Y, Eickbush TH (1990) Origin and evolution of retroelements based upon their reverse transcriptase sequences. EMBO J 9: 3353–3362

3 Transposable Elements in Baculoviruses: Their Possible Role in Horizontal Gene Transfer

J.A. Jehle and J.M. Vlak

Department of Virology, Wageningen Agricultural University, Binnenhaven 11, 6709 PD Wageningen, The Netherlands

Abstract

Baculoviruses are a primarily insect-pathogenic group of viruses which have received considerable attention because of their use as powerful expression system of eucaryotic genes and their potential in biological control of insect pests. Recent progress in the molecular analyses of baculoviruses also has indicated that they are able to capture transposable elements originated from host genomes during infection of cultured insect cells or host larvae. These observations raised questions on the possible involvement of baculoviruses in the horizontal spread of transposable elements among different insect species. In this contribution we present an overview of the present knowledge of transposon-mediated mutagenesis of baculoviruses.

Introduction: Basics of Baculovirus Biology

Baculoviruses are double-stranded DNA-viruses with a genome size ranging between 80–230 kb. Their enveloped virions are occluded in a crystalline matrix mainly consisting of a single protein with a MW of 25–33 K, the so-called polyhedrin or granulin (Rohrmann 1986). Corresponding to morphological characteristics baculoviruses are subdivided into two genera (Murphy et al. 1995): (1) the genus *Nucleopolyhedrovirus* (NPV) forms polyhedral occlusion bodies with many virions each containing a single or multiple nucleocapsids; (2) the genus *Granulovirus* (GV) has granule-like occlusion bodies containing one virion with a single enveloped nucleocapsid.

The replication of the baculoviruses takes place in the nucleus of the host cell (for reviews, see Granados and Williams 1986; Blissard and Rohrmann 1990). During baculovirus replication, which is most intensely investigated for *Auto-*

E.R. Schmidt/T. Hankeln (Eds.)
Transgenic Organisms and Biosafety

grapha californica nucleopolyhedrovirus (AcMNPV), two different viral phenotypes are generated. In early stages of replication extracellular viruses are produced. They consist of nucleocapsids, which have budded through the cell membrane into the hemocoel and thereby acquired a membrane. The extracellular viruses are responsible for secondary infections of other host tissues. Late in infection the nucleocapsids are enveloped in the cell nucleus by a protein layer and occluded into polyhedra. Finally, the host cells disintegrate and the polyhedra are released.

Since the polyhedrin gene is not necessary for virus replication this gene can be replaced by another one allowing the expression of foreign genes under the control of the strong polyhedrin promoter (Smith et al. 1983; Miller 1988). The exploitation of the promoters of polyhedrin or p10 (another very late, overexpressed gene) is the key to the baculovirus expression system for the production of a foreign protein (Luckow and Summers 1988; Vlak et al. 1990; Vlak and Keus 1990; Luckow 1991). By the expression of insect specific toxins or hormones, it is also possible to cut the time needed to kill host insects (Vlak 1993). This strategy is used to generate recombinant baculoviruses with enhanced insecticidal properties which are expected to substitute some chemical insecticides in the future.

At present, the specific insertion of genes by using the baculovirus expression vectors is an artificial way to introduce foreign genetic information into the virus genomes. Recent observations suggest that in nature another way exists: the spontaneous accommodation of transposons, which are derived from the insect host genome, into the baculovirus genome.

Transposable Elements

Transposons are mobile genetic elements and are present in procaryotes and eucaryotes (Berg and Howe 1989). They can cause extensive mutations such as deletions or insertions and are considered as "genomic parasites". In short, two different classes of transposons are distinguished (Lewin 1990): (1) *(DNA-) transposons*, such as the P element of *Drosophila* and the Tc1 element of *Caenorhabditis elegans*, contain terminal inverted repeats and encode a transposase gene; (2) *retrotransposons*, such as the *copia* and *gypsy* elements of *Drosophila*, are characterized by the presence of long terminal repeats (LTR). Retrotransposons encode a reverse transcriptase and, in contrast to transposons, they transpose by a RNA intermediate. A common and characteristic feature of transposon and retrotransposon activity is the duplication of the target sequence (2–13 bp) flanking the transposons to the right and to the left after insertion.

Transposons in *Nucleopolyhedroviruses*

In NPVs, members of both transposon classes have been observed (Table 1). Most of these were detected when plaque morphology mutants of NPVs were analyzed and found to carry host cell insertions of variable length. These so-called "few polyhedra" (FP) mutants differed from wild type virus in generating plaques in cell culture monolayers with a significantly reduced number of polyhedra per infected cell (Fraser and Hink 1982; Fraser 1986).

The AcMNPV FP-D isolate, for example, was obtained after serial passages in cell cultures of the cabbage looper *Trichoplusia ni*. It carried a 7.5-kb retrotransposon named TED, which was apparently derived from the host genome (Miller and Miller 1982). TED was inserted at map unit 86.4–86.6 of the virus genome, where it disrupts a gene encoding a 94-kDa protein (Friesen and Miller 1987). Sequence analysis indicated that TED is a member of the *gypsy* family of retrotransposons (Friesen and Nissen 1990). It contains LTRs of 273 bp and

Table 1. Comparison between different *Nucleopolyhedrovirus* (NPV) transposons (modified from Friesen 1993)

Transposon	Origin	Length (bp)	Target sequence	Inverted terminal repeat (bp)	Terminal triplet	Open reading frame	References
Retrotransposons							
TED (AcMNPV)	*Trichoplusia ni*	7510	AATG	LTR[a]	–	+	Miller and Miller 1982; Friesen and Nissen 1990
Transposons							
IFP2 (GmMNPV)	*T. ni*	2745	TTAA	13 + 17	CCC/GGG	+	Cary et al. 1989
TFP3 (AcMNPV)	*T. ni*	782	TTAA	15	CCC/GGG	–	Wang et al. 1989
TFP3 (GmMNPV)	*T. ni*	830	TTAA	15	CCC/GGG	–	Wang et al. 1989
IFP2.2 (AcMNPV)	*Spodoptera frugiperda*	2164	GTTTTTC	–	–	?	Beames and Summers 1990
IFP1.6 (AcMNPV)	*S. frugiperda*	1565	TTAA	14	CCT/AGG	?	Beames and Summers 1990
E (AcMNPV)	*S. frugiperda*	630	TTAA	14	CCT/AGG	–	Schetter et al. 1990
M5 (AcMNPV)	*S. frugiperda*	290	TTAA	13	CCG/CGG	–	Carstens 1987

[a]Retrotransposon with long terminal repeat (LTR)

internal open reading frames (ORFs) similar to the *gag*, *pol* and *env* genes of retroviruses. The activity of these genes in AcMNPV FP-D infected cell culture was confirmed by transcription and expression studies. It appeared that this element is active after integration into the AcMNPV genome since retrovirus-like empty particles were observed (Friesen and Nissen 1990; Lerch and Friesen 1992; Friesen 1993).

In numerous other FP mutants of AcMNPV/MNPV and *Galleria melonella* MNPV transposable elements which belong to the class of (DNA)-transposons were found (Table 1). These transposons were derived from *T. ni* or *S. frugiperda* host cells. Although their sizes vary between 290–2475 bp, they share some common structural characteristics, e.g. short inverted terminal repeat sequences of 13–15 bp and a duplicated 4-bp target sequence TTAA, suggesting that they all belong to the same lepidopteran transposon family (for reviews, see Fraser 1986; Friesen 1993). Nearly all of these transposons were found to be inserted into the locus (map unit 36.0–37.0) of a 25-kDa protein gene which is assumed to be involved in the switch from the production of extracellular viruses to polyhedra derived viruses during the replication cycle (Jarvis et al. 1992). The block of polyhedra formation by disrupting the gene for the 25-kDa protein favored the production of extracellular viruses. Since extracellular viruses are about three orders of magnitude more infectious for cultured cells than polyhedra derived viruses, it is likely that transposon-carrying FP mutants have a selective advantage over wild type viruses during the specific conditions of serial passaging in cultured cells. With the exception of transposon IFP2, none of the transposons contain ORFs which could encode a transposase gene (Fraser et al. 1995). Thus, NPV transposons identified so far represent probably defective transposons that are not able to transpose autonomously.

Transposons in *Granuloviruses*

In granulosis viruses, transposable elements have been found in *Cydia pomonella* granulovirus (CpGV) mutants which were isolated from a mixed infection of *Cryptophlebia leucotreta* larvae with both CpGV and *C. leucotreta* granulovirus (ClGV) and by subsequent *in vivo* cloning of the virus progeny in *C. leucotreta* and *C. pomonella* (Jehle et al. 1995; Jehle 1994). Out of 194 *in vivo* cloned CpGV and ClGV isolates two CpGV mutants were identified which carried transposon-like insertions. Both transposable elements, designated TCl4.7 and TCp3.2, were isolated and completely sequenced. Their structural characteristics clearly differ from those found in NPVs and resemble the Tc1-like elements found in other invertebrates (Table 2). Transposon TCl4.7 is 4726 bp long and has inverted terminal repeats of 29 bp. It hybridized to middle repetitive sequences of *C. leucotreta* DNA suggesting that it is derived from *the C. leucotreta* genome. In contrast, TCp3.2 is assumed to originate from the genome of *C. pomonella*. It is

Table 2. Comparison of TCl4.7 and TCp3.2 with different Tcl/mariner-like transposons

Transposon (Virus)	Origin	Length (bp)	Target sequence	Inverted terminal repeat (bp)	Terminal triplet	Open reading frames	References
TCl4.7 (CpGV)	*Cryptophlebia leucotreta*	4726	TA	29	CAG/CTG	+	Jehle 1994; Jehle et al. 1995
TCp3.2 (CpGV)	*Cydia pomonella*	3239	TA	756	CAG/CTG	+	Jehle 1994
Tcl	*Caenorhabditis elegans*	1610	TA	54	CAG/CTG	+	Rosenzweig et al. 1983
Tc3	*C. elegans*	2335	TA	462	CAG/CTG	+	Collins et al. 1989
Tc4	*C. elegans*	1605	TNA	774	CTA/TAG	–	Yuan et al. 1991
Tc6	*C. elegans*	1603	TA	765	CAG/CTG	–	Dreyfus and Emmons 1991
uhu	*D. heteroneura*	1646	TATA	46	CAG/CTG	+	Brezinsky et al. 1990
minos	*Drosophila hydei*	1775	?	255	ACG/CGT		Franz and Savakis 1991
HB1	*D. melanogaster*	1652	TA	31	CAG/CTG	+	Brierley and Potter 1985
mariner	*D. mauritiana*	1286	TA	28	CCA/TGA	+	Hartl 1989
MLE	*Hyalophora cecropia*	1255	?	38	?	+	Lidholm et al. 1991

3239 bp long and has extended inverted terminal repeats of 756 bp (Jehle 1994). Transposons TCl4.7 and TCp3.2 were inserted at a TA dinucleotide into non-translated regions at 9.5 and 28.3 map unit of the CpGV genome, respectively (Figs. 1, 2).

Both transposons carry at least two ORFs, which potentially represent two exons of a spliced gene. The amino acid sequences of these ORFs show significant similarity to transposases of Tc1- and mariner-like transposons found in *Drosophila* and *Caenorhabditis* species (Jehle 1994). The most homologous amino acid domains encompass highly conserved aspartic (D) and glutamic acid (E) residues forming a "D35E" motif. This motif was shown to be present in transposases of a many eucaryotic and procaryotic transposons, which were recently combined into a IS630-Tc1 superfamily (Doak et al. 1994). Many members of this transposon superfamily insert at a TA dinucleotide, similar to TCl4.7 and TCp3.2. It is not known whether the putative transposases of TCl4.7 and TCp3.2 are transcribed and expressed in the CpGV mutants. Molecular analyses have been initiated to study this aspect in more detail.

The presence of TCl4.7 and TCp3.2 transposon in CpGV mutants adds new significance to the phenomenon of host DNA transposition into baculoviruses. First, this observation demonstrated that within the phylum Arthropoda Tc1-like elements are not restricted to the family Diptera, but are also present in Lepi-

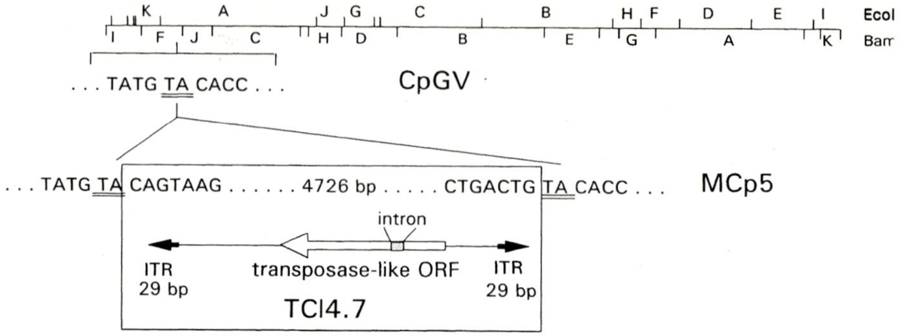

Fig. 1. Structural organization of the transposable element TCl4.7 found in the mutant in *Cydia pomonella* granulovirus (CpGV)-MCp5 (Jehle et al. 1995). The sequences of the integration site (dinucleotide TA) in the genome of CpGV and the border sequences of TCl4.7 in MCp5 are given. Inverted terminal repeats (*ITR*) are indicated by *black arrows*, the transposase-like open reading frame (*ORF*) by a *white arrow*

doptera. Second, the presence of these elements is an unequivocal *in vivo* conformation of the host transposition previously observed only with NPVs in cell culture system.

Host Transposons: A Source of Baculovirus Genetic Diversity?

By the insertion of host transposable elements the baculovirus genome is able to acquire new genetic information. This process can result in an insertion of new functional genes or at least of regulatory-like sequence affecting the transcription and expression patterns of a given virus. Both happened e.g. in the case of the retro-element TED in AcMNPV (Friesen et al. 1986; Friesen and Nissen 1990; Lerch and Friesen 1992). Transposon insertion may also cause disruption of viral genes rendering these non functional, as seen for the 25-kDa or 94-kDa proteins, and may result in a change of viral gene function. Finally, transposon activity in the virus genome can possibly lead to extensive genome rearrangements, similar to those known from many procaryotic and eucaryotic transposons. A joint cooperation of two neighboring transposons could result in a deletion or translocation of a block of genes located in between (Carstens 1987). It seems therefore conceivable that together with the transposon host-specific genes may also be transferred from the host into the baculovirus genome, or between different baculoviruses. Although it can be expected in most cases that these mutations will reduce the competitiveness of an altered virus as compared to the wild-type virus, the isolation of the FP mutants and the transposon carrying CpGV mutants have demonstrated that these viruses are viable and increase the genetic heterogeneity and diversity of baculoviruses.

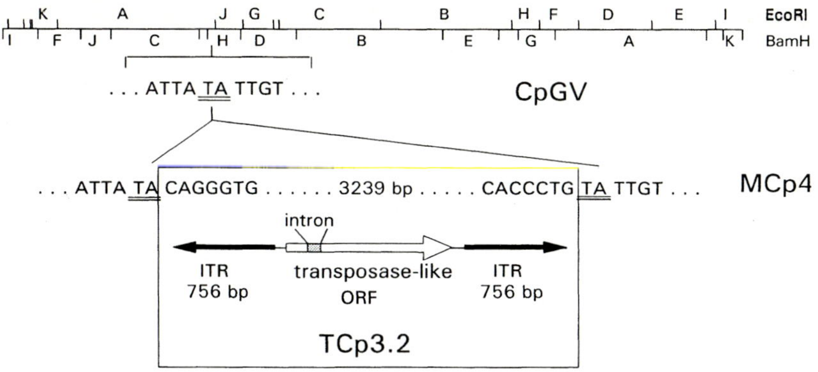

Fig. 2. Structural organization of the transposable element TCp3.2 found in the mutant in *Cydia pomonella* granulovirus (CpGV)-MCp4 (Jehle 1994). *ITR,* inverted terminal repeat; *ORF,* open reading frame. For details, see legend to Fig. 1

Baculoviruses: A Possible Vector for Insect Transposons?

The distribution of many transposon families in distantly related taxa strongly supports the concept that transposable elements are not only vertically transmitted from generation to generation. They are also able to cross species barriers and invade new gene pools (Hurst et al. 1992; Robertson 1993). The mechanisms of this horizontal transmission of transposons are not yet understood, but shared pathogens or parasites were supposed to act as possible vectors (Houck et al. 1991; Hurst et al. 1992). The potential role of baculoviruses as a vector of transposons among insect species was already suggested when the transposon TED was discovered (Miller and Miller 1982), but was never experimentally validated.

In addition to the acquisition of a transposon by a baculovirus genome, three further prerequisites need to be met for a successful transmission of a transposon by baculoviruses into other organisms: (1) the maintenance of transposon mobility after integration into the baculovirus genome, (2) an abortive infection of a non-susceptible host species during which the transposons can jump back into an insect genome, and (3) the transposon insertion into the germ line of the new host to establish a hereditable position within the new gene pool.

The transcriptional activity of the retroelement TED in AcMNPV mutants is circumstantial evidence for the mobility of transposons after incorporation into the baculovirus genome (Lerch and Friesen 1992). For TCp3.2 and TCl4.7, similar analyses are under way. Several studies have demonstrated that an infection may proceed without lethality, when less susceptible instars of a host or when non-susceptible host insect cells are infected (for review Gröner 1986; Jehle et al. 1993). For example, AcNPV transcribes and partially translates a number of early and late genes in non-permissive lepidopteran or dipteran cells (Carbonell et al.

1985; Carbonell and Miller 1987; Guzo et al. 1992). Consequently, it is conceivable that a transposon can be mobilized during an abortive infection and thereby invade a new host genome. The final step, the establishment of the transposon within the germ line of an insect without killing it, is pivotal but experimentally most difficult to prove. However, it can be envisaged that a physical interaction between baculovirus and reproductive organs occurs during latent or carry-over transovarial infection of the host insect. Smits (1987) demonstrated that, as a result of carry-over, about 15% of the newly hatched larvae from eggs derived from insects that survived a sublethal dose of baculovirus during their last instar stage, died from polyhedrosis. The presence of latent NPVs in host insects without apparent infection was demonstrated (Jukovicova 1979; Wood et al. 1986; Hughes et al. 1993). In these cases the latent viruses could be activated by superinfection with heterologous baculoviruses (Jukovicova 1979; Hughes et al. 1993). It is not clear whether baculovirus transmission is transovarial or intraovarial. The transposition into germ line cells may be a low-frequency event but, if it happens, it may entail far-reaching consequences in evolutionary terms. In conclusion, the data gathered so far support the view of a possible symbiotic interaction between baculoviruses and host transposons (Blissard and Rohrmann 1990; Friesen 1993). The Tc1-like transposons found in CpGV mutants are thus an important tool to substantiate this hypothesis.

Acknowledgements. Studies performed by J.A. Jehle and reported herein emerged from the project "Risiken bei der Freisetzung von Baculoviren: Abschätzung der Möglichkeiten einer unkontrollierten Ausbreitung gentechnischer Veränderungen durch Bestimmung der Rekombination zwischen Baculoviren mit heterologem Wirtsbereich", on which E. Fritsch, H. Backhaus und J. Huber participated. This project was financially supported by the German Ministry of Research and Technology (no. 03 19306A). J.A. Jehle is presently a fellow of the Human Capital and Mobility Programme of the European Union.

References

Beames B, Summers MD (1990) Sequence comparison of cellular and viral copies of host cell DNA insertions found in *Autographa californica* nuclear polyhedrosis virus. Virology 174, 354–363

Berg DE, Howe MM (eds) (1989) *Mobile DNA*. American Society of Microbiology, Washington, DC

Blissard GW Rohrmann GF (1990) Baculovirus diversity and molecular biology. Annu Rev Entomol 35, 127–155

Brezinsky L, Wang GVL, Humphreys T, Hunt J (1990) The transposable element Uhu from Hawaiian *Drosophila* – member of the widely dispersed class of Tc1-like transposons. Nucleic Acids Res 18, 2053–2059

Brierley HL, Potter SS (1985) Distinct characteristics of loop sequences of two *Drosophila* foldback elements. Nucleic Acids Res 13, 485–500

Carbonell LF, Klowden MJ, Miller LK (1985) Baculovirus-mediated expression of bacterial genes in dipteran and mammalian cells. J Virol 56, 153–160

Carbonell LF, Miller LK (1987) Baculovirus interaction with nontarget organisms: a virus-borne reporter gene is not expressed in two mammalian cell lines. Appl Environ Microbiol 53, 1412–1417

Carstens EB (1987) Identification and nucleotide sequence of the regions of *Autographa californica* nuclear polyhedrosis virus genome carrying insertion elements derived from *Spodoptera frugiperda*. Virology 161, 8–17

Cary LC, Goebel M, Corsaro BG, Wang H-G, Rosen E, Fraser MJ (1989) Transposon mutagenesis of baculoviruses: analysis of *Trichoplusia ni* transposon IFP2 insertion within the FP-locus of nuclear polyhedrosis virus. Virology 172, 156–169

Collins J, Forbes E, Anderson P (1989) The Tc3 family of transposable genetic elements in *Caenorhabditis elegans*. Genetics 120, 621–627

Doak TG, Doerder FP, Jahn CL, Herrick G (1994) A proposed superfamily of transposase genes: Transposon-like elements in ciliated protozoa and a common "D35E" motif. Proc Natl Acad Sci USA 91, 942–946

Dreyfus DH, Emmons SW (1991) A transposon-related palindromic repetitive sequence from *C. elegans*. Nucleic Acids Res 19, 1871–1877

Franz G, Savakis C (1991) Minos a new transposable element from *Drosophila hydei*, is a member of the Tc1-like family of transposons. Nucleic Acids Res 19, 6646

Fraser MJ, Hink WF (1982) The isolation and characterization of the MP and FP plaque variants of *Galleria mellonella* nuclear polyhedrosis virus. Virology 117, 366–378

Fraser MJ (1986) Transposon-mediated mutagenesis of baculoviruses: transposon shuttling and implications for speciation. Ann Entomol Soc Am 79, 773–783

Fraser MJ, Cary L, Boonvisudhi K, Wang H-GH (1995) Assay for movement of lepidopteran transposon IFP2 in insect cells using a baculovirus genome as a target DNA. Virology 211, 397-407

Friesen PD (1993) Invertebrate transposable elements in the baculovirus genome: characterization and significance. In: Parasites and Pathogens of Insects, Vol. 2: Parasites, pp 147–178. Academic, London

Friesen PD, Rice WC, Miller DW, Miller LK (1986) Bidirectional transcription from a solo long terminal repeat of the retrotransposon TED: symmetrical RNA start sites. Mol Cell Biol 6, 1599–1607

Friesen PD, Miller LK (1987) Divergent transcription of early 35- and 94-kilodalton protein genes encoded by the *Hin*dIII genome fragment of the baculovirus *Autographa californica* nuclear polyhedrosis virus. J Virol 61, 2264–2272

Friesen PD, Nissen MS (1990) Gene organization and transcription of TED, a lepidopteran retrotransposon integrated within the baculovirus genome. Mol Cell Biol 10, 3067–3077

Granados RR, Williams KA (1986) *In vivo* infection and replication of baculoviruses. In: The Biology of Baculoviruses, Vol. I, Biological Properties and Molecular Biology, pp 89–108. Granados RR and Federici BA (eds) CRC Press, Boca Raton

Gröner A (1986) Specificity and safety of baculoviruses. In: The Biology of Baculoviruses, Vol.I, Biological Properties and Molecular Biology, pp 177–202. Granados RR and Federici BA (eds) CRC Press, Boca Raton

Guzo D, Rathburn H, Guthrie K, Dougherty EM (1992) Viral and host cellular transcription in *Autographa californica* nuclear polyhedrosis virus-infected gypsy moth cell lines. J Virol 66, 2966–2972

Hartl DL (1989) Transposable element mariner in *Drosophila* species. In: Mobile DNA. DE Berg and MM Howe (eds.), pp 485–502. American Society of Microbiology, Washington DC

Houck MA, Clark JB, Peterson KR, Kidwell MG (1991) Possible horizontal transfer of *Drosophila* genes by the mite *Proctolaelaps regalis*. Science 253, 1125–1129

Hughes DS, Possee RD, King LA (1993) Activation and detection of a latent baculovirus resembling *Mamestra brassicae* nuclear polyhedrosis virus in *M. brassicae* insects. Virology 194, 608–615

Hurst GDD, Hurst LD, Majerus EN (1992) Selfish genes move sideways. Nature 356, 659–660

Jarvis DL, Bohlmeyer DA, Garcia Jr A (1992) Enhancement of polyhedrin nuclear localization during baculovirus infection. J Virol 66, 6903–6911

Jehle JA, Fritsch E, Huber J (1993) Aspects of the stability of the host-pathogen interaction of baculoviruses. In: Transgenic organisms, pp 175–191. K Wöhrmann and J Tomiuk (eds.) Birkhäuser, Basel

Jehle JA (1994) Sicherheitsaspekte der Gentechnologie: Verwandtschaft und Variabilität der Genome des *Cryptophlebia leucotreta* Granulosevirus und des *Cydia pomonella* Granulosevirus. PhD thesis, TU Braunschweig

Jehle JA, Fritsch E, Nickel A, Huber J, Backhaus H (1995) TCl4.7: a novel lepidopteran transposon found in *Cydia pomonella* granulosis virus. Virology 207, 369-379

Jukovicova M (1979) Activation of latent virus infections in larvae of *Adoxophyes orana* (Lepidoptera: Tortricidae) and *Barathra brassicae* (Lepidoptera: Noctuidae) by foreign polyhedra. J Invertebrate Pathol 34, 213–223

Lerch RA, Friesen PD (1992) The baculovirus-integrated retrotransposon TED encodes gag and pol proteins that assemble into viruslike particles with reverse transcriptase. J Virol 66, 1590–1601

Lewin B (1990) Genes IV. Oxford University Press, Oxford

Lidholm D-A, Gudmundsson GH, Bomann HG (1991) A highly repetitive, mariner-like element in the genome of *Hyalophora cecropia*. J Biol Chem 266, 11518–11521

Luckow VA, Summers MD (1988) Trends in the development of baculovirus expression vectors. Bio/Technology 6, 47–55

Luckow VA (1991) Cloning and expression of heterologous genes in insect cells with baculovirus vectors. In: Recombinant DNA Technology abd Applications, pp. 97–153. A. Prokop, R.K. Bajpai and C.S. Ho (eds.) McGraw-Hill, New York

Miller DW, Miller LK (1982) A virus mutant with an insertion of a copia-like transposable element. Nature 299, 562–564

Miller LK (1988) Baculoviruses as gene expression vectors. Annu Rev Microbiol 42, 177–199

Murphy FA, Fanquet CM, Bishop DHL, Ghalrial SA, Jarvis AW, Martelli GP, Mayo MA, Summers MD (eds) (1995) Virus Taxonomy. Classification and Nomenclature of Viruses. Springer, Wien

Robertson HM (1993) The mariner transposable element is widespread in insects. Nature 362, 241–245

Rohrmann GF (1986) Polyhedrin structure. Journal of General Virology 67, 1499–1513

Rosenzweig, B., Liao, L.W., Hirsh, D. (1983) Sequence of the *C. elegans* transposable element Tc1. Nucleic Acids Res 11, 4201–4209

Schetter C, Oellig C, Doerfler W (1990) An insertion of insect cell DNA in the 81-map-unit segment of *Autographa californica* nuclear polyhedrosis virus. J Virol 64, 1844–1850

Smith GE, Summers MD, Fraser MJ (1983) Production of the human beta interferon in insect cells infected with a baculovirus expression vector. Mol Cell Biol 3, 2156–2165

Smits PH (1987) Nuclear polyhedrosis virus as biological control agent of *Spodoptera exigua*. PhD thesis, Wageningen Agricultural University

Vlak JM, Schouten A Usmany M. Belsham GJ, Klinge-Roode EC, Maule AJ, Van Lent JWM, Zuidema D (1990) Expression of cauliflower mosaic virus gene I using a baculovirus vector based upon the baculovirus p10 gene and a novel selection system. Virology 179, 312–320

Vlak JM, Keus JAR (1990) Baculovirus expression vector system for production of viral vaccines. In: Advances in Biotechnological Processes, Vol. 14 Viral Vaccines, pp 97–127. A. Mizrahi (ed.) Liss, New York

Vlak JM (1993) Genetic engineering of baculoviruses for insect control. In: Molecular Approaches to Fundamental and Applied Entomology, pp 90–127. J Oakeshott and MJ Whitten (eds.) Springer-Verlag, New York

Wang HH, Fraser MJ, Csiszar Cary L (1989) Transposon mutagenesis of baculoviruses: analysis of TFP3 lepidopteran transposon insertions at the FP locus of nuclear polyhedrosis. Gene 81, 97–108

Wood HA, Burand JP, Hughes PR, Flore PH Gettig RR (1986) Transovarial transmission of *Lymantria dispar* nuclear polyhedrosis virus. In: Fundamental and Applied Aspects of Invertebrate Pathology, p405. RA Samson, JM Vlak and D Peters (eds) Wageningen

Yuan J, Finney M, Tsung N, Horvitz HR (1991) Tc4, a *Caenorhabditis elegans* transposable element with an unusual fold-back structure. Proc Natl Acad Sci USA 88, 3334–3338

4 From Gene Transfer to Risk Assessment: Experiences with Genetic Variability in Baculoviruses and New Approaches to Analyse the Reaction of Microbial Communities to Stress Factors

H. Backhaus[1], A. Nickel[1], E. Fritsch[2], J. Huber[2], K. Wendt-Potthoff[3], B. Engelen[1], A. Felske[1], U. Nübel[1], and J.A. Jehle[4]

[1]Institut für Biochemie und Pflanzenvirologie, Biologische Bundesanstalt für Land- und Forstwirtschaft, Messeweg 11/12, 38104 Braunschweig, Germany
[2]Institut für biologischen Pflanzenschutz, Heinrichstr. 243, 64287 Darmstadt, Germany
[3]GKSS Institut für Gewässerforschung, Am Biederitzer Busch 12, 39114 Magdeburg, Germany
[4]Department of Virology, Wageningen Agricultural University, Binnenhaven 11, P.O.B. 8045, 6709 PD Wageningen, The Netherlands

Abstract

Two different approaches to an experimentally based risk assessment of deliberate releases of genetically modified micro-organisms (GMO) are presented. First, we investigate the idea of biological containment as a tool for safety assurance in the particular case of genetically modified baculoviruses used for insect control. Molecular studies on the variability of two granulosis virus genomes demonstrate the processes of recombination, transposition and duplication as mechanisms of gene movement and variability. Thus, a modified gene in a baculovirus genome will be modified and exchanged with other viruses to some extent, the rate of which is difficult to determine with any precision. Information on potentials and directions of gene movement in the environment is obligatory for a sound risk assessment but any improvements on the estimation of its frequencies might only marginally contribute to the biosafety assessment of modified baculoviruses. An investigation of the potential effect of the new gene on their behaviour, competiveness and population dynamics is considered to be a more sensible and effective choice. The application of GMOs in the environment will potentially result in an establishment of modified genes from different sources in natural populations of microorganisms. Then, a refinement of methods to investigate the effects on natural communities resulting from this establishment, may be seen as a challenge for a sensible approach to these biotechnology applications. This is the rationale of our second approach, to

E.R. Schmidt/T. Hankeln (Eds.)
Transgenic Organisms and Biosafety

evaluate and develop new methods for monitoring of the structure and functionality of microbial populations. Their structure may be analysed by determining the abundance of sequences coding for the small subunit ribosomal RNA of microbial species. We obtain fingerprints, representing the heterogeneity of the community, by separating PCR products of different sequences from these target molecules in a temperature gradient gel electrophoresis (TGGE). Features of the method are described together with some data indicating our goal, to use this kind of analysis to monitor changes in the composition of communities due to any kind of stress factor, including the introduction of GMOs. The use of the BIOLOG system for investigating the oxidative catabolism of a set of carbon sources by microbial communities is presented as a tool for differentiating their functionality. Data from single strain cultures and some communities demonstrate basic features of the data obtained and some determinants of their variation and inferences on the analysis by multivariate statistics are mentioned.

Introduction

With respect to the general subject of BIOSAFETY, it is the kind and structure of the relation of natural processes like fate, movement, replication and expression of newly introduced genes to any adverse effects on organisms and the environment as legally protected goods, that has to be investigated in the course of any consequence assessment. The problem might be put into the following question: Which are the characteristics of organisms and genes that have to be predicted and controlled for the sake of an adequate risk assessment or safety assurance in particular cases of introduction of GMOs into the environment?

Questions of safety assurance generally have to be answered on a case by case basis, but some generalizations are possible from our viewpoint. We put some of these into the following theses, presented also with the intention of indicating some inherent bracketing of the heterogeneous subjects of this contribution.

1. Once released from inside the laboratory, the safety of GMOs cannot be established by demonstrating the efficiency of natural or artificial biological barriers supposed to prevent their functional and genetic interaction with the natural environment. – For replicating entities any low chance of "escape" might be considered too high and unacceptable, if the consequences are unknown (and assumed as being potentially disastrous).
2. The establishment of GMOs in the environment will have some effect on other organisms and the development and function of managed or natural ecosystems. – Risk assessment of gene technology applications may be seen as a challenge to refine the methods of analysis of the structure and function of natural communities for an improvement in our ability to classify any interferences as being beneficial or detrimental.

3. Our predictive capabilities will remain limited as will our potential for performing adequate assessments. Science can professionally contribute to the value loaded process of risk balancing only by providing testable criteria.

Leaving aside any comments on the last statement as a kind of ethical position, we substantiate the other points in the following contribution together with some conclusions derived from them. It will comprise two aspects of the risk assessment of the deliberate release of genetically engineered micro-organisms: First, the genetic variability of two granulosis viruses is investigated in a molecular approach, in order to evaluate the predictability of the fate of new genes as a parameter related to risk assessment. Second, we describe our methods used for a determination of the structure and functionality of microbial communities as a tool to investigate their reaction to stress factors, potentially useful for an assessment of any effects resulting from the deliberate release of GMOs.

Results

Genetic Variability and Biological Safety of Modified Baculoviruses for Insect Control

Together with biodegradation of environmental pollutants, growth promotion of agricultural crops and new vaccines, the improvement of the applicability of biocontrol agents of insect pests is seen to be one of the most promising fields for the use of genetically modified micro-organisms in the environment. Biological control of insect pests looks back on a long history and different strategies have been developed to control insect pests by pathogens, parasites and predators. Biological control tries to combat pest species, which compete with humans for the use of agricultural production, by introducing or stimulating their natural enemies. This endeavor can be subdivided into four broad categories (Table 1).

Classical biological control has frequently used the introduction of new enemies of insect pests to field or greenhouse cultures. The safety record of this type of handling so-called "exotic" species seems to be superior to other types of introductions of non-indigenous organisms, which were less well planned in most cases (Ehler 1991). Manipulation of the environment means to provide niches for the persistence and propagation of the control species, for example by providing adequate life conditions, or in the case of parasites, to provide refuges for their normal or alternate hosts. Inoculative augmentation means to enhance the infection pressure during times of excessive welfare of the target insect pest, when it is flourishing on the rich table provided by our agricultural, horticultural, and forest practices. By inundative augmentation as many biological control organisms are released as are necessary to immediately infect the target

Table 1. Types of biological control of insect pests by pathogens (parasites, predators)

Control type	Comments
Introduction and establishment of new pathogens	–
Environmental manipulation	Conserve or enhance the pathogen
Inoculative augmentation	"Booster shot" of indigenous pathogens
Inundative augmentation	Comparable to chemical control No replication of pathogen required to enhance infective dose for target pest

pest and to keep it under a damage threshold. This strategy has a rapid effect and does not depend on replication or establishing the control agent, as inoculative augmentation does.

Baculoviruses as Biological Control Agents

Baculoviruses can be used for both, for inoculative and for inundative augmentation. This virus group is specific for arthropods (mainly insects) and to date more than 600 isolates originating from different hosts have been described. A brief introduction to the taxonomy, morphology and biology of baculoviruses is given in the contribution of J.A. Jehle and J.M. Vlak. In the following we will focus on granulosis viruses (GVs), which are less well studied than nuclear polyhedrosis viruses (NPVs), which both represent the main groups of baculoviruses. GVs are characterized by a huge protein matrix of granulin protein (homologous to the polyhedrin of nuclear polyhedrosis viruses) forming an occlusion body of about 200×400 nm, which contains a single virion. The occlusion body protects the virion from environmental decay (e.g. from exposure to UV light) and allows the virions to stay infective for a certain time outside the host. It is easily visible by light microscopy, facilitating the detection or titer determination by the experimentalist. The coding sequence of granulin conveniently serves as a molecular taxonomic marker (Jehle and Backhaus 1994b).

The host infection and virus replication of GVs are similar to those of NPVs. In general, however, GVs infect only species of the Ordo Lepidoptera. They also show a higher tissue specificity than NPVs, and they replicate in the nucleus as well as in some cytoplasmic compartments whereas replication of NPVs is restricted to the nucleus (Crook 1991). The pronounced host specificities of baculoviruses mainly reside in the control of gene transcription and expression and DNA replication, whereas the first steps of infection – as virus uncoating or transport of viral DNA to the host nucleus – are not host- specific and even partly occur in vertebrate cells.

Since infected larvae release a huge amount of viruses, an initial virus application can result in a heavy epizootic resulting in a complete breakdown of a pest population. The high virus multiplication in a host larvae is widely used for an easy in vivo production of baculovirus insecticides, simply by rearing and infecting host larvae, whose cadavers are then homogenized and sprayed. However, there are some pitfalls in the use of baculovirus insecticides which confine their competitiveness to chemical insecticides. Besides their higher production costs there are also some specific drawbacks originating from their character as biologicals:

1. Only insect larvae can get infected by baculoviruses, eggs and adults are not susceptible to infection. The limited persistence of the virus in the environment then requires a careful timing of virus application and usually a multiple repetition is necessary.
2. After virus ingestion it takes about 7–14 days until larvae stop feeding and die from infection. During this delay damage to the crop plants continues.
3. The narrow host specificity, which is one of the most prominent biological features of baculovirus insecticides and a reason for their safety to non-target organisms, is of economic disadvantage in all those crop cultures where more than one predominant pest species has to be controlled.

Altogether, baculovirus insecticides are endowed with a certain duality of ecological benefits and application constraints. The latter are the main reasons stimulating efforts to improve some properties of baculovirus insecticides by using gene technology. With the establishment of cell culture techniques and the development of the baculovirus expression systems, it became possible to introduce (through homologous recombination with vector constructs containing the foreign sequence embedded within a segment of the baculovirus genome) and to express foreign genes in the baculovirus genome. The field release of an *Autographa californica* nuclear polyhedrosis virus (AcNPV) containing an 80 bp marker sequence was the first "deliberate release" experiment of a micro-organism in Europe (Bishop et al. 1988). In recent years most efforts were concentrated on accelerating the speed of kill of insect larvae by the introduction of insect-effective genes into the baculovirus genome. One of the most promising approaches so far seems to be the insertion of an insect specific toxin gene from the North African scorpion *Androctonus australis* (Cory et al. 1994). One major concern of the deliberate release of genetically engineered baculoviruses is the possibility that recombination with another naturally occurring baculovirus could result in a host-range expansion of the genetically engineered virus or in a transfer of the introduced gene to another baculovirus genome (Fuxa 1990, Jehle et al. 1993).

For that reason, risk assessment of baculoviruses demands a well-grounded knowledge of the genetic stability and variability as well as on the possibility of recombination of these viruses. In a model study on genetic recombination between baculoviruses with differing host range, we performed mixed infection experiments in larvae of the false codling moth, *Cryptophlebia leucotreta*, using

Table 2. Infectivities of *Cydia pomonella* granulosis virus (CpGV) and *Cryptophlebia leucotreta* granulosis virus (ClGV, two strains) for different host insect larvae

Host	LC_{50} (L4)		
	CpGV	ClGV–CV3	ClGV–CV4
Cryptophlebialeucotreta	75 000	15	32
Cydia pomonella	47	> 300 000	–

Parameters for infectivity comparisons: stage of infected instar larvae (L4); time of evaluation (14 days); percentage of killed larvae; slope of dose-response curve (virus dose versus percentage of killed larvae).

two granulosis viruses, the *Cryptophlebia leucotreta* granulosis virus (ClGV) and *Cydia pomonella* granulosis virus (CpGV) (Jehle et al. 1992b; Jehle et al. 1994c).

C. leucotreta, which causes severe damage in plantings of citrus fruits or avocado, is susceptible for both, ClGV and CpGV. The susceptibility for CpGV, however, is significantly lower (three orders of magnitude) than for its homologous virus, the ClGV (Table 2). CpGV is a registered control agent in Germany and a few other European countries for larvae of the codling moth *C. pomonella.* In the mixed infection experiments with *C. leucotreta* larvae both viruses were used in doses approximately corresponding to their lethal doses. The infection was simply performed by adding the viruses to the artificial diet of the larvae. The characterization of the progeny included restriction analysis to determine the relative amount of virus types. Virus clones were obtained by an in vivo cloning procedure, infecting larvae of *C. leucotreta* or *C. pomonella* with a very

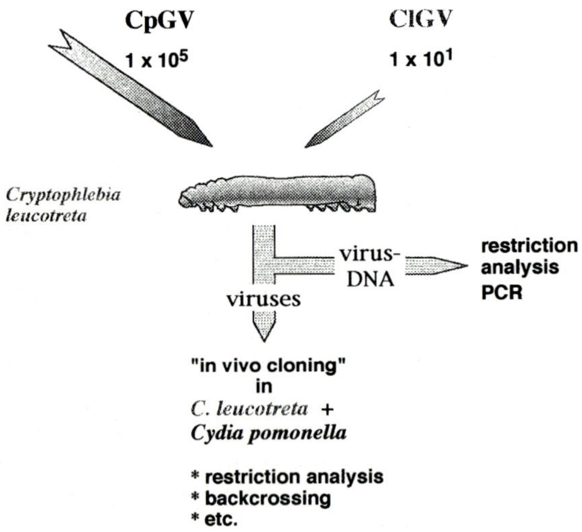

Fig. 1. Mixed infection experiments and their analysis. *PCR*, polymerase chain reaction

low virus dose. The low number of killed larvae then results from replication of a single virus genotype per infected larvae. In so far, the in vivo cloning can be compared to an endpoint dilution done in living host larvae (Fig. 1).

Using the in vivo cloning procedure more than 300 ClGV and CpGV genotypes were isolated and characterized by DNA restriction analysis. The results of these experiments were: (a) most of the isolated genotypes of ClGV and CpGV were similar to the original virus inoculum used in the mixed infection experiment, (b) none of the isolated genotypes was a recombinant between ClGV and CpGV, and (c) among the progeny viruses, there was one ClGV genotype and three CpGV genotypes carrying insertions of 0.9–4.7 kb.

Recombination Studies

A possible reason for our failure to isolate recombinants between ClGV and CpGV became evident when the DNA homology between both viruses was examined. Using high stringent and low stringent Southern hybridization, only 2–4 regions of the complete genomes were characterized to comprise highly conserved sequences. Among these are the granulin gene and the DNA binding protein gene region, which were identified in more detail (Jehle et al. 1992a; Jehle and Backhaus 1994a,b). The latter showed a well-preserved similarity in sequence and gene arrangement to the respective region of the more distantly related *Autographa californica* nuclear polyhedrosis virus (AcNPV) (Jehle and Backhaus 1994a). Thus, the limited degree of sequence homology between the two viruses tested in our experiments could have been one reason for a low rate of recombination and the failure to isolate recombinants. The specific conditions of the characterization of the virus progeny from crossings, potentially selecting against a putative recombinant with a lower fitness against the wild type inoculum by the in vivo cloning, might have been a second reason.

In order to become independent of the biological activity of potential recombinants and to enhance the sensitivity of the detection of recombination a PCR approach was chosen, which allowed the analysis of recombination events on the molecular level. The principle of these PCR experiments is shown in Fig. 2. Using two primers (one specific for ClGV, the other one specific for CpGV), complementary to sequences on both sides of the granulin gene, only those DNA sequences are amplified which contain a new linkage of sequences by recombination within the granulin gene. An artificial recombinant plasmid containing the 5' end of the ClGV granulin gene and the 3' end of the CpGV granulin gene was used as a positive control to optimize the PCR reaction.

Using this PCR strategy, viral DNA obtained from the progeny of different virus crosses and used as a template resulted in the generation of signals with differing intensities. Simple mixtures of ClGV and CpGV DNA in a test tube containing a higher template concentration (4x) served as a negative control and did not generate a PCR signal. The observed intensities of PCR signals could not

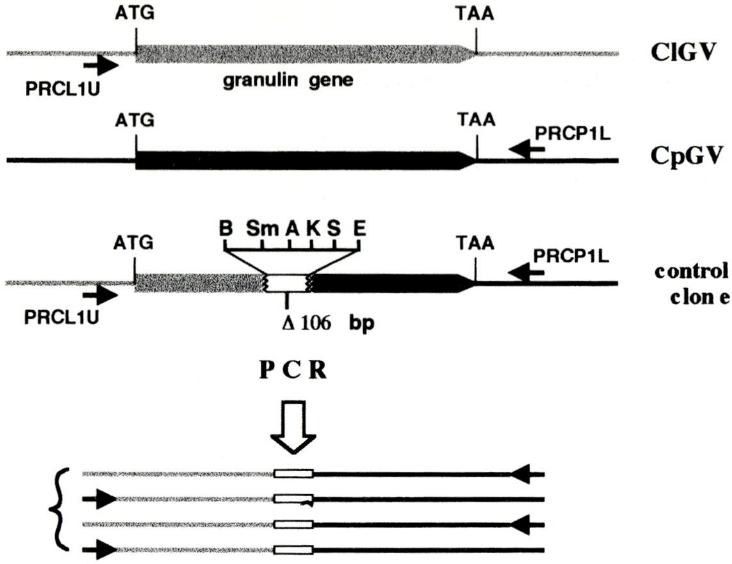

Fig. 2. Principle of polymerase chain reaction (*PCR*) analysis to demonstrate recombinant virus DNA in the progeny of mixed infection experiments

be correlated to the respective experimental conditions of the different crosses (infecting dose, relative amounts of virus types). In addition, the sequencing of one cloned PCR product demonstrated the presumed recombinant structure.

We interpret our results as being suggestive of the occurrence of recombination events within regions of sequence homology, even though the type of PCR analysis performed lacks the complete stringency of a proof. A quantitative interpretation of our data would indicate that recombination events may occur once every 1000 molecules under similar infection conditions. Thus, we conclude from these experiments that a foreign gene which is inserted into a region of shared homology between related viruses will have a significant likelihood of being exchanged by recombination, when both viruses coinfect the same host.

Among the progeny viruses some variant ClGV and CpGV genotypes with differing restriction patterns were isolated. These variants were analyzed in more detail and seen to carry insertions which showed structural features of a highly repetitive DNA and of transposons, respectively.

Repetitive Sequences

In the genome of ClGV a highly variable region was identified within the restriction fragment *Eco*RI-H. As indicated by a ladder of Southern hybridization signals, this genome region shows a regular variation in length generated by a

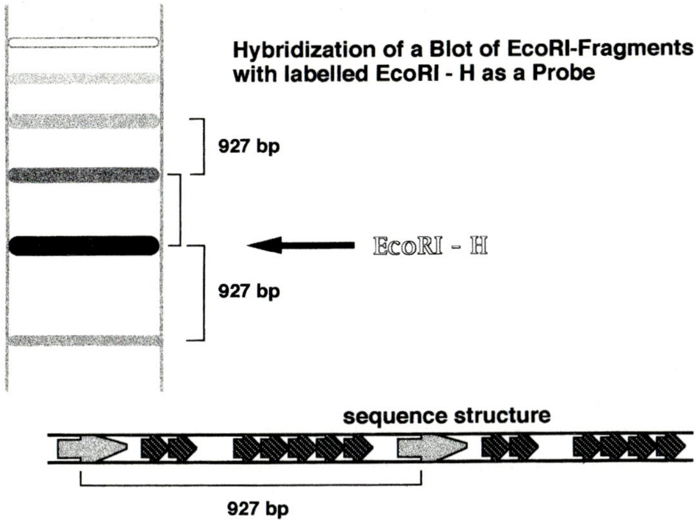

Fig. 3. Detection of the variability of a repetitive region in the genome of some ClGV clones. The repetitive sequence structure is indicated

concatemeric multiplication of a fragment of about 927 bp (Fig. 3). Ladders with up to more than seven steps were observed in the DNA from different virus progenies suggesting their formation during replication in the host. In contrast, some ClGV genotypes never showed any heterogeneity within this region.

By sequencing a fragment containing two repetitive 927 bp units, a substructure consisting of smaller repetitive units of 27 bp was identified. Except for 4 bp both 927 bp units were completely identical. A head-to-tail duplication or multiplication can be supposed as the possible mechanism of their formation. These repetitive structures resemble repetitious sequences found in other viral and eucaryotic genomes but they do not share any homology to hr regions or the repeat structures found in defective interfering particles of other NPVs (e.g. AcNPV) (Strauss et al. 1991). Of two clones not showing any variability of this region, the first contained only one of the 927 bp unit, the other demonstrated some rearrangements of the sequence, lacking the small repeats altogether. Further investigations were initiated to analyze whether sequences are genuine for the ClGV genome or whether preliminary indications of their origin from the host genome are correct. With its peculiar repeat structure, the heterologous variability of sequence variants in different clones and a potential relation to host DNA sequences the detection and further analysis of this region is prone to extend our knowledge on the role of repetitive sequences in the variability of baculovirus genomes.

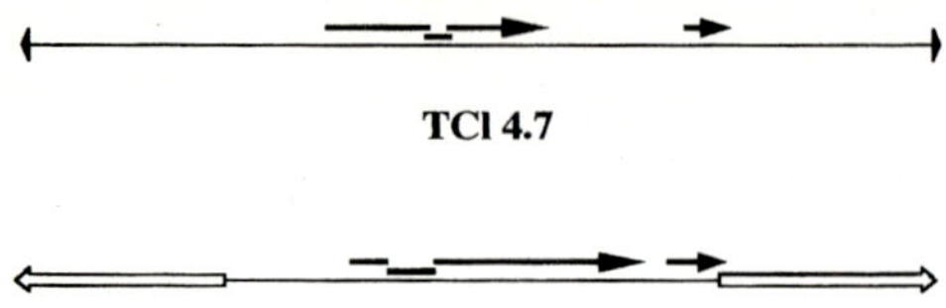

TCl 4.7

TCp 3.2

Fig. 4. Structure of TCl4.7 and TCp3.2 transposons. End terminal inverted repeats are indicated by *arrows*, open reading frames by *thick lines* with the transcription direction indicated

Host Transposable Elements

A further result of our investigations was the unambiguous demonstration of the horizontal transmission of host transposons into the baculovirus genome during the infection of the host larvae. Two different transposable elements were identified in the CpGV insertion mutants originating from the mixed infection experiments and in vivo cloning procedure (Fig. 4, Table 3). A detailed characterization of both of these transposable elements is presented in the contribution of J. A. Jehle and J. M. Vlak. As indicated by Southern hybridization the transposon TCl4.7 apparently originated from the genome of *C. leucotreta*. It is evident that it transposed into the CpGV genome during our mixed infection experiments in *C. leucotreta* larvae because CpGV will not normally encounter this host in the laboratory or in its natural environment (Jehle et al. 1994c, 1995). The second transposon (termed TCp3.2) was also found in a CpGV mutant, but it hybridized to genomic DNA of *Cydia pomonella* indicating that this one was most likely mobilized during the in vivo cloning procedure in the homologous host (Jehle et al. 1994c). Both transposons have ORFs whose translated amino acid sequence share significant homology to transposases of Tc1/mariner-like transposons from different *Caenorhabditis* and *Drosophila* species. Both elements were inserted into non-coding regions of the CpGV genome and comparative infection experiments gave no indications that the insertion of either of the transposons interferes significantly with the biological activity of the viruses.

Table 3. Summary of characteristic features of the transposons detected in the granulosis virus genomes

	TCl4.7	TCp3.2
Length	4726 bp	3238 bp
Site duplication	TG[TA]CA	TA
Inverted repeat	29 bp (4 mismatches)	756 bp (7 mismatches)
Host	Cryptophlebia leucotreta	Cydia pomonella
Copy number	$\gg 10$	4–5
Transposase coding region	343 aa	284 aa

aa, amino acid; bp, base pairs.

Our mixed infection experiments with ClGV and CpGV demonstrated a considerable variability and plasticity of baculovirus genomes comprising intergenomic recombination, intragenomic duplication of repetitious sequences and the horizontal transfer of transposable elements from the host to the virus genome. The detection of highly variable and repetitive sequence structures as well as the identification of host transposable elements in the genome of granulosis viruses may illustrate two different mechanisms of a natural variability of baculovirus genomes which both result in an increase of genome complexity and a possible acceleration of virus evolution (Strauss et al. 1991).

We summarize our interpretation of the results as follows.

1. The exchange of genetic information among baculoviruses as well as between host and baculoviruses interferes with the stability of a (modified) gene within a baculovirus genome. An expression of the stability or variability of a baculovirus genome depends on the details of the analysis and the time scale of observation. The isolation of two different transposon-carrying CpGV mutants out of about 300 analyzed virus genotypes suggests that under certain circumstances (e.g. our experimental conditions), the transposon-mediated mutagenesis of a baculovirus genome can reach surprisingly high frequencies.
2. Although extrapolations from our experiments to frequencies occuring in the environment are far from being straightforward, we consider that a foreign gene introduced into the baculovirus genome by genetic engineering will have a fair chance of being exchanged with different virus populations. Transposons might contribute to gene exchange in secondary processes by transposition functions or by providing sites for homologous recombination.
3. The conclusion that the probability of gene exchange by homologous recombination will depend on homologies and chances of mixed infections in the first place, seems to be a trivial one. The frequencies of primary events will also increase with the period of time which the new gene/modified baculovirus will survive in the environment, and the amount of viruses which will be released.

Some inferences from these results and their interpretation for the risk assessment of the deliberate release of genetically engineered baculoviruses or GMO in general are presented in the discussion.

We might take the potential establishment of a new gene in the environment as the endpoint of our considerations, assuming that this is defined as an undesired event, which may be true at first glance for the scorpion toxin gene in baculoviruses. Sometimes the considerations on risk assessment of GMO applications are not extended any further. – But with other introductions, an establishment of GMOs or their genes in the environment may be an accepted or even desired outcome, as e.g. with the establishment of new agricultural crops or growth promoting bacteria in agricultural settings or bacteria for bioremediation in other managed systems. Approaches to an effects assessment in these cases are presented below.

Monitoring Effects on the Structure and Function of Microbial Communities as a Tool in Risk Analysis

During experiments performed to monitor the fate of a genetically modified phytopathogenic bacterium and its genes in a simple model system, its potential to survive and be established was noted (Wendt-Potthoff et al. 1992, 1994; Wendt-Potthoff and Backhaus 1994). It was not only at this time that we realized that the crucial point in any prerelease model system research with microorganisms might be the capability of monitoring any effects on a microbial community resulting from the release of a GMO, and correlating this effect to "exposure" parameters like survival, replication and gene transfer.

Analysis of the Structure of Microbial Communities

A microbial community can be defined by its species composition and the functions relating its components and determining its interactions with the environment. A variety of conventional methods of determining species composition of microbial communities rely on cultivation methods by which only a small fraction of the population is tested. We become increasingly aware of the fact that frequently cultivation methods do not represent the spectrum of organisms active in the habitat of choice. Among modern methods trying to determine the components of communities, the direct sampling of DNA or RNA and cloning and sequencing of characteristic sequences is an established method. It is a labor intensive approach and has some other inherent pitfalls (Gutell et al. 1985, Pace et al. 1986).

We set out to modify such types of analysis with the goal of getting fingerprints characteristic for the genetic structure of communities, which could be used to monitor any stress factors on their diversity (Atlas 1984; Muycer 1993). Target of our approach is the most important molecule for bacterial taxonomists today, the small subunit 16S ribosomal RNA or the genes coding for it, respectively. The molecule may be folded into a secondary structure depending on the arrangement of inverted repeats within the sequence. Sequences which are well conserved between bacterial species are surrounded by less well conserved and hypervariable regions. First, primers for PCR amplifications were chosen within well conserved regions. Amplification of DNA from some population will yield fragments of more or less identical length, not separable by conventional electrophoresis techniques. By superimposing a temperature gradient, varying denaturing conditions are applied to these molecules in our system. At a certain position determined by the temperature, melting domains of the molecule induce a drastic retardation of mobility (Rosenbaum and Riesner 1987).

Time and degree of mobility shifts as seen in a gel with a temperature gradient perpendicular to the direction of electrophoresis depend on details of the sequence structure of the melting domain of the separated PCR product. Even

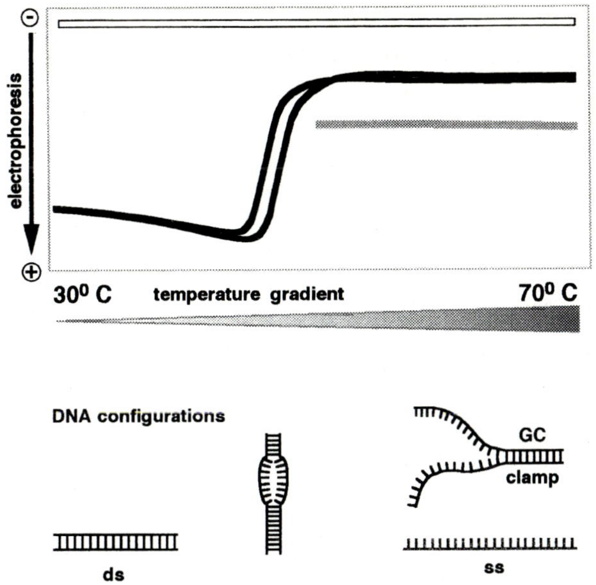

Fig. 5. Representation of the banding pattern of polymerase chain reaction (PCR) products from 16S RNA genes of two bacterial strains (*Pseudomonas paucimobilis* and *Comamonas testosteroni*) in a temperature gradient gel electrophoresis (TGGE) with the temperature gradient applied perpendicular to the direction of electrophoresis. The DNA configurations forming the observed bands are symbolized below

a single base pair difference in a melting domain is sufficient for a separation in such an analysis. In this gel the separation of PCR products from a mixture of two bacterial strains, *Pseudomonas paucimobilis* and *Comamonas testosteroni* is demonstrated (Fig. 5).

If the direction of electrophoresis is superimposed on the temperature gradient in the same direction, single bands from the molecules retarded by melting domains are observed by choosing the adequate conditions, which can be calculated from the results of a perpendicular melting analysis.

Figure 6 combines the analysis of some single strain PCR products with the patterns obtained from different community DNAs. One unequivocal result of our analysis so far is the observation that sometimes the ribosomal RNA genes of a single strain yield a pattern of heterogeneous bands (e.g. lanes 5, 6). It is evident that this finding contradicts the simplest assumptions we had made at the beginning. Obviously, the genetic heterogeneity resulting from this type of analysis is somewhat greater than the heterogeneity of the community at the species level. Bacteria may contain an heterogeneous set of ribosomal RNA genes. We are in the process of investigating this phenomenon in more detail, because referencess to it in the literature are only marginal (Carbon et al. 1979; Sogin 1990).

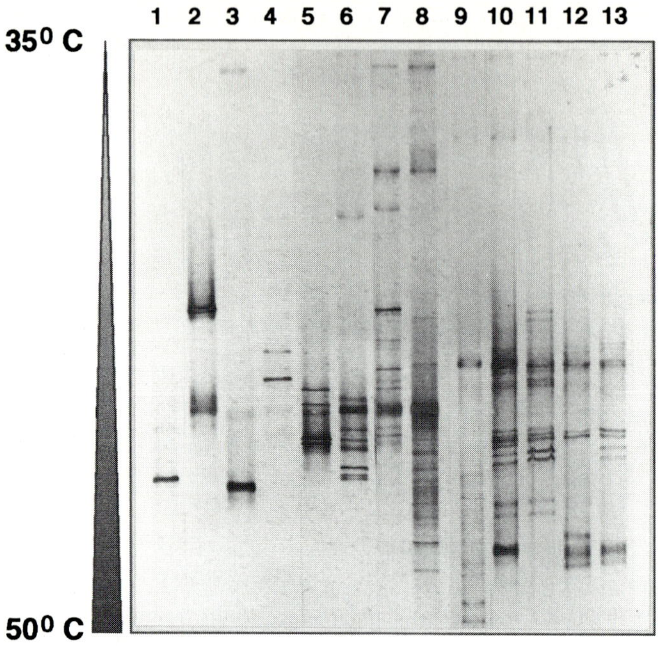

Fig. 6. Banding patterns of polymerase chain reaction (PCR) products from bacterial strains and microbial communities using a set of primers complementary to conserved regions of the 16S-RNA (968–984; 1330–1346, *Escherichia coli* position numbering). *Lane 1*, *Agrobacterium tumefaciens*; *2*, *Acinetobacter johnsonii*; *3*, *Pantoea agglomerans*, *4*, *Erwinia carotovora*; *5*, *Clavibacter michiganense*; *6*, *Bacillus polymyxa*; *7*, manure slurry; *8*, agricultural soil; *9*, agricultural soil BBAI, without cultivation; *10*, BBAI – bacterial community as cultivated on agar media; *11*, BBAII, parallel sample of the same soil, cultivated; *12*, result of cultivation from sample BBAI after 4 days; *13*, sample cultivated identically at the same time from sample BBAI. See the text for more details on the source and isolation of DNAs

Community DNAs were isolated from bacteria separated from their habitat matrix. Patterns derived from DNAs of bacterial communities from different sources like manure slurry and different agricultural soils are presented. Profiles in lanes 10–13 (Fig. 6) are derived from bacteria grown on agar plates after a cultivation of the community. In addition to their apparent distinction from DNA extractions of separated bacterial communities without precultivation, these lanes clearly hint at problems of reproducibility which are among a variety of technical questions which have to be solved in the course of our work.

Our approach has only very briefly been introduced here. Its principal goal is to get meaningful fingerprints of the structure of bacterial communities, in order to monitor any induced changes. In principal the methodology will enable any detailed analysis of missing or additional bands by their isolation and sequencing. Directions of our present work with this system can be summarized as follows:

1. The complexity of the fingerprints will be increased by using different sets of selected primers in order to get more detailed patterns of the members of complex communities in different lanes. Nevertheless we are aware of the fact that the type of analysis will not allow us to get a complete picture of complex communities. Similar to other approaches it opens some windows of observation with certain limits inherent in the method employed.
2. We complement our analysis of PCR products from DNA sampling by direct isolation of ribosomes, which seem to be a convenient readily extractable source for the respective target sequences. The quantitative response of this type of analysis will differ from using the ribosomal genes as target sequences, depending on the correlation of relative ribosome numbers to the respective bacterial activity in the community.
3. We will also extend our comparison of results between DNA extractions from bacterial fractions directly separated from the habitat matrix with DNA from cultivated communities as demonstrated in Fig. 6.

Analysis of the Functionality of Bacterial Communities

Conventionally microbial activity can be analyzed by determining the kinetics of elementary processes such as CO_2 production, nitrate reduction or nitrogen fixation. These parameters are also used in testing protocols, e.g. for pesticide registrations. Elementary processes certainly can be performed by communities of different composition and their functionality will be redundant with respect to the diversity of species involved. We looked for a simple and more detailed methodology and got a valuable hint from a publication in which an identification system for microbial species was used for the characterization of bacterial communities (Garland and Mills 1991). The system consists of microtitre plates with a set of 95 carbon sources. Their degradation by respiring bacteria is measured by the colour change of a tetrazolium dye. Colour change follows some kinetics, in a first approximation depending upon the number of cells in the well of the microtiter plate. To compensate for different inoculum densities, the colour development in each well may be normalized with respect to the total colour development of the plate. Using multivariate statistics, the comparison of the complex set of data from different communities then can be visualized by calculating for each plate one point in a two dimensional net of two principal components. Principal components mathematically combine the most significant differences within a set of compared data – here the set of carbon sources, the utilization of which differed most between the microbial communities tested. In the publication of Garland and Mills (1991), the data from three communities (soils, freshwater, hydroponic) formed well separable clusters in the plane spanned by two principal components.

We shall briefly describe a few investigations on the system coming then to its potential value for analysing effects of any kind of stress factor on microbial communities (Zak et al. 1994). For an analysis of the sensitivity of the system we

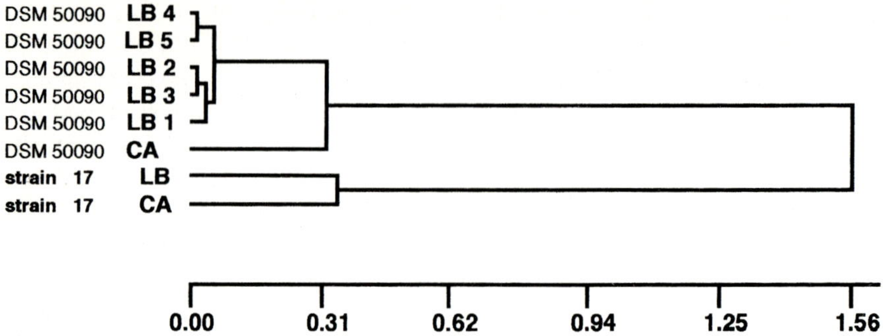

Fig. 7. Clustering of BIOLOG data from two bacterial strains of the same species (*Pseudomonas fluorescens*) precultivated in two different media. *LB*, Luria Broth; *CA*, Casamino Acids medium

again used pure cultures. The system is able to register shifts in the physiological state of cells as seen in a conventional cluster analysis (Fig. 7).

Bacterial cultures of one strain, grown on the same medium in parallel, cluster very close together, stressing a good reproducibility of the system. The same strain, precultivated on a different medium gave results more distantly related to the previous ones. An environmental isolate of the same species demonstrates an identical distance when precultured on the different media. Overall, this strain is only distantly related to the reference strain in this comparison.

Challenging the system by reducing the number of cells per well through serial dilutions showed almost identical results in the final colour distribution

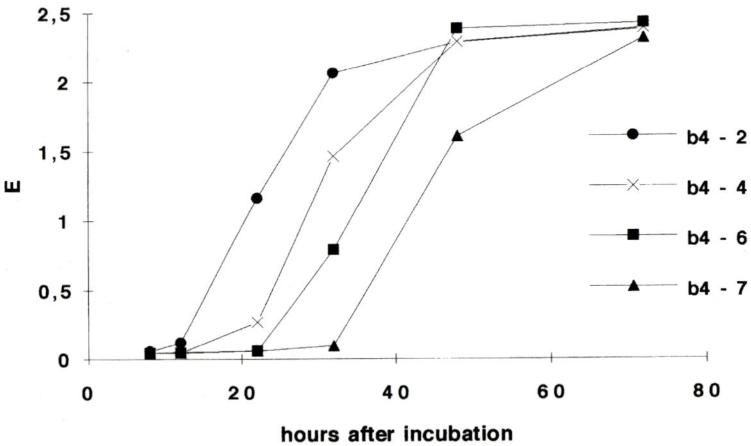

Fig. 8. Kinetics of colour development in one well of the BIOLOG plates (b4, D-galactose) with serial dilutions of a bacterial culture. The different dilutions are indicated. The highest dilution (10^{-7}) resulted in an inoculum of six colony-forming units (CFU) per well

within the wells down to a number of six colony-forming units (CFU) per well. However, the rate of colour development is significantly different, as shown in Fig. 8. This figure clearly hints to potential pitfalls in the analysis, if it is based on the evaluation at a predetermined timepoint. To include some simple representation of the kinetics of colour development in the statistical analysis might be adviseable for an improvement of its sensitivity. With an average cell number of 0.6 CFU/well the signal distribution gets variable due to the Poisson distribution of productively inoculated wells. This again is a factor which has to be taken into account in adjusting inoculum densities, especially for the analysis of communities.

Figure 9 shows a visual representation of the results with the system used for the analysis of soil communities. Data from a mixed randomized probe from a fallow field with sugarbeets grown in the previous year are compared to one from a field with sugarbeets as growing crops. Significant differences as well as si-

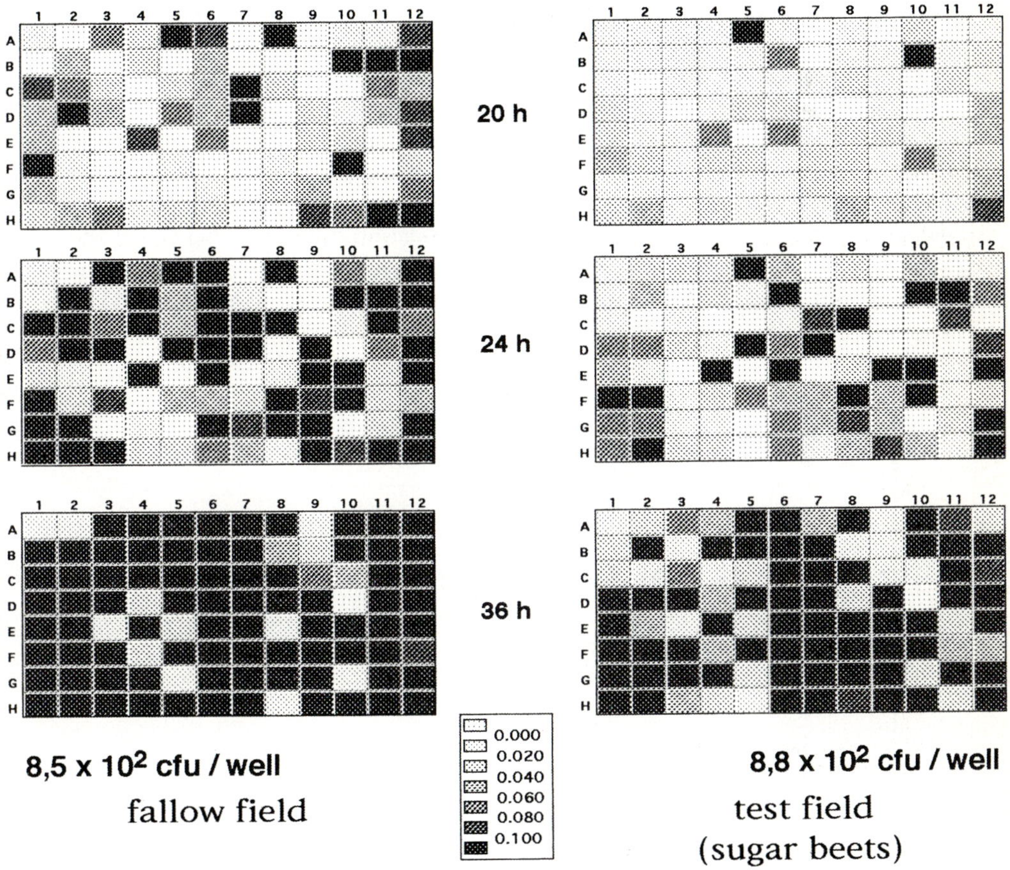

Fig. 9. Gray scale representation of the colour development of BIOLOG microtitre plates incubated with microbial communities from different agricultural fields

milarities in the patterns are easily identified by simple visual inspection. Different probes of either field demonstrated a good reproducibility of the results.

Figure 10 demonstrates a cluster analysis of bacterial communities from the model system work designed to monitor the fate of a genetically modified *Pseudomonas syringae* and briefly mentioned at the beginning of this chapter (Wendt-Potthoff and Backhaus 1994). In the analysis of this model system we tried to use the BIOLOG system for the first time for a functional characterization of bacterial communities and its potential shifts by experimentally introduced parameters. We were not able to distinguish the model system with and without the GMO or with respect to other experimental parameters. However, we got some indication of shifts in community functions due to the time course of the experiments (the age of the model system) and a rather clear differentiation of communities from leaves and the rhizosphere was represented in samples from this simple closed system, as demonstrated in the figure.

Figure 10 is also used to visualize our goal in using this type of analysis, as indicated. We want to elaborate the BIOLOG system (or some variation of it) with respect to its sensitivity and discriminatory power, in order to differentiate communities with the presence of a GMO as the only variant parameter. An increase in the sensitivity clearly is desirable in order to monitor more subtle variations than between the communities which can be separated so far (e.g. Figs. 9, 10). Also, the conditions for getting reproducible results have to be carefully tested. To include the dynamics of colour development in the data analysis and to get a precise mathematical formulation of the significance of the separation of different communities clearly will be a future task for more complex multivariate statistics. By improving the methodology in using this kind of

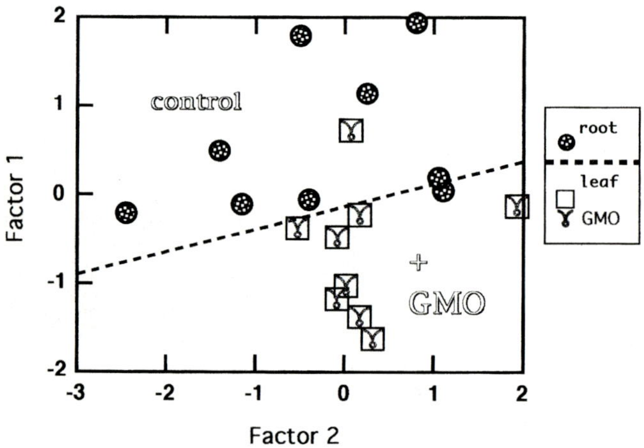

Fig. 10. Principal component analysis of bacterial communities from a simple model system with bean plants. Separation of root and leaf communities. Sample points from leaf communities are superimposed with a symbol for the presence of a genetically modified (micro-)organism (*GMO*) to indicate simply the ultimate goal of this type of analysis

"observation window" as indicated, we will potentially be able to register the application of different types of GMO as a kind of stress factor influencing the functionality of microbial communities. We are now investigating the BIOLOG-system with the application of a herbicide as a challenge for the functioning of soil and epiphytic bacterial communities, envisaging that such an effect might be more easily identified than more subtle changes potentially induced by the presence of established GMOs.

Discussion

Scientific risk assessment has to extrapolate present knowledge about causal relations of the application of a (potential) risk source to any undesired effects on organisms or the environment as legally protected goods. The generation of scientific data through experimentation can be used to expand the available knowledge base in order to improve the inference from the database to the consequences of particular applications. Figure 11 tries to indicate schematically the structure of some causal relations between characteristics of GMO or their modified genes and any consequences of their application in the environment, together with the potential for inferences from their knowledge or control to an assessment of undesired effects. We shall briefly try to exemplify the significance of our data and approaches, intended as a contribution to biosafety research, by referring to the scheme presented in Fig. 11.

Constructing a GMO from a recipient and a "new" gene, e.g. a modified baculovirus by inserting a scorpion toxin gene, consequences of its application for the biological control of insect pests have to be considered within two domains:

1. *The new phenotype of the GMO.* Parameters essential to assess the quality and safety of this organism with respect to its application may be tested by laboratory and model system experimentation, e.g.:
   a) Infectivity parameters, persistence
   b) Host specificity, target versus nontarget effects
   c) Mode of action of the new gene (product) in its new background From such data the potential of the GMO to establish itself in the environment, as well as any effects resulting from it, may be deduced.
2. *The motility and biological features of the new gene.* This is influenced by its new host and the ecological context into which it is introduced.

A relatively low initial (primary) frequency of gene exchange in the environment will generally not infer any negative effects on organisms or the environment directly. But in considering the properties of replicating entities, such a low rate may not be easily translated into a negligible rate of "exposure" on an extended time scale. Then, a risk analysis may and should be based on the assumption of some type and extent of gene exchange occuring within the new context of that gene. Data requirements for a risk assessment consequently would have to stress more explicitly the evaluation of the chance of establishment of the gene, as a prerequisite to exerting any unwanted effects in natural or managed systems. This means primarily to ask for the potential selective value of the gene in different contexts, which seems a difficult task at a first glance.

At least some type of inference is possible from the properties and phenotype of the GMO itself. For example, the introduction of the toxin gene into the baculovirus genome enhances its speed of action with respect to target organism killing, but it also results in a reduced virus multiplication and virus yield, indicative of an inferior effect on the potential of the virus to multiply and establish in nature (Cory et al. 1994). Such data are indicative of the inference that these viruses may be suitable for inundative biological control but not for inoculative control, requiring replication, effective transmission, and a limited potential of establishment in the environment (Table 1). Conclusions on risk or application related issues can be substantiated by analysing other parameters influencing population dynamics, like persistence, host specificity, fate of infected hosts etc. (Fuxa 1990; Hochberg and Waage 1991). – Inference from such data to the selective value of the gene in different virus contexts is inevitably incomplete and assumes that some generalizations are possible, despite the fact that the selective value in detail will depend on specific organism and environ-

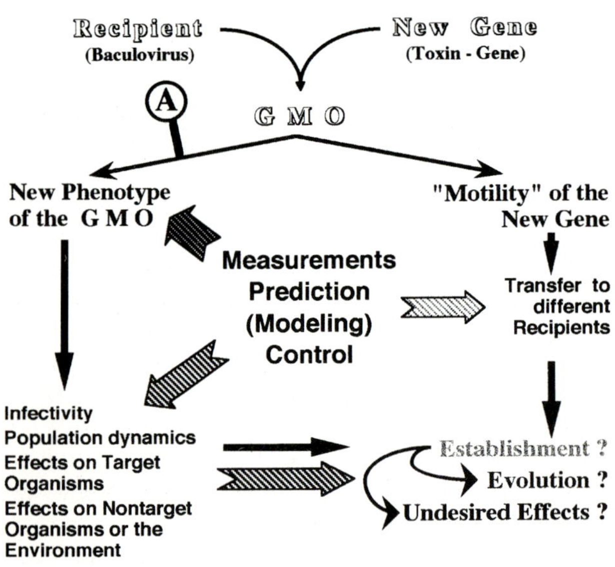

Fig. 11. Causal relations between genetically modified (micro-)organism (*GMO*) applications and effects on organisms or the environment. *A*, new gene experimentally introduced into alternate recipients. *Black arrows* indicate causal connections; *shaded arrows* indicate the potential for inference and significance of knowledge and monitoring results for risk analysis and assessment of GMOs (potentials symbolized by the intensities of the shading)

mental characters, where they are (naturally) tested. Thus, the quality of inference could also be improved by testing phenotypic effects of the genetically modified gene in different contexts, e.g. by its introduction into the genome of a different host (A in Fig. 11).

An inspection of our data on the variability of baculovirus genomes and reflections on their potential improvement and extension suggests that we cannot predict the chance of the fixation of genetic modifications in natural populations from laboratory data. First, the result of transfer events is more sensitive to any changes in selective values than to transfer frequency for most realistic parameter values (Levin and Stewart 1980; Levin 1986; Kim et al. 1991). Secondly, we may miss the correct extent of gene exchange by orders of magnitude. This is the case because our knowledge about mechanisms is incomplete, and especially the ignorance of the heterogeneity of the environment with respect to parameters influencing movement rates is tremendous. And it will remain so, despite all efforts to test lots of variables. (How to determine or estimate chances of mixed baculovirus infections in the environment?).

We will briefly summarize our conclusions following these arguments. It is well conceivable that their implications for a risk assessment and the relation to evolutionary concepts is valid for most types of environmental introductions of genetically modified organisms (Morozov et al. 1989; Greene and Allison 1994):

1. GMOs will be established in managed systems in our environment by their introduction on a commercial scale. Inevitably this will have some consequence for the structure and functioning of our environment. Direct effects on organisms or the environment resulting from the GMO phenotype can be tested in an experimental phase. The ability of the GMO to establish itself in natural systems without human intervention will be different from that of the host organism.
2. By testing the parameters of GMOs, contributing to competitiveness, some sound inferences on the effect of the new gene on its establishment potential and environmental consequences of this establishment are possible.
3. It is rational to expect the rate of transfer of engineered genes to be governed by their new host background and ecological context in the first place. Any "biological containment" of a gene in an organism will depend more on its contribution to the competitiveness and function of the new and other potential hosts than on barriers to its movement. Experience with the risk analysis of chemicals stresses the notion that we should not rely on any simple assumptions about the distribution of potential "harmful agents" in the ecosystem (Marco et al. 1987; Rowland and Molina 1994).
4. In so far as inferences from low initial gene transfer rates to a limitation of effects on natural systems are invalid for replicating and evolving entities on an extended time scale, an environmental assessment of a gene must include a prediction of the effects of the new gene on competitiveness of organisms and of its presence in a new ecological context on the viability and functioning of organisms and communities.

Our experimental approaches described in the second part of this contribution relate to the notion that an ability to monitor any effects of an establishment of new organisms or genes in our environment is imperative, if the biological properties do not preclude such an establishment or if it is a desired consequence of the application of a particular GMO. With the intention to elaborate sensitive systems for the structural and functional analysis of bacterial communities for a risk analysis of environmental application of GMOs, some questions may be raised which relate to their interpretation and to a sensible risk balancing. These questions certainly include the following: Is the observed effect significant (statistical separation of cases)? How do we have to evaluate the effect (positive – harmful; reversible – irreversible; immediate – long term)?

Risk balancing depends on our present definition of "protected goods" or may contribute to its improvement. It must be done by sensible comparisons to other variables, being natural and induced by human intervention. It will influence any decision about a consent (or denial) to the application of a GMO together with potential management strategies.

Risk assessment has to make predictions, conventionally, if it is taken as a tool to determine the frequency of occurence of well known dangers, extrapolating from their realizations in the past – much more, if it shall serve as an instrument of prophylaxis from risks hypothetical so far. The identification of any hazards potentially resulting from the introduction of the GMO itself or from consequences of a natural spread of its genetic modifications is the most challenging step in any risk assessment of environmental introductions of GMOs. As one means of orientation within a multitude of parameters, relations and processes, the intended goal of the application could be taken as one focal point for a structuring of the analysis (Sagoff 1992). Safety assurance finally has to include the statement of a lack of control over a wealth of natural processes.

We would like to stress that any efforts to get a predictability on processes like the establishment of genes in natural populations is ambitious in so far as we try to forecast evolutionary processes. Ecology and population genetics should make their contributions case by case, but the process will suffer from uncertainties as do all of our inferences on the consequences of actions in the real world.

Acknowledgements. We thank O. Büttner for performing the principal component analysis on some data sets, and H. Heuer and K. Smalla for providing samples of bacterial community DNA isolations. The work has been supported by the Ministry of Science and Technology in different projects.

References

Atlas RM (1984) Use of microbial diversity measurements to assess environmental stress. In: Klug MJ, Reddy DA (eds) Current perspectives in microbial ecology. American Society for Microbiology, Washington, pp 540–545

Bishop DHL, Entwistle PF, Cameron IR, Allen CJ, Possee RD (1988) Field trials of genetically-engineered baculovirus insecticides. In Sussman M, Collins CH, Skinner FA, Stewart-Tull DE (eds) The release of genetically engineered micro-organisms. Academic, London, pp 143–179

Carbon P, Ehresmann C, Ehresmann B, Ebel J-P (1979) The Complete Nucleotide Sequence of the Ribosomal 16-S RNA from Escherichia coli. Experimental Details and Cistron Heterogeneities. Eur.J Biochem 100, 399–410

Cory JS Hirst ML, Williams T, Halls RS, Goulson D, Green BM, Carty TM, Possee RD, Cayley PJ, Bishop DHL (1994) Field trial of a genetically improved baculovirus insecticide. Nature 370, 138–140

Crook NE (1991) Baculoviridae: Subgroup B. In: Kurstak E (ed) "Viruses of Invertebrates. Dekker, New York, pp 73–110

Ehler LE (1991) Planned Introductions in Biological Control. In: Lev R Ginzburg (Ed) Assessing Ecological Risks of Biotechnology. Butterworth-Heinemann, Boston, London etc, pp 21–40

Fuxa JR (1990) Fate of released entomopathogens with references to risk assessment of genetically engineered microorganisms. Bulletin of the Entomological Society of America 35, 12–24

Garland JL, Mills AL (1991) Classification and characterization of heterotrophic microbial communities on the basis of patterns of community level sole-carbon utilization. Appl Environm Microbiol 57, 2351–2359

Greene A, Allison RF (1994) Recombination Between Viral RNA and Transgenic Plant Transcripts. Science 263, 1423–1425

Gutell RR, Weiser B, Woese CR, Noller HF (1985) Comparative anatomy of 16-S-like ribosomal RNA. Prog Nucleic Acids Res 12, 5837–5852

Hochberg ME, Waage JK (1991) Control engineering. Nature 352, 16–17

Jehle JA, Backhaus H., Fritsch E, Huber J (1992a) Physical map of the *Cryptophlebia leucotreta* granulosis virus genome and its relationship to the genome of *Cydia pomonella* granulosis virus. J Gen Virol 73: 1621–1626.

Jehle JA, Fritsch E, Backhaus H, Huber J (1992b) Risk assessment of genetic engineering of baculoviruses: In vivo integration of host DNA into the genome of granulosis viruses. In: Casper R, Landsmann J (Eds) Proceedings of the 2nd International Symposium on Biosafety Results of Field Tests of Genetically Modified Plants and Microorganisms, Goslar. Biologische Bundesanstalt für Land- und Forstwirtschaft, Braunschweig, pp 250–256

Jehle JA, Fritsch E, Huber J (1993) Aspects of the stability of the host-pathogen interaction of baculoviruses. In: Wöhrmann K and Tomiuk J (Eds.) Transgenic Organisms. Birkhäuser Verlag Basel, pp 175–191

Jehle JA, Backhaus H (1994a) Genome organization of the DNA-binding Protein gene region of *Cryptophlebia leucotreta* granulosis virus is closely related to that of nuclear polyhedrosis viruses. J Gen Virol 75, 1815–1820

Jehle JA, Backhaus H (1994b) The granulin gene region of *Cryptophlebia leucotreta* granulosis virus: sequence analysis and phylogenetic considerations. J Gen Virol 75, 3667–3671

Jehle JA, Fritsch E, Backhaus H, Huber J (1994c) Sicherheitsforschung zur Gentechnik bei Baculoviren: Rekombination und Transposition als mögliche Ursachen eines horizontalen Gentransfers. In: Projektträger Biologie, Energie, Ökologie; KFA Jülich (ed)

Biologische Sicherheit Band 3. Bundesministerium für Forschung und Technologie, Bonn, pp 833–856

Jehle JA, Fritsch E, Nickel A, Huber J, Backhaus H (1995) TCl4.7: a novel lepidopteran transposon found in Cydia pomonella granulosis virus. Virology, in press

Kim J, Ginzburg LR, Dykhuizen DE (1991) Quantifying the Risks of Invasion by Genetically Engineered Organisms. In: Ginzburg LR (Ed) Assessing Ecological Risks of Biotechnology. Butterworth-Heinemann, Boston, London etc, pp 193–214

Levin BR (1986) The Maintenance of Plasmids and Transposons in Natural Populations of Bacteria. In: Levy SB, Novick RP (eds) Antibiotic Resistance Genes: Ecology, Transfer and Expression. Cold Spring Harbor Laboratory, New York, pp 57–70

Levin BR, Stewart FM (1980) The population biology of bacterial plasmids: a priori conditions for the existence of mobilizable nonconjugative factors. Genetics 94, 425–443

Marco GJ, Hollingworth RM, Durham W (1987) Silent Spring Revisited. American Chemical Society, Washington DC

Morozov SY, Dolja VV, Atabekov JG (1989) Probable Reassortment of Genomic Elements among Elongated RNA-Containing Viruses. J Mol Evol 29, 52–62

Muycer G, De Waal EC, Uitterlinden AG (1993) Profiling of complex microbial populations by denaturing gradient gel electrophoresis analysis of polymerase chain reaction-amplified genes coding for 16S rRNA. Appl Environ Microbiol 59, 695–700

Pace NR, Stahl DA, Lane DJ, Olsen GJ (1986) The analysis of natural microbial populations by ribosomal RNA sequences. Adv Microbiol Ecol 9, 1–55

Rosenbaum V, Riesner D (1987) Temperature-gradient gel electrophoresis. Thermodynamic analysis of nucleic acids and proteins in purified form and in cellular extracts. Biophysical Chemistry, 26, 235–246

Rowland FS, Molina MJ (1994) Ozone depletion: 20 years after the alarm. Chemical and Engineering News, pp 8–13

Sagoff M (1992) On Making Nature Safe for Biotechnology. In: Ginzburg LR (Ed) Assessing Ecological Risks of Biotechnology. Butterworth-Heinemann, Boston, London etc, pp 341–365

Sogin ML (1990) Amplification of Ribosomal RNA Genes for Molecular Evolution Studies. In: Innis A, Gelfand DH, Sinsky JJ, White TJ (Eds) PCR Protocols. A Guide to Methods and Applications. Academic Press Inc., San Diego, pp 307–314

Strauss EG, Strauss JH, Levine AJ (1991) Virus evolution. In: Fields BN, Knipe DM (Chief Eds.) *"Fundamental Virology"*, Raven, New York, pp 167–190

Wendt-Potthoff K, Zock S, Smalla K, Niepold F, Backhaus H (1992) Colonization of Bean Plants by Genetically Engineered Pseudomonas syringae pv. syringae: Effects of Inoculation Technique. In: Casper R, Landsmann J (eds) Proceedings of the 2nd International Symposium on The Biosafety Results of Field Tests of Genetically Modified Plants and Microorganisms. Biologische Bundesanstalt, Braunschweig, pp 275–280

Wendt-Potthoff K, Backhaus H (1994) Monitoringverfahren für Mikroorganismen und Gene in Modellsystemen. In: Projektträger Biologie, Energie, Ökologie. Forschungszentrum Jülich (ed) Biologische Sicherheit Band 3. Bundesministerium für Forschung und Technologie, Bonn, pp 542–562

Wendt-Potthoff K, Backhaus H, Smalla K (1994) Monitoring the fate of engineered bacteria sprayed on the phylloplane of bush beans and grass. FEMS Microbiol Ecol 15, 279–290

Zak JC, Willig MR, Moorhead DL, Wildman HG (1994) Functional Diversity of Microbial Communities: A Quantitative Approach. Soil Biol Biochem 26, 1101–1108

5 Examination of the Putative Horizontal Gene Transfer from Transgenic Plants to *Agrobacteria*

I. Broer[1], W. Dröge-Laser[2], and M. Gerke[1]

[1]Universität Bielefeld, Fakultät für Biologie, Lehrstuhl für Genetik, Universitätsstrasse, Postfach 100131, 33501 Bielefeld, Germany
[2]*Current address*: The Salk Institute for Biological Studies, Plant Biology Laboratory, 10010 North Torrey Pines Road, La Jolla, CA 92037, USA

Abstract

The putative horizontal transfer of foreign genes from transgenic plants to infecting wildtype agrobacteria has been examined. As a first step, the ability of *Agrobacteria* to take up foreign DNA via transformation has been analysed using different conditions. It could be confirmed that neither the presence of plant material nor the induction of virulence genes influenced the transformation frequency which was found to be very low in any case. Accordingly, in spite of the close contact of plants and *Agrobacteria* during the infection, no transfer of foreign genes could be traced. In conclusion, no wide-spread distribution of transgenes from plants to *Agrobacteria* in the free living can be expected.

Introduction

The usage of transgenic plants in agriculture is of public concern and intensively discussed by several groups. Many questions have been raised and only few of them can be answered to date. Hence, scientists all around the world try to analyse potential risks connected with this field. Horizontal gene transfer is one of the most mentioned problems associated with the deliberate release of transgenic plants. Although gene transfer between microorganisms has already been demonstrated in the soil (Stotzky et al. 1990), the only example of gene transfer between prokaryotes and eukaryotes discovered in free living is the *Agrobacterium* mediated gene transfer. Other cases of 'trans-kingdom' gene transfer have been detected in the laboratory (e.g. Bryngelson et al. 1988; Heinemann and Sprague 1989). Preliminary evidence for horizontal gene transfer between higher plants and *Aspergillus niger* has been reported by Hoffmann et al. (1992). Nevertheless, transfer from plants to soil bacteria has not been described

E.R. Schmidt/T. Hankeln (Eds.)
Transgenic Organisms and Biosafety

to date. However, since soil bacteria are capable of transferring genes to other microorganisms, and, in the case of *Agrobacterium*, even to the plant, this seems to be an important step in the hypothetical distribution of transgenes in the environment. Hence we tried to analyse the potential horizontal gene transfer between transgenic plants and *Agrobacteria* present in the intercellular spaces of plant tumors. The main assumption for this is that the close contact between the plant and the microorganism supports gene transfer. We chose the soil bacterium *Agrobacterium tumefaciens* as recipient for the gene transfer for several reasons. Firstly, the transfer of DNA from *A. tumefaciens* to the plant is well known and analysed. Nevertheless, to date there is no evidence that the transfer system can only be used in one direction. Secondly, *Agrobacteria* carry sequences homologous to transgenes transferred to the plant via *Agobacterium* mediated gene transfer (e.g. T-DNA borders). This may facilitate the establishment of transgene-carrying DNA fragments from the plant in the recipient by homologous recombination. Thirdly, the tumor induced by Agrobacteria represents a natural habitat for the inducers living in the intracellular spaces. The analysis of these tumors in the laboratory is therefore very close to natural conditions. In addition, the integration of plant-derived transgenes into the T-DNA of wildtype Agrobacteria presents the possibility of further transfer of the genes to other dicot plant cells.

Results and Discussion

Although the transformation frequency of *Agrobacteria* induced by an artificial transformation technique is very low (Holsters et al. 1978), it cannot be excluded that the differing conditions in the tumor and the induction of virulence (*vir*) genes might increase this frequency significantly. We therefore decided to analyse the influence of plant material and *vir*-inducing substances on in vitro transformation frequency in the laboratory. A binary vector was constructed carrying the *aac*I gene conferring gentamicin (Gm) resistance to bacteria, the *npt*II coding region from Tn*5* fused to the nos-promoter conferring kanamycin resistance to plants (Herrera-Estrella et al. 1983) and the luziferase coding region (*luc*) of the North American firefly *Photinus pyralis* (Ow et al. 1986) driven by the CaMV35S promoter, which leads to expression in plants and bacteria. All three were inserted in between the T-DNA borders on a replicon carrying the *npt*II gene conferring kanamycin resistance to bacteria (Bevan 1984). The freeze-thaw technique used for transformation according to Holsters et al. (1978) using ccc-plasmid DNA led to frequency comparable with that published in the literature. In order to induce virulence genes, the pH was changed to condition optimal for the *vir* induction and acetosyringon was added as inducing substance. Transformants were selected on media containig Gm or Km and afterwards tested for luziferase activity. Only Gm^r or Km^r and luc^+ clones were considered to be transformants. Under these conditions, no increase in transformation frequences could be observed.

Under natural conditions, genes originating from the plant chromosome will not be present as ccc plasmid DNA. Therefore, we used linearised plasmid DNA under the same conditions described above. However, transfer frequency of the competent as well as the *vir*-induced, competent bacteria decreased dramatically.

Vir proteins do not only act as endonucleases, but also bind as single strand binding proteins on the T-strand (for a review, see Zambrysky 1992). Therefore, *vir* induction might only support the transfer of single stranded DNA from plants to *Agrobacteria*. However, no transformants could be detected using single stranded DNA in the transformation assay described.

The fact that no influence of *vir* induction could be observed may be due to the lack of other components present in the plant. We therefore added homogenized plant material to the competent bacteria in the presence or absence of ccc-plasmid-DNA. Transfer frequency in the presence of plasmid DNA was the same with and without plant material, no positive influence of plant specific substances was detectable.

Nevertheless, this might depend on the destruction of the plant tissue. Therefore, transgenic plants carrying the *aac*I gene conferring gentamicin (Gm) resistance to bacteria, the *npt*II coding region from Tn5 (Herrera-Estrella et al. 1983) fused to the nos-promoter conferring kanamycin resistance to plants and the luziferase coding region (*luc*) of the North American firefly *Photinus pyralis* (Ow et al. 1986) driven by the CaMV35S promoter, which leads to expression in plants and bacteria, were constructed as donors for the transfer experiment. Transformation was carried out via *Agrobacterium* mediated gene transfer (strain LBA4404/IB19).

It has already been demonstrated that *Agrobacteria* used for the transformation are still present in the regenerated plant (e.g. van der Hoeven et al. 1991; Mogilner et al. 1993). In accordance with this, in spite of the addition of bacteriostatic agents to the culture medium, LBA4404/IB19 could be reisolated from tumors induced by a different wildtype strain. This could be proven as well by the analysis of the *Agrobacterium*-specific plasmid content as by the presence of the pIB19 replicon that is not transferred to the plant cell. The amount of LBA4404/IB19 in the tumor was surprisingly high, while only a few bacteria could be reisolated from homogenized plants which were not infected with a wildtype strain. It may be speculated that the *Agrobacteria* still present in the plant start to grow and divide in the presence of opines induced by the wild-type T-DNA.

In order to prevent gene transfer from remaining *Agrobacteria* used for transformation, only S1 plants, from which no *Agrobacteria* could be reisolated, were used in the experiment. As recipient, the *Agrobacterium* wildtype strain A281 (Jin et al. 1987), which is Gm^s, luc^- and can be distinguished from LBA4404 by its plasmid content, was chosen. Transformants were selected on media containig Gm and afterwards tested for luziferase activity. Only Gm^r and luc^+ clones were considered to be transformants. Nevertheless, Gm^r, luc^- clones were analysed for the presence of the *aac*I gene by Southern analysis. The analysis of numerous tumors and reisolated bacteria revealed Gm^r clones which proved to be A281. None of them were *luc*$^+$. Southern analysis of representative clones proved that these strains did not contain the *aac*I gene, but were spontanously resistant to Gm. We therefore conclude that the induction of virulence genes and the presence of plant material do not improve the capacity of *Agrobacteria* to take up foreign DNA. At least in the case of *Agrobacterium*-plant interaction the frequency of DNA transfer to the bacterium is below 10^{-12}. Hence, supposing this mechanism is actually working, it should not play an essential role in the distribution of foreign genes into the environment, although this bacteria-plant interaction seems to represent optimal conditions for an event that has not been observed up to now.

References

Bevan M (1984) Binary *Agrobacterium* vectors for plant transformation. Nucleic Acid Res 12: 8711–8721

Bryngelsson NT, Gustafson M, Green B, Cannon F (1988) Uptake of host-DNA by the parasitic fungus *Plasmodiophora brassicae.* Physiol Mol Plant Pathol 33: 163–171

Heinemann JA, Sprague GF (1989) Bacterial conjugative plasmids mobilize DNA-transfer between bacteria and yeast. Nature 340: 205–209

Herrera-Estrella L, De Block M, Messens E, Hernalsteens JP, van Montagu M, Schell J (1983) Chimaeric genes as dominant selectable markers in plant cells. EMBO J 2: 987–995

Hoffmann T, Glotz C, Schieder O (1992) Preliminary evidence for horizontal gene transfer between higher plants and *Aspergillus niger.* In: Casper R, Landsmann J (eds) The biosafety results of field tests of genetically modified plants and microorganisms. Biologische Bundesanstalt für Land und Fortwirtschaft, Braunschweig, Germany, pp 247–249

Holsters M, de Waehle D, Depicker A, Messens E, van Montagu M, Schell J (1978) Transfection and transformation of *Agrobacterium tumefaciens.* Mol Gen Genet 163: 181–187

Jin S, Komari T, Gordon MP, Nester EW (1987) Genes responsible for the supervirulence phenotype of *Agrobacterium tumefaciens* A281. J Bact 10: 4417–4425

Mogilner N, Zutra D, Gafny R, Bar-Joseph M (1993) The persistence of genetically engineered *Agrobacterium tumefaciens* in agroinfected plants. MPMI 6–5: 673–675

Ow D, Wood K, DeLucca M, DeWett J, Helinski D, Howell S (1986) Transient and stable expression of the firefly luciferase gene in plant cells and transgenic plants. Science 342: 856–859

Stotzky G, Devanas MA, Zeph LR (1990) Methods for studying bacterial gene transfer in soil by conjugation and transduction. Adv Appl Microbiol 35: 57–169

van der Hoeven C, Dietz A, Landsmann J (1991) Latente Agrobakterien in transgenen Pflanzen nachgewiesen. Nachrichtenblatt Deutscher Pflanzenschutz 43: 249–251

Zambrysky PC (1992) Chronicles from the *Agrobacterium*-plant cell DNA transfer story. Annu Rev Plant Physiol Plant Mol Biol 43: 465–490

6 Can Agrobacteria Be Eliminated from Transgenic Plants?

J. Landsmann, E. Graser, A. Riedel-Preuβ, and C. van der Hoeven

Biologische Bundesanstalt für Land- und Forstwirtschaft, Institut für Biochemie und Pflanzenvirologie, Messeweg 11/12, 38104 Braunschweig, Germany

Introduction

Plant transformation with *Agrobacterium tumefaciens* is a routine tool of genetic engineering. It was generally taken for granted that the endobiotic relationship of the disarmed agrobacteria with the transgenic plant is a transient one and that no agrobacteria survive and persist after early counterselection steps. However, occasionally bacteria can be isolated from transformed plants many months after infection (van der Hoeven et al. 1991; van der Hoeven 1992). Here we report on the characterization of reisolated agrobacteria and show that they can persist within the plant even under counterselection pressure. Implications for release into the environment of genetically engineered plants are discussed and experiments to eliminate agrobacteria from infected plants are presented.

Leaf Disk Transformation of Tobacco

For comparison of conditions for plant transformation and survival of the agrobacteria, the protocol for leaf disk transformation used (Horsch et al. 1985) is outlined below.

Leaves from sterile grown *Nicotiana tabacum* var. W38 were cut into strips vertical to the middle vein and soaked for 10 min in 2.5×10^8 log phase *Agrobacterium tumefaciens* LBA4404 per ml MS-medium (Murashige and Skoog 1962). The agrobacteria contained the binary vector PARGUSH resp. PARBASTA resp. PARPET which are derivatives of pLX222 (Landsmann et al. 1988). The leaf pieces were placed on sterile paper towel to rid them from excess water, then put on MS9-medium agar petri dishes at 26 °C in the dark. After 48 h of cocultivation the leaves, now visibly surrounded by agrobacteria, were gently washed in MS-medium and placed onto MS9-medium agar containing 100 μg/ml kanamycin

E.R. Schmidt/T. Hankeln (Eds.)
Transgenic Organisms and Biosafety

and 250 μg/ml Claforan (Hoechst). They were incubated at 24 °C at a 16 h light/ 8 h dark period and transferred to fresh plates once a week. No agrobacteria grew on this selection/counterselection medium. After 4 weeks the emerging shoots were transferred to MS-medium agar containing kanamycin and Claforan. Another 4 weeks later the rooted plants were isolated into MS-medium agar jars containing just the selecting antibiotic kanamycin. Shoot tips comprising at least five leaves were transferred to new jars in 2- to 3-week intervals. After six passages, the antibiotic kanamycin was also omitted.

Analysis of Transformed Plants and Reisolation of Bacteria

Those regenerating shoots which rooted on hormone free MS-medium containing 100 μg/ml kanamycin were classified as being transformed. NPT activity, GUS activity (for PARGUSH plants), PAT activity (for PARBASTA plants) and CAT activity (for PARPET plants) were determined in leaves and roots.

During shoot tip propagation, i.e. more than 5 weeks after release of Claforan counterselection and more than nine weeks after transformation, occasionally bacteria could be detected growing out of the cut stem site of the majority of 140 transgenic PARGUSH plants. We especially made sure that accidental contamination of the cutting site could be excluded.

From two other PARGUSH transformation experiments two out of 26 plants were heavily contaminated with agrobacteria more than two years after cocultivation. From one transformation with PARBASTA seven out of 30 transgenic plants released agrobacteria into the culture medium three years after transformation. From two transformations with PARPET three out of 20 transgenic plants regularly released agrobacteria.

Identification of the Reisolated Bacteria

All PARGUSH bacteria isolated six months after plant transformation from 6 individual transformants were positive in a ketolactose test (Bernaerts and De Ley 1963), indicating they were *A. tumefaciens.* Two PARGUSH and the three PARPET isolates were tested in a BIOLOG test (Biolog Inc., Hayward, CA, USA) where they showed slightly different fermentation capabilities. However, they all could be clearly identified as *Agrobacterium* spp.

The antibiotic resistance markers of eight PARGUSH isolates and seven PARBASTA isolates were tested. The isolates grew well on LB-medium agar plates containing 100 μg/ml rifampicin plus 5 μg/ml tetracyclin and 100 μg/ml kanamycin, respectively. Two of the three PARPET isolates had lost the tet-resistance marker.

Plasmid preparations from all isolates resulted in plasmid DNA of the expected size, although the vectors were generally not completely stable in *A. tumefaciens* LBA4404. Restriction with EcoRI or HindIII endonucleases showed the restriction pattern of the original binary vectors. Three DNA preparations from individual PARGUSH colonies out of 60 digested with HindIII showed deviating fragment lengths. Southern hybridization with a PARGUSH probe detected the correct fragments although additional rearrangements had taken place.

Functional Integrity of the Reisolated Agrobacteria

In order to check the integrity of the constructs two of the reisolated PARGUSH *Agrobacterium* strains were used to transform tobacco with the leaf disk method. Six regenerated plants of each transformation experiment were grown to the rooting stage. The level and variability of GUS activity of the regenerated plants proved to be comparable to the primary set of transformants.

Protoplast Culture May Exclude Agrobacterial Contaminations

As the agrobacteria used for plant transformation can obviously survive within the transgenic plants, methods to eliminate these bacteria need to be developed. *Agrobacterium* is supposed to persist predominantly between plant cells (Lehoczkky 1971; Stellmach 1990; Mogilner et al. 1993). Thus protoplast culture offers a means of regenerating bacteria-free plants; it is, however, time consuming and not easily adaptable to recalcitrant crop species. Preliminary evidence from such experiments with tobacco are reported on the poster. Simpler methods for elimination of agrobacteria have been tested with varying success (see below).

Cefotaxim (Claforan) Is Not Sufficient for Elimination of Agrobacteria

Pieces of leaves from transgenic in vitro tobacco plants, still harbouring recombinant agrobacteria, were incubated on MS-medium to regenerate shoots. During the regeneration process 250 μg/ml and 500 μg/ml Claforan, respectively, were applied. Fifty shoots of each experiment were analysed. One of the shoots developed agrobacteria when transferred to antibiotic-free medium. However, the majority of the shoots were positive in polymerase chain reaction (PCR) analyses with primers deduced from the agrobacterial chromosomal ros gene (Cooley et al. 1991; Dong et al. 1992; Dsouzaault et al. 1993; Zweigerdt 1993), even after high dosage application of Claforan.

Shoot Meristem Culture Reduces Agrobacterial Contaminations

Shoot tips were prepared from transgenic in vitro tobacco plants and used to regenerate calli and shoots (Graser 1994). A total of 50% of the regenerated calli from 57 meristem preparations greater than 0.5 mm in diameter and 85% of calli regenerated from 139 meristems less than 0.5 mm in diameter did not develop agrobacterial colonies on the medium. DNA preparations of these calli were submitted to PCR analyses including hybridization with the ros gene as hybridization probe. A total of 71% of the calli showed negative results in the ros hybridization of the PCR gels. Primers deduced from the tet gene of the binary vector pLX222 could not detect visible amplification products in PCR analyses with the bacteria-free candidates. The tet gene is located outside the T-DNA region and should not be transferred to the plant chromosome upon plant transformation.

Tobacco Seeds Seem to Exclude Agrobacteria

A total of 525 seeds from three individual tobacco transformants were aseptically germinated on MS-medium; 375 of the emerging plantlets were cut into pieces and incubated at 28°C on LB-medium agar. After 10 days still no agrobacteria were visible. The remaining plantlets were grown for further 4 months, with no bacteria appearing on the MS medium.

Seedlings from different transgenic tobacco lines were submitted to PCR analyses with the ros and the tet primers. A total of 100% of 46 tet-PCR analyses of two transgenic lines (comprising two to three pooled seedlings each) and all 20 tet-PCR analyses of leaves from individual plantlets of another transgenic line were negative.

Summary and Discussion

Engineered agrobacteria residing within transgenic plants in the field are capable of gene transfer to other bacteria in the environment. An uncontrolled spread of genes from non-bacterial kingdoms to the microflora of natural and agricultural ecosystems is not desired to date. Thus plants released into the environment should be free from engineered agrobacteria.

It should be noted, however, that in the vast majority of cases no significant impact and thus no risk for the environment would be connected with such a horizontal gene transfer.

Persisting agrobacteria after the plant transformation appear to be defying even severe attempts for elimination (Gould and Smith 1989). Nevertheless, seed transmission of agrobacteria can probably be denied at least for tobacco. However, tests proving the bacteria-free state of a plant depend upon the detection limit (Ponsonnet and Nesme 1994) and need to be standardized for general use. The detection limit for visible PCR products on ethidium bromide stained gels lay at <100 bacteria per reaction tube. Additional hybridization could detect even 1 bacterium. However, when calculating the necessary dilutions (PCR does not work in concentrated extracts) and extrapolating to the plant material assayed it came down to a minimum of 750 agrobacteria which escape detection.

Special attention has to be given to false positive results due to contamination in laboratories where the agent to be screened for in trace amounts is prevalent.

Persistence of free Ti-plasmid DNA in transgenic plants has not yet been investigated.

Acknowledgements. We would like to thank Ute Köllner for excellent technical assistance and Robert Zweigerdt for his help and for some of the ros primers. This work was supported by BMFT grant 0310042A.

References

Bernaerts MJ, De Ley J (1963) A biochemical test for crown gall bacteria. Nature 197: 406–407

Cooley MB, D'Souza MR, Kado CI (1991) The virC and virD operons of the *Agrobacterium* Ti plasmid are regulated by the ros chromosomal gene: analysis of the cloned ros gene. J Bacteriol 173: 2608–2616

Dong I-C, Sun C-W, Thies KL, Luthe DS, Graves JCH (1992) Use of polymerase chain reaction to detect pathogenic strains of *Agrobacterium*. Pytopathology 82: 434–439

Dsouzaault MR, Cooley MB, Kado CI (1993) Analysis of the ros repressor of *Agrobacterium* virC and virD operons – molecular intercommunication between plasmid and chromosomal genes. J Bacteriol 175: 3486–3490

Gould JH, Smith RH (1989) A non-destructive assay for GUS in the media of plant tissue cultures. Plant Mol Biol Rep 7: 20 9–216

Graser E (1994) Eliminierung von Agrobakterien aus gentechnisch veränderten Pflanzen. Thesis, Göttingen University

Horsch RB, Fry JE, Hoffmann NL, Eichholtz D, Rogers SG, Fraley RT (1985) A simple and general method for transferring genes into plants. Science 227: 1229–1231

Landsmann J, Llewellyn D, Dennis ES, Peacock WJ (1988) Organ regulated expression of the *Parasponia andersonii* haemoglobin gene in transgenic tobacco plants. Mol Gen Genet 214: 68–73

Lehoczkky J (1971) Further evidence concerning the systematic spreading of *Agrobacterium tumefaciens* in the vascular system of the grapevines. Vitis 10: 215–221

Mogilner N, Zutra D, Gafny R, Barjoseph M (1993) The persistence of engineered *Agrobacterium tumefaciens* in agroinfected plants. Mol Plant Microbe Interaction 6: 673–675

Murashige T, Skoog F (1962) A revised medium for rapid growth and bio assays with tobacco tissue cultures Physiol Plant 15: 473–497

Ponsonnet C, Nesme X (1994) Identification of Agrobacterium strains by PCR-RFLP analysis of pTi and chromosomal regions. Arch Microbiol 161: 300–309

Stellmach G (1990) Experiences with the isolation of latent tumorgenic agrobacteria from grapevines. Nachrichtenbl Dtsch Pflanzenschutzd 42: 151–153

van der Hoeven C (1992) Untersuchungen zur Variabilität von gentechnisch veränderten Tabakpflanzen mit Resistenz- und Reportergenen. Ph.D. dissertation, Hannover University

van der Hoeven C, Dietz A, Landsmann J (1991) Agrobacteria shown to reside in transgenic plants. Nachrichtenbl Dtsch Pflanzenschutzd 43: 249–251

Zweigerdt R (1993) Nachweis persistierender Agrobakterien in transgenem Kartoffelmaterial. Thesis, University of Braunschweig

7 Preliminary Findings of DNA Transfer from Transgenic Plants to a Wild-Type Strain of *Aspergillus niger*

T. Hoffmann[1,2], C. Golz[1,3], and O. Schieder[1]

[1]Institute of Applied Genetics, Free University of Berlin, Albrecht-Thaer-Weg 6, 14195 Berlin, Germany
[2]*Current address:* Environmental State Office of Brandenburg, Department of Gene Technology (I 5), POB 601061, 14410 Potsdam, Germany
[3]*Current address:* Federal Environmental Agency, POB 330022, 14191 Berlin, Germany

Abstract

Nine different field isolates of bacteria and fungi were cocultivated with four species of transgenic plants or leaf material expressing the hygromycin B resistance gene (hph) in sterile microcosms to determine the possibility of gene transfer events of the resistance gene to the microorganisms. Significantly higher resistance levels were detected only for the filamentous fungus *Aspergillus niger* after coculture, reisolation and selection of propagated fungus material. The hph-gene, other vector sequences and plant specific sequences were detected in the DNA of some resistant colonies and in a recloned plasmid. However, the mitotic stability seemed to be low. Results of transformation experiments in the fungus with plant transforming vectors gave rise to the assumption that a high gene dose of the 35S-promoted hph-gene was necessary to resist the selection pressure. High cotransformation efficiencies and the recovering of altered vector plasmids from uncut DNA of transformed clones indicated a high recombination activity in fungal cells. The possible uptake of foreign DNA might have occured during the growth of fungal hyphae on the plant material. A special role of hyphal tips as the location of active growth and protein secretion could be assumed.

Introduction

Gene exchange among microorganisms, especially in bacteria, is thought to be a powerful tool for the evolutionary development of microorganismal populations and ranks among the issues considered in risk assessment with respect to the release of GMOs. There is growing evidence that such a transfer may occur even between distantly related species by conjugation, transduction and transformation. A species-dependent competence state, the adsorption, uptake and in-

E.R. Schmidt/T. Hankeln (Eds.)
Transgenic Organisms and Biosafety

tegration of homologous and heterologous DNA, a longtime integrity and protection of free DNA in natural habitats, sequence similarities of transposable elements and enzyme genes are observed phenomena and supposed pointer hints for natural transformation events. The rarely reported supposed events of horizontal transmission of genetic material among fungi (Bryngelsson et al. 1988, Kellner et al. 1993) or other eukaryots seem to depend on the close natural contact of organisms. In a range of experiments combining several transgenic plants and field isolates of bacteria and fungi we were able to detect foreign DNA sequences in a wild type strain of *Aspergillus niger* (a fungus imperfectus) after close coculture with transgenic plants in microcosms. Often the fungus killed the plants by colonizing the plants' surface indicating a nutritional function of plant material for the fungus. The existence of strains of filamentous fungi that produce very high levels of secreted enzymes awoke interest in the use of fungi as hosts for the expression of recombinant proteins also for industrial applications (Peberdy 1994, Nyyssönen et al. 1993). Protein secretion in filamentous fungi is supposed to be associated with the process of hyphal tip growth. Conidial- and mycelial-bound exopectinases of *Aspergillus sp.* (Aguilar and Huitron 1993) and the glucoamylase secretion predominantly at the growing hyphal tips in *A. niger* (Wösten et al. 1991) give evidence for the high external enzymatic activity. Therefore several strains of *A. niger* were classified into the group of optional phytopathogenic organisms causing fruit and plant damages by the secretion of pectolytic, cellulolytic and proteolytic enzymes (Mehta et al. 1993; Jarai and Buxton 1994). The movement of such large molecules through the cell wall is paradox and the porosity of cell walls for the passage of macromolecules is assumed to vary in different fungi and even different strains of a particular species (Peberdy 1994). Possibly the reverse way of macromolecule transmission including proteins and nucleic acids also occurs and may serve as an explanation for the uptake of the observed foreign DNA sequences.

Material and Methods

Several transgenic hygromycin B resistant plants of *Brassica napus, Brassica nigra, Datura innoxia* and *Vicia narbonensis* generated by *A. tumefaciens*-mediated or direct gene transfer were cocultured with *A. niger* mycelial material in aseptic microcosms with 20 g moist soil for four weeks under light. Cut leaf material was also exposed to the fungus. Fungal progeny was reisolated in 0.5% NaCl solution and plated onto nutrition agar supplemented with 750 mg/l hygromycin B to select for resistant fungal colonies. A large array of examination was choosen to allow statistical interpretations of resistance events. Resistant fungal colonies were propagated via conidial spores under selection pressure and DNA was isolated, separated on agarose gels and blotted for Southern analysis to demonstrate the presence of the hph-gene. The DNA of positive clones was cut, self-ligated and offered the *E. coli* strain DH5α to clone the detected foreign DNA sequences. Several Southern blot, PCR and sequence analyses were carried out in such a clone (Plasmid 2/5–7) to evaluate the origin and composition of the DNA. Mycelial protoplasts were produced to transform the fungus strain with several plasmids and study the 35S-promoted expression of marker genes like hph and GUS and to compare with the expression by a fungal promotor. A quantitative GUS-assay based on the fluorescence emission of the substrate 4-methyl-umbelliferyl-β-D-glucoronide was used to measure expression values.

Results and Discussion

Among nine species of bacteria and fungi tested for the production of a higher resistance level after coculture with transgenic, hph gene carrying plant material, only in *A. niger* we were able to find an increased number of resistant colonies as compared with the control of delivered wild type plants. The ratios varied between 4:1 and 36:1. A statistical interpretation of the values by the *t*-test showed high significance ($\alpha = 0.01$) of detected differences for used leaf material of *Brassica napus* and *B. nigra*. Lower significance level ($\alpha = 0.025$–0.05) or no significance was obtained in the use of whole plants and for *D. innoxia* and *V. narbonensis* (Hoffmann et al. 1994). On the other hand, we found the presence of the hph-gene in only 5% of nearly 200 resistant colonies examined by Southern analysis. Furthermore, in a large proportion of the hph-gene carrying clones we were unable to find the gene again after prolonged subculture despite the maintenance of selection pressure (Fig. 1).

However, the assumption that the presence of the hph-gene was due to earlier gene acquisition was supported by additional findings as the simultaneous presence of other sequences of plant transforming vectors (nptII and amp) and even a plant specific sequence (Fig. 2).

To give an explanation for the observed apparent instabilities of transgenic clones derived from the coculture with transgenic plants we transformed mycelial protoplasts of the fungal strain with plant transforming vectors and compared expression and stability of the integrated resistance gene with the properties of gene inserts expressed by a fungal promotor (gpd-promotor of *A. nidulans*). In the case of common fungal transforming vectors a slightly higher transformation efficiency as compared with the one described by Punt et al. (1987) and an enormous increase of transformation frequency as well as a very high cotransformation efficiency using a combination of expression (GUS-gene) and selection (hph-gene) vectors were found. These findings may indicate a high recombinational activity. A high degree of homologous and nonhomologous recombination and duplication in transforming DNAs before integration in filamentous fungi has been demonstrated previously (Aleksenko 1994). The transformation efficiencies of plant transforming vectors (with 35S-promotor) were very low but also enhanced in the case of cotransformation.

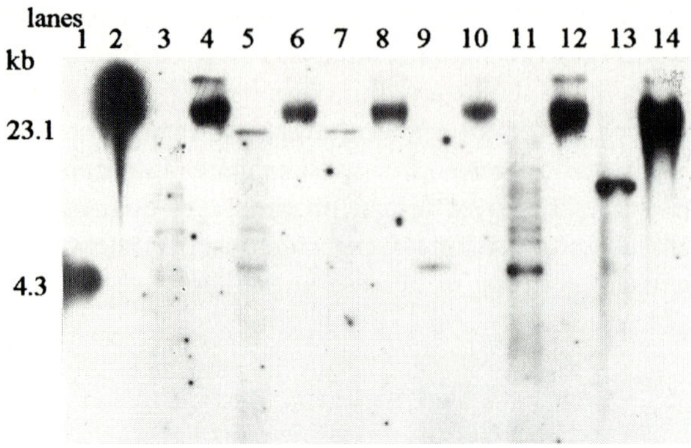

Fig. 1. Southern hybridization in DNA of resistant fungal colonies for the detection of the hph-gene after coculture with transgenic plants (^{32}P-probe: 1.1 kb hph-gene sequence). *Lanes 1,2,* EcoRV cut and uncut DNA of *A. tumefaciens* used to transform *B. napus* plants; *3–12,* EcoRV cut and uncut DNA, respectively, of resistant fungal colonies (*lanes 5–10,* DNA of a clone from three subcultures); *13, 14,* EcoRV cut and uncut DNA of the co-cultured transgenic line of *B. napus;* wild-type DNAs of *B. napus* and *A. niger* did not hybridize to the hph-gene sequence

The expression of a 35S-promoted GUS gene was three magnitudes lower than a gpd-promoted one, whereas the patterns of transient expression were comparable (Fig. 3) indicating a low expression activity of the 35S-promotor in the fungus.

The Southern analysis of protoplast-derived transformed clones gave rise to the assumption of a predominant effect of the gene dose for the level of GUS-expression and probably also for the origin of resistance to hygromycin in the case of a 35S-promoted hph-gene (Fig. 4).

Clones with a lower copy number of the resistance gene might be lost due to the selection pressure. The only colony which maintained the transgenic character after a coculture experiment with a high hph-copy number containing plant of *D. innoxia* also showed several copies of the hph-gene in the genome. From this fungal DNA we were able to recover a plasmid with an altered composition and an additional sequence as compared with the plant transforming vector pGL2. In PCR experiments and Southern hybridizations we found pieces of the sequence both in wild type plant and wild type fungal DNA indicating sequence similarities. In PCR experiments with different primer pairs after partial sequencing of the DNA we could show the fungal origin of the sequence and the presence of the original sequence composition of the recloned plasmid in the transgenic fungal clone (Hoffmann et al. 1994). Probably the recloned plasmid is the product of a phenomenon frequently observed in transgenic filamentous fungi: the recovering of altered transforming vectors partially even with fungal genome sequences offering uncut fungal DNA to *E. coli* $RecA^-$ host strains. Such findings indicate posttransformational events that might be responsible for the observed transgenic curing (Ward et al. 1993) or for mitotic instabilities

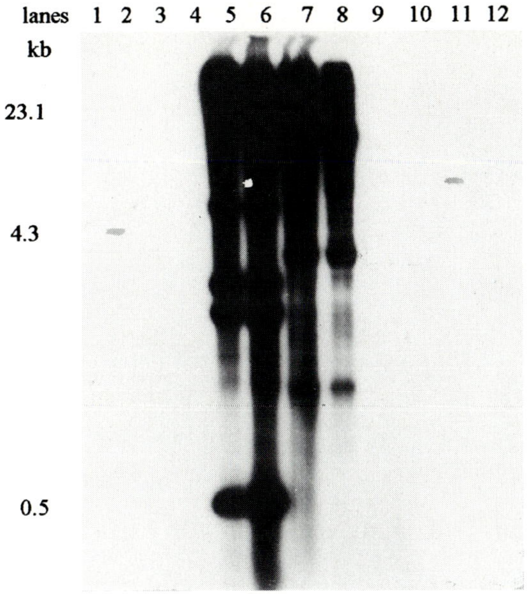

Fig. 2. Southern hybridization of *B. nigra* DNA and fungal DNA after coculture with transgenic *B. nigra* and selection against hygromycin (^{32}P-probe of a repetitive sequence from *B. nigra* specific for the B genome; Gupta et al., unpublished results). *Lanes 1–6,* BamH1 cut DNA; *7–12,* EcoRV cut DNA; *1, 12,* fungal wild-type DNA; *1–4, 9–12,* DNA of resistant fungal colonies; *5, 8,* DNA of the plant wild type; *6,7,* DNA of the transgenic plant

(Rhounim et al. 1994). The discovery of fungal sequences supporting an unexpected autonomous replication of transforming vectors in filamentous fungi enhancing transformation efficiencies enormously (Randall and Reddy 1992, Woods and Goldman 1993, Gems and Clutterbuck 1993) demonstrates that further research is necessary to increase knowledge about fungal genetics. Although we did not observe mitotic genetic instabilities, there could be other reasons to explain the apparent genetic instabilities after coculture. First, the wild type fungus was able to adapt to the high hygromycin concentration, and second

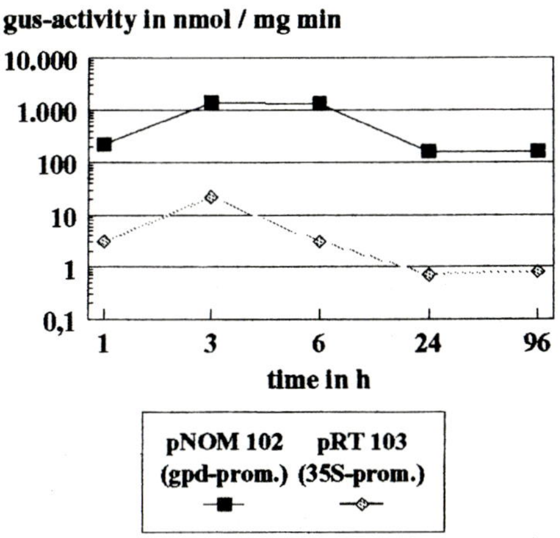

Fig. 3. Transient GUS-activity of *Aspergillus niger* protoplasts after transformation with different plasmids carrying a fungal (gpd)-or a 35S-promoted GUS-gene

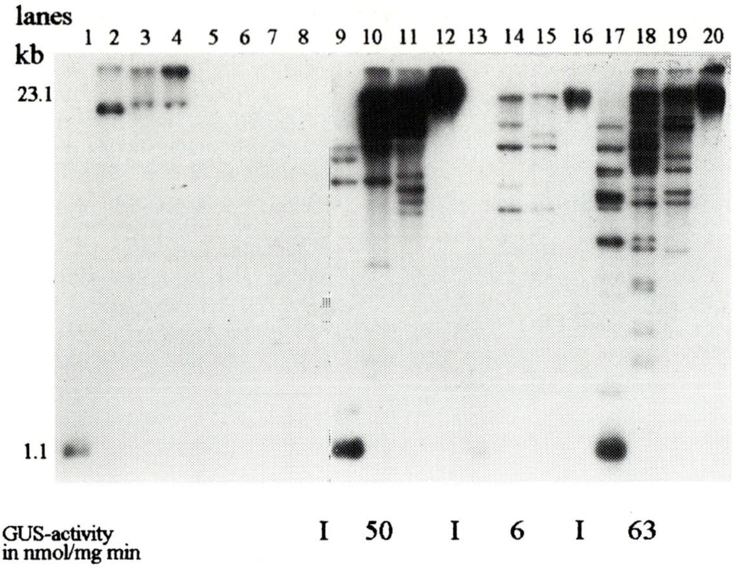

Fig. 4. Southern hybridization in protoplast derived, resistant fungal clones transformed with the plasmid pGSC 1704/GUS with 35S-promoted hph-and GUS-genes (^{32}P-probe: hph-gene sequence of 1.1 kb). Corresponding GUS-activities are shown, probably indicating a gene dose-response effect. *Lanes 1,5,9,13,17,* BamH1 cut DNA; *2,6,10,14,18,* EcoRV cut DNA; *3,7,11,15,19,* HindIII cut DNA; *4,8,12,16,20,* uncut DNA; *1–4,* plasmid pGSC 1704/GUS; *5–8,* DNA of the fungal wild type; *9–20,* DNA of transformed fungal clones

taking into account such properties as a low expression of the 35S-promoted hph-gene and the presence of multinucleolar hyphal cells and probably also conidial spores, there is no guarantee of maintaining the transgenic character. The occurrence of abortive transformants up to 90% was reported for *N. crassa* and *A. nidulans* (reviewed by Fincham 1989). The selection in the protoplast system may also be more efficient than in conidial spores or mycelial material, where a much higher selection pressure is necessary to avoid escapes. However, questions regarding the mode of action of the supposed gene transfer to *A. niger* remain unresolved, although there is also evidence for the uptake of free DNA in soil detected with a GUS-assay and Southern hybridizations. The assumed porosity of cell walls especially in the growing and actively protein secreting hyphal tips might be the reason for the passage of macromolecules also in the reverse direction. For filamentous fungi transformation was shown to occur also without degradation of cell walls in the presence of alkali ions (reviewed by Fincham 1989). Transformation efficiencies were enhanced by linearised DNA implicating that the same property might be valid also for chromosomal DNA. The integration of foreign DNA might be facilitated by sequence homologies and similarities as shown for bacteria with the ability of natural transformation (reviewed by Lorenz and Wackernagel 1994). We detected sequence similarities of wild type DNA sequences found in the recovered plasmid and furthermore unexpected hybridizations of fungal wild type DNA with the sequence of the maize transposable element AC and with an AT rich repetitive sequence isolated and cloned from *B. nigra* (Gupta et al. 1990) and assumed to be specific for this plant.

Acknowledgements. The authors express their gratitude to U. Bernd for invaluable technical assistance. We also acknowledge the support by the laboratory group of Prof. T. Börner (Humboldt-Universität, Department of Biology) and Prof. J. Wöstemeyer (IGF Berlin). The strains of *A. niger* and other microorganisms were isolated by Prof. M. D. Sacristan. The work was supported by a grant from the Ministry of Research and Technology of Germany, project no. 0319285A.

References

Aguilar G, Huitron C (1993) Conidial and mycelial-bound exo-pectinases of *Aspergillus sp.* FEMS Microbiol Lett 108: 127–132

Aleksenko AY (1994) Cointegration of transforming DNAs in *Aspergillus nidulans*: a model using autonomously-replicating plasmids. Curr Genet 26: 352–358

Bryngelsson T, Gustaffsson M, Green B, Lind C (1988) Uptake of host DNA by the parasitic fungus *Plasmodiophora brassicae*. Physiol Mol Plant Pathol 33: 163–171

Fincham JRS (1989) Transformation in fungi. Microbiol Rev 53: 148–170

Gems DH, Clutterbuck AJ (1993) Co-transformation with autonomously-replicating helper plasmids facilitates gene cloning from an *Aspergillus nidulans* gene library. Curr Genet 24: 520–524

Gupta V, Jagannathan V, Lakshmikumaran MS (1990) A novel AT rich tandem repeat of *Brassica nigra*. Plant Sci 68: 223–229

Hoffmann T, Golz C, Schieder O (1994) Foreign DNA sequences are received by a wild-type strain of *Aspergillus niger* after co-culture with transgenic higher plants. Curr Genet 27: 70–76

Jarai G, Buxton F (1994) Nitrogen, carbon, and pH regulation of extracellular acidic proteases of *Aspergillus niger*. Curr Genet 26: 238–244

Kellner M, Burmester A, Woestemeyer A, Woestemeyer J (1993) Transfer of genetic information from the mycoparasite *Parasitella parasitica* to its host *Absidia glauca*. Curr Genet 23: 334–337

Lorenz MG, Wackernagel W (1994) Bacterial gene transfer by natural genetic transformation in the environment. Microbiol Rev 58: 563–602

Mehta A, Chopra S, Mehta P (1993) Production of cell-wall degrading enzymes by three isolates of *Aspergillus niger* under different cultivation conditions. J Basic Microbiol 33: 193–199

Nyyssönen E, Penttilä M, Harkki A, Saloheimo A, Knowles JKC, Keränen S (1993) Efficient production of antibody fragments by the filamentous fungus *Trichoderma reesei*. Bio/Technology 11: 591–595

Peberdy JF (1994) Protein secretion in filamentous fungi – trying to understand a highly productive black box. Tibtech 12: 50–57

Punt PJ, Oliver RP, Dingemanse MA, Pouwels POH, van den Hondel CAMJJ (1987) Transformation of *Aspergillus* based on the hygromycin B resistance marker from *Escherichia coli*. Gene 56: 117–124

Randall TA, Reddy CA (1992) The nature of extra-chromosomal maintenance of transforming plasmids in the filamentous basidiomycete *Phanerochaete chrysosporium*. Curr Genet 21: 255–260

Rhounim L, Gregoire A, Salama S, Faugeron G (1994) Clustering of multiple transgene integrations in highly unstable *Ascobolus immersus* transformants. Curr Genet 26: 344–351

Ward M, Wilson LJ, Kodama KH (1993) Use of *Aspergillus* overproducing mutants, cured for integrated plasmid, to overproduce heterologous proteins. Appl Microbiol Biotechnol 39: 738–743

Woods JP, Goldman WE (1993) Autonomous replication of foreign DNA in *Histoplasma capsulatum*: role of native telomeric sequences. J Bacteriol 175: 636–641

Wösten HAB, Mouhka SM, Sietsma JH, Wessels JGH (1991) Localization of growth and secretion of proteins in *Aspergillus niger*. J Gen Microbiol 137: 2017–1023

8 Gene Dispersal from Transgenic Crops

P.J. Dale and J.A. Scheffler

John Innes Centre, Colney Lane, Norwich NR4 7UJ, United Kingdom

Introduction

Before transgenic plants are released in small-scale field experiments or for widespread commercial use, it is necessary to complete a risk or safety assessment. There is international agreement on the need for risk assessment because a greater diversity of genes can be introduced by transformation than by conventional plant breeding methods. Risk assessment involves addressing a range of questions including: the nature of the transgene, its affect on the plant's phenotype, and how the plant may interact with the environment. When assessing the interaction between a transgenic plant and the environment, one important aspect to consider is the likelihood and consequences of the movement of the transgene from the transgenic crop plant to other organisms. Usually a risk assessment considers the possibility of transgene movement by pollen to other plants, movement to microorganisms associated with the transgenic plant or to microorganisms found in the gut of animals that consume the transgenic plant. In this paper we will concentrate on the movement of transgenes to other plants by pollen.

Opportunities for Gene Transfer by Pollen

For transgenes to be dispersed by pollen, the donor and recipient plants must be present in the same location, flower at the same time and be sexually compatible. To have a potential environmental consequence, the F1 hybrid embryo must be capable of producing a hybrid plant which can self or cross pollinate, or reproduce by vegetative means. Because of the desire of plant breeders over past decades to incorporate genes into new plant varieties, there is considerable previous experience of cross-pollination between our crop plants and their re-

E.R. Schmidt/T. Hankeln (Eds.)
Transgenic Organisms and Biosafety

lated species and genera. As a consequence there is a large body of data on sexual compatibility for many of our crop plants.

In discussing the issues surrounding transgene dispersal by pollen to related species, oilseed rape (*Brassica napus*) will be used as a model. This crop is likely to be one of the first transgenic crop plants to be in widespread agricultural use in Europe. Within our research group, we have carried out a series of studies to determine the distance pollen can travel and successfully cross pollinate other oilseed rape plants, and an extensive literature review to evaluate the sexual compatibility of *B. napus* with its related species.

Distance of Transgene Dissemination by *Brassica napus* Pollen

In an attempt to determine the frequency of cross-pollination at different distances, transgenic oilseed rape containing a Basta herbicide resistance transgene, was surrounded by non-transgenic oilseed rape plants. Both transgenic and non-transgenic plants were of the spring variety Westar. The transgenic plants were present in a 9-m diameter plot in the centre of the experimental site and the non-transgenic plants in an approximately one square hectare (105 m × 105 m)

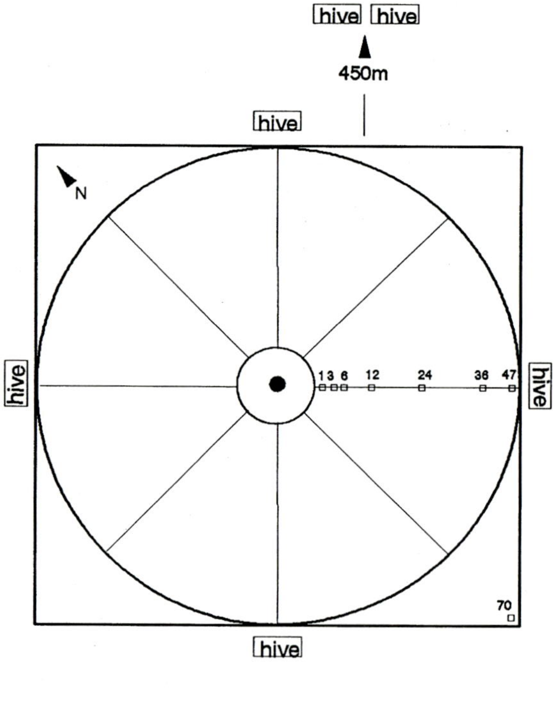

Fig. 1. Plan of the oilseed rape field experiment. The *dark centre circle* (diameter, 1 m) contained non-transgenic plants. The *surrounding circle* (diameter, 9 m) contained transgenic plants. The remaining area within the square contained non-transgenic plants. Samples were harvested from each of the eight sections at 1, 3, 6, 12, 24, 36 and 47 m from the edge of the transgenic circle. The location of the honey bee hives are shown

surrounding the transgenic plot. In the centre of the transgenic plot was a 1 m diameter circular plot of non-transgenic plants.

The principal pollinating insects of oilseed rape are honey bees and bumble bees. To ensure that there were adequate numbers of pollinating insects, six honey bee hives were introduced into the experimental area. Four hives were placed at the edge of the experimental plot, and an additional two hives were positioned 450 m to the northeast of the plot. There was good flowering synchrony throughout the plot and a high level of honey bee activity. Indigenous bumble bees were also observed to be moving from plant to plant during the flowering period. The experimental design is illustrated in Fig. 1. Samples of seeds were harvested from the plot and screened for resistance to Basta either in a glasshouse or in a 2.5-hectare field plot. The Basta resistant plants were expected to be hemizygous (i.e. with no corresponding allele on the homologous chromosome; Tt, heterozygous; T–, hemizygous) for the transgene, because they would be hybrids between the transgenic pollen donor plants and the non-transgenic recipient plants. Progeny of the Basta resistant plants were evaluated for Mendelian segregation of the resistance transgene, and samples of the resistant plants were also analyzed by Southern blotting to confirm the presence of the transgene. The proportion of progeny sampled at the different distances that contained the transgene is shown in Fig. 2. The frequency of progeny carrying

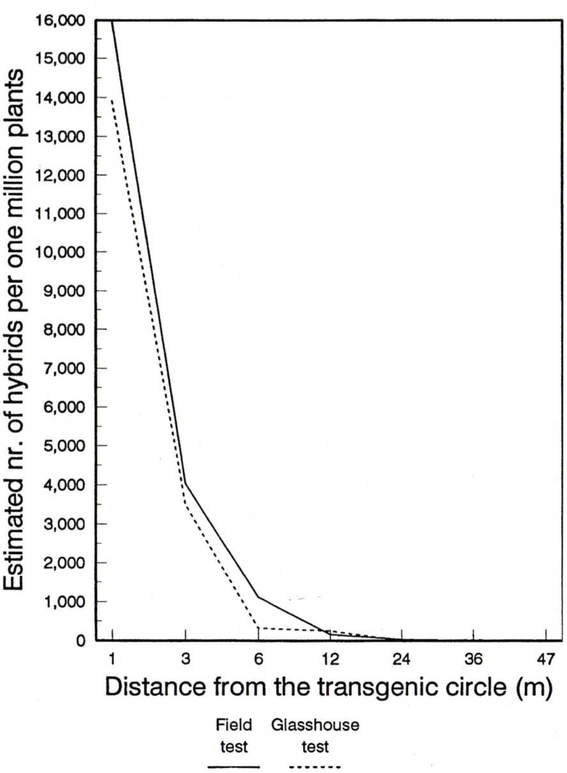

Fig. 2. Estimated number of hybrids produced by cross-pollination per one million progeny, at different distances from the pollen source

the transgene fell sharply up to 12 m from the transgenic plants and the frequency at 47 m was estimated as three per million seeds (Scheffler et al. 1993).

These data are now being used to make more informed judgements during the risk assessment process. Similar experiments were carried out at INRA in France (Rennes) and at Plant Genetic Systems in Belgium (Gent) and the frequencies of pollination were similar in the three locations. Because the size of the donor plot of transgenic plants and the number of non-transgenic recipient plants might affect the amount of cross-pollination, further experiments have been carried out to obtain additional estimates of pollen dispersal. These will be reported elsewhere.

Sexual Compatibility of *Brassica napus* with its Related Species

The extent of pollen dissemination will depend on the degree of sexual compatibility between the crop and any species receiving the pollen. In a survey of sexual compatibility between *B. napus* and its related species, we reviewed approximately one thousand publications and attempted to estimate the degree of sexual compatibility with a wide range of species. In reviewing the work of others, it is necessary to rely on the integrity and judgement of the different authors. However, many of the hybridizations have been attempted by several workers, so if there are reports of successful hybridization from different laboratories, it is more likely that hybridization is possible. If a hybridization is difficult or has limited success, the number of reports in the literature may not reflect the amount of effort put into attempting the cross, because of the tendency to under-report negative results. In interpreting the data, a distinction was made between hybridizations that have been observed under field conditions, those that have required manual pollination and those requiring ovary culture or in vitro embryo rescue. In several cases, the crosses involved the use of male sterile *B. napus* which increases the opportunity for cross-pollination.

Data were included on hybrids that have been produced only with the aid of manual pollination because it is possible that once transgenic plants are in widespread commercial production, they will be exposed to a range of environments and multiple ecotypes of a related species, and rare hybrids may form. A consideration of the success of hybridization following manual cross-pollination makes it possible to assess the extent of any post-hybridization genetic barriers, such as the fertility and fitness of the F1, F2 or the backcross progeny. A full discussion of sexual compatibility between *B. napus* and its relatives is presented elsewhere (Scheffler and Dale 1994) but is represented diagramatically in Fig. 3.

Sexual compatibility in *B. napus* can be summarised as follows: *B. rapa* (=*B. campestris*) and *B. juncea* will cross pollinate with *B. napus* under field condi-

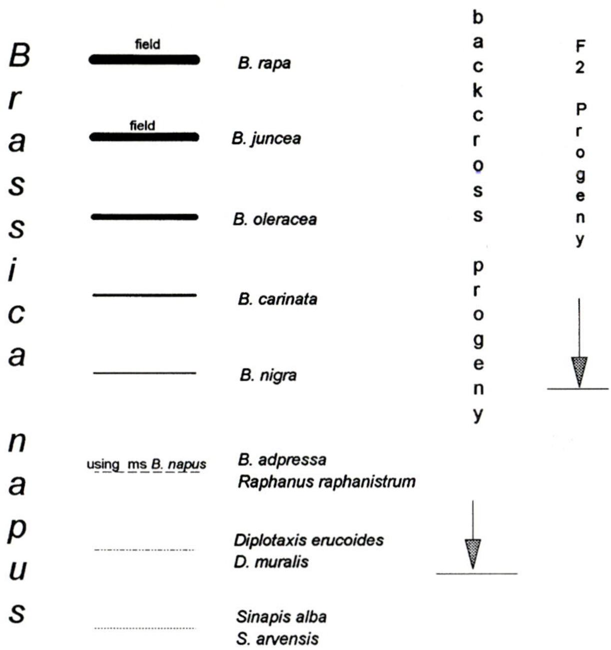

Fig. 3. Species reported to hybridize with *Brassica napus* are listed in order, with the most sexually compatible species at the *top*. Hybrids reported to produce F_2 or backcross progeny are indicated by the *arrows*. *Field* indicates that hybrids were formed by open pollination under field conditions. *B. adpressa* and *Raphanus raphanistrum* formed hybrids in the field when sterile male (*ms*) *B. napus* was used

tions; *B. adpressa* and *Raphanus raphanistrum* have been observed to cross pollinate with male sterile *B. napus* under field conditions. All other species were either incompatible or required manual pollination and sometimes *in vitro* embryo rescue to be successful.

Significance of Pollination Distance and Sexual Compatibility Data for Risk Assessment

Information on the frequency of pollination events at different distances is useful for the initial small-scale evaluation of novel transgenic plants (Scheffler et al. 1993; McPartlan and Dale 1994). Researchers and breeders are introducing a wide array of transgenes into plants which are assessed initially in the laboratory and glasshouse (Dale et al. 1993). Following this, selected transgenic plants are evaluated on a small scale under agricultural field environments. During this initial evaluation it may be desirable to isolate them genetically from other sexually compatible plants. A guarantee of total genetic isolation can rarely be ensured, so the implications of this should be considered in the risk assessment.

Once plants are in commercial production, genetic isolation by distance, use of barrier crops, asynchronous flowering or removal of flowers, is usually impractical. Therefore, information on sexual compatibility, rather than pollination distance, is most relevant for determining the likelihood and consequences of transgene dispersal. Where sexual hybridization is believed to be possible, the focus of risk assessment on transgene dispersal should be on an analysis of the *consequences* of hybridization between the transgenic plant and the related species.

Consequences of Transgene Dispersal

It is reasonable to assume that once transgenic plants are in widespread commercial use, hybridization will occur between crop plants and other sexually compatible species. There can be no scientifically justifiable general comment on the consequences of transgene dispersal (e.g. genetic pollution), each transgene needs to be considered on its merits and requires a case by case risk assessment. Some classes of transgenes are considered briefly below.

Herbicide Tolerance

The transfer of herbicide tolerance to a wild plant population should be of no consequence for natural plant populations because herbicides are not used in these environments. There should also be no impact in semi-natural habitats, such as road verges, because the vegetation is usually controlled mechanically rather than chemically. If a potential weed population receives the herbicide tolerance gene, then that weed could not be controlled by the corresponding herbicide. If that herbicide is important for weed control in a particular crop and there are no alternatives, then weed control could become a problem. A risk assessment in preparation for the commercial release of a herbicide tolerant crop would need to address these issues.

Consideration of a proposal to release a variety resistant to one particular herbicide is relatively straightforward, because there are usually several herbicides available for weed control. With the potential to introduce several different herbicide tolerance genes into different varieties of the same crop species, there is the possibility that weeds could be produced that have multiple resistance to different herbicides. If there are submissions to register transgenic varieties carrying different herbicide tolerance genes in the future, we need to debate what kind of strategic planning is necessary and who should be responsible for its implementation.

Pest and Disease Resistance

The impact of the transfer of pest or disease resistance to sexually compatible species will depend on the extent to which particular natural populations of plants are restrained or influenced by insect pests or diseases. If there are wild plant populations that are limited by the activity of a pest or disease, then the transfer of a corresponding resistance transgene could confer a selective advantage to the wild plant and potentially make it more persistent or invasive.

The transfer of resistance to wild relatives of crop plants would be expected to expose the corresponding pathogen or insect to a greater selection pressure to overcome the resistance. This is especially likely if the same or similar resistance strategies are used in a range of different crop plants. Throughout the history of traditional plant breeding there has been a continuing struggle between breeders producing new resistant plant varieties and pathogens or insects breaking down that resistance. Some of the resistance strategies possible using transgenic plants may be particularly difficult to overcome by the pathogen or insect, but some form of strategic planning may be important for the application of more general defence mechanisms (e.g. chitinase against fungal pathogens).

Other Characters

Many other kinds of transgenes are being introduced into crop plants (Dale et al. 1993). For example, there are significant scientific developments in modifying fatty acid composition of oilseed rape to produce oils for industrial use and manufacturing (Murphy 1994). There is also progress in the use of plants for the production of pharmaceutical chemicals. Another important research goal is to gain a better understanding of the genetics of stress tolerance in plants (cold, drought, heat, salt, metals) so that transgenic plants might be made more tolerant of environmental stress. The possibilities of the transfer of genes of this kind to wild relatives would have to be considered in a risk assessment, to estimate their possible effect on the ecology and distribution of those species.

Transgenic Plant Releases in the Centres of Origins of Crop Plant Species

Risk assessments are generally made with reference to the country or region where the release is being carried out, often with little reference to surrounding geographic areas. With the move to commercialise transgenic crop plants it will become increasingly important to consider the wider impact of their commercial use. With the ease and frequency of intercontinental travel it is likely that transgenic plants and seeds will intentionally or unintentionally (e.g. eaten fruits containing viable seeds passing into sewerage systems) be taken from one part of the world to another. This is an important reason why it is necessary to adopt internationally harmonised standards for the release and use of transgenic plants. Transgenic plants could be carried to the centre of origin (which often corresponds to an area of great genetic diversity) so it is important to consider the possible impact of transgene movement to wild relatives that may not be present in the primary country or continent of release.

Consequences of Gene Transfer from Conventionally Bred Plant Varieties

Most of the transgenic plants produced to date fall into classes familiar to the traditional plant breeder eg. pest resistance, disease resistance, herbicide tolerance and changes in quality of plant product. It is reasonable to ask, therefore, what evidence there is of gene movement from conventionally bred plants to wild plant populations. A number of reports have addressed this issue and there are several well quoted examples of gene flow between crop and sexually compatible relatives. Some examples include rice (*Oryza sativa*) and perennial rice (*Oryza perennis*), maize (*Zea mays*) and Teosinte (various *Zea* species), sugarbeet (*Beta vulgaris*) and wild beet (*B. vulgaris* spp. unknown) (Dale 1994). In some cases this has contributed to the evolution of more troublesome weeds. So far as we are aware, there is no evidence of gene transfer having a significant impact on plants in natural (non-agricultural) habitats.

Conclusions

With the development of transformation methods for most of our important crop plants and the rapid advancement in gene isolation and analysis, new techniques are available to continue improvement of our crops for the future. With new opportunities there are inevitably new responsibilities that need to be considered carefully by scientists and by society as a whole. The general public demands more information on developments in science and technology, and we as scientists have an increasing responsibility to explain, to educate and to justify what we do. It is important, during our discussions of risk assessment, that we do not lose sight of the positive opportunities available using transgenic plants in breeding programmes to develop new varieties suited to more sustainable agricultural systems, both here in Europe and in developing countries.

Acknowledgements. We thank the European Union BRIDGE Programme for supporting our collaboration on risk assessment with other European laboratories. Plant Genetic Systems (Gent, Belgium) for supplying the original transgenic seeds. We also acknowledge support from PROSAMO (Planned Release of Selected and Modified Organisms) a programme established by ten industrial companies and the UK Department of Trade and Industry and the Agricultural and Food Research Council, to investigate the environmental safety of the genetic modification of plants and micro-organisms for agricultural benefit.

References

Dale PJ (1994) The impact of hybrids between genetically modified crop plants and their related species: general considerations. Mol Ecol 3: 31–36

Dale PJ, Irwin JA, Scheffler JA (1993) The experimental and commercial release of transgenic crop plants. Plant Breeding 111: 1–22

McPartlan HC, Dale PJ (1994) An assessment of gene transfer by pollen from field-grown transgenic potatoes to non-transgenic potatoes and related species. Transgenic Res 3: 216–225

Murphy DJ (1994) Designer oil crops breeding, processing and biotechnology. VCH, Weinheim

Scheffler JA, Parkinson R, Dale PJ (1993) Frequency and distance of pollen dispersal from transgenic oilseed rape (*Brassica napus*). Transgenic Res 2: 356–364

Scheffler JA, Dale PJ (1994) Opportunities for gene transfer from transgenic oilseed rape (*Brassica napus*) to related species. Transgenic Res 3: 263–278

9 Evaluation of Potential Gene Transfer from Transgenic Plants

D. Mitten[1], K. Redenbaugh[2], and J. Lindemann[3]

[1]Mitten Consulting, 720 Borchard, Woodland, CA 95695, USA
[2]Calgene, Inc., 1920 Fifth St., Davis, CA 95616, USA
[3]Lindemann Consulting, El Cerrito, CA 94530, USA

Introduction

Calgene began commercial sales of its genetically engineered FLAVR SAVR tomato in 1994, after concluding that the FLAVR SAVR tomato was as safe as other tomatoes (FDA 1994a) and after FDA issued a food additive regulation for APH(3′)II, the product of the transformation marker gene kan^r. The FDA concluded that "use of APH(3′)II as a processing aid in the development of new varieties of tomato, oilseed rape, and cotton is safe" (FDA 1994b), amending the food additive regulations to permit the use of APH(3′)II in food for human consumption and in feed and drinking water of animals.

Calgene's Request for Food Additive Regulation for APH(3′)II (Calgene 1993) contained a complete Environmental Assessment, in which the potential for gene transfer from the transgenic crops, among many other topics, was evaluated. Two critical questions concerning gene transfer were asked: "Will gene transfer occur?" and "What would be the consequences of gene transfer?" There are two scenarios for gene transfer that are scientifically relevant: plant–to plant transfer and plant-to bacteria transfer. (Potential transfer to human epithelial cells lining the intestinal lumen was deemed to be of no concern since they are terminally differentiated – do not divide – and continually shed.) This paper summarizes the methods and results of the evaluation, using the FLAVR SAVR tomato as an example.

Potential for Plant to Plant Transfer

For plant to plant transfer, Calgene relied upon publicly available knowledge and expert opinion concerning the pollination biology of the crop. Since tomato is almost exclusively self-pollinating in the United States, and only minimal spatial

E.R. Schmidt/T. Hankeln (Eds.)
Transgenic Organisms and Biosafety

isolation is needed to maintain the purity of genetic lines, Calgene concluded that there is very little probability of the FLAVR SAVR gene moving into other varieties of cultivated tomato. However, even if this should occur, there would be no adverse effects since the only phenotypic changes in tomato are those directly associated with reduction in the polygalacturonase enzyme (Redenbaugh et al. 1993).

The issue of potential plant to plant gene transfer and the lack of adverse consequences thereof was addressed extensively by both Calgene and U.S. regulatory agencies for the FLAVR SAVR tomato. USDA concluded for example that "There is no likelihood that the FLAVR SAVR tomato will increase the weedy potential of another plant, the FLAVR SAVR tomatoes exhibit no plant pathogenic properties, are no more likely to become a weed than nonengineered tomato varieties, do not cause damage to processed agricultural commodities and are unlikely to harm other organisms that are beneficial to agriculture" (USDA APHIS 1992). Further, Calgene concluded that there is no probability of tomatoes outcrossing with any other native or introduced plant species in the United States, since there are no other sexually compatible species that occur in the United States. Two key conclusions by the FDA (FDA 1994b) were the following: (1) introduction of the kan^r gene will not confer a competitive advantage upon a plant receiving it, and (2) transfer of the kan^r gene to other crops or related weeds will have no significant adverse environmental effects.

On February 22, 1994, USDA concluded that there would be no significant environmental impact from gene transfer from Calgene's BXN cotton and issued a Determination on Nonregulated Status (USDA APHIS 1994). USDA has reached the same conclusion for Laurate canola and a Determination was issued on November 4, 1994 (USDA 1994).

Potential for Plant to Bacteria Transfer (Horizontal Gene Transfer)

For plant to bacteria transfer, publicly available knowledge and expert opinion indicated that there would be some potential that bacteria could acquire and express the kanamycin resistance gene from a transgenic crop plant.

Three questions were identified as pertinent to the issue of plant-to-bacteria transfer of the kan^r gene from transgenic crop plants:

1. What is the estimated frequency of *kan*^{r} gene transfer from transgenic crop plants to bacteria in the soil or the gut?
2. Would transfer of the *kan*^{r} gene from crops to bacteria result in a significant increase in the number of kanamycin resistant bacteria in the soil or the gut?
3. If the answer to question (2) were yes, could this potentially compromise antibiotic therapy in humans or animals?

Calgene addressed these questions by constructing models of gene transfer, using data from the published literature on in vitro transformation events and reasoned scientific judgment. Calgene did not attempt to directly measure gene transfer into wild-type strains of bacteria under simulated field or gut conditions because background levels of kanamycin resistance in bacteria are fairly high and transformation events would be so rare as to be undetectable relative to the background. Calgene consulted with experts who agreed that the existing scientific database was sufficient to evaluate the potential for horizontal gene transfer.

Based on the results of the evaluations, Calgene concluded and the FDA (FDA 1994b) concurred that:

- There will be no significant increase in exposure to kanamycin-resistant bacteria from consumption of the FLAVR SAVR tomato: at most one new *kan*^{r} bacterial cell will be produced for every 750 billion that are already present in the human gastrointestinal (GI) tract.
- There will be no significant increase in kanamycin-resistant bacteria in agricultural soils: at most one new *kan*^{r} bacterial cell will be produced for every 10 million that are already present in soil.

Gene Transfer Model: Gut

To construct the scenario for gene transfer in the human GI tract, first the estimated daily intake of DNA containing the *kan*^{r} gene that would be available for transformation was calculated. No DNA fragment large enough to contain an intact *kan*^{r} gene can be detected to survive digestive conditions for even 10 min. Based upon evidence from the scientific literature, the probability of a functional gene sequence surviving intestinal digestion may be considered zero (Berkowitz 1990). However, Calgene conservatively assumed that 0.1% of ingested DNA would be available for transformation for the sake of constructing a worst case scenario.

After 10 min each in the low pH environment of the stomach and the pancreatic secretions of the upper small intestine essentially no DNA fragments greater than 1 kb survive. Normally stomach contents are exposed to these conditions for 3–5 h rather than 10 min. Pancreatic secretions contain DNA degrading enzymes capable of breaking down both double and single stranded DNA. Upon complete digestion with pancreatic DNAse I, DNA is reduced to polynucleotides of an average length of four nucleotides. In contrast, the *kan*^{r} gene is 1000 nucleotides in length.

Intake was calculated for the average consumer of fresh fruits and vegetables and for the 90th percentile consumer, using published data from the USDA Consumer Nutrition Center. The DNA concentration available for transformation in the gut was calculated assuming that transformation would occur only in the lower small intestine and large intestine, since conditions in other parts of the GI tract appear particularly non-conducive to natural transformation.

Protein degrading enzymes in the stomach and upper small intestine would likely destroy the protein competence factor excreted by transformable *Streptococcus* species and required in high concentrations for natural transformation.

For the average consumer of fresh fruits and vegetables, a maximum of 6.7×10^{-6} mg DNA per ml of intestinal fluid would be available for transformation. This figure is increased by a factor of 2.5 for the 90th percentile consumer.

To estimate transformation in the gut, *Streptococcus* was used as the model organism since it is the only genus with species indigenous to the gut that has been demonstrated to have a natural transformation system in at least some species. Further, most of the other bacteria that colonize the lower small intestine are obligate anaerobes and as such are less capable of aminoglycoside uptake.

It should be noted that natural transformation has not been demonstrated for *Streptococcus* or other bacteria either in the human GI tract or under simulated GI tract conditions. In particular, *E. coli* is not able to be transformed under the conditions of the human gut.

Three assumptions were made concerning the conditions for gene transfer. First, Calgene assumed that all of the nonanaerobic bacteria would be transformable, even though none of the resident bacterial species are known to be transformable. Second, Calgene assumed that none of the kan^r genes would be disrupted during recombination, and third, Calgene assumed that all kan^r genes incorporated into a bacterial genome would be expressed. Although these three assumptions are extremely unlikely, they can be used to make the worst possible case.

Since the gut contains approximately 10^{12} *Streptococcus* bacteria, and 75%–86% of these are resistant to kanamycin already, the worst case scenario concludes that one new kanamycin resistant cell would be produced for every 750 billion that are already present in the GI tract. Thus, the margin of safety in the model is at least 11 orders of magnitude. The estimated transformation frequency is also five orders of magnitude less than the frequency of natural mutation to kanamycin resistance per bacterial replication.

In their final ruling, the FDA "determined that the probability of transfer of the kan^r gene to gut microflora is remote and that even under worst-case conditions, the number of microorganisms that would be converted to kanamycin resistance is negligible when compared to the reported prevalence of gut microflora that are already resistant to kanamycin" (FDA 1994b).

Gene Transfer Model: Soil

Calgene also evaluated the potential for transformation of soil bacteria, once again by constructing a hypothetical model and using available published literature and reasoned scientific judgment. Calgene concluded and the FDA concurred that at most one new *kan*^{r} bacterium would be produced via transformation for every 10 million that are already present in the soil.

Scenarios were considered for two bacterial species: the gram-positive *Bacillus subtilis* which has a natural transformation system, but no homology to the inserted DNA, and the gram-negative *Agrobacterium tumefaciens*, which has no known natural transformation system, but has homology to the inserted T-DNA border regions in the subject transgenic crops.

The models estimated the worst case and a more likely case. For example, the worst case scenario assumed that 100% of soil bacteria were transformable, which is extremely unlikely. The more likely case assumed that 10% would be transformable. The worst case assumed that every fragment of crop DNA in the soil contained an intact *kan*^{r} gene. This is extremely unlikely given that the *kan*^{r} gene constitutes less than one 100 thousandth of the tomato genome. Finally, for the worst case assumption that following recombination that the gene would always be expressed and that the gene product would be both active and stable in the bacterial cell. Once again, these events are extremely unlikely.

When put together, these factors yielded a range of values for a *Bacillus*-type transformation system:

- Two transformants per hectare in the more likely case
- 9×10^5 transformants per hectare in the worst case scenario

It follows from the calculation that, at worst, kanamycin resistant *Bacillus* transformants resulting from uptake of DNA from crop debris left in the field would constitute not more than one 10 millionth of the existing kanamycin resistant soil bacteria. Thus the margin of safety in the model is seven orders of magnitude.

For a bacterial species such as *Agrobacterium tumefaciens* that is lacking a transformation system, the estimates are three orders of magnitude less. In that case there would be at most one new *kan*^{r} bacterial cell for every 10 billion that are already present in the soil. The margin of safety in this model is ten orders of magnitude.

The FDA agreed with Calgene that "there would be no increase in kanamycin-resistant soil microorganisms because it is highly unlikely that the *kan*^{r} gene could move from the plant genome into soil microorganisms via horizontal gene transfer. Further, the agency has determined that, even if such transfer could occur, the rate at which it could occur is such that it would not result in a detectable increase over the existing background population of kanamycin-resistant bacteria" (FDA 1994b).

The gene transfer models illustrate the kind of reasoning and analysis that Calgene thinks is sufficient to evaluate the potential for gene transfer from genetically engineered plants.

References

Berkowitz DB (1990) The food safety of transgenic animals. Bio/Technology 8: 819–825

Calgene (1993) Request for food additive regulation for aminoglycoside 3′-phosphotransferase II. Food Additive Petition No. 3A4364, Docket no. 91A-0330

Redenbaugh K, Berner T, Emlay D, Frankos B, Hiatt W, Houck C, Kramer M, Malyj L, Martineau B, Rachman N, Rudenko L, Sanders R, Sheehy R, Wixtrom R (1993) Regulatory issues for commercialization of tomatoes with an antisense polygalacturonase gene. In Vitro Cell Dev Biol 29: 17–26

FDA (1994a) Calgene Inc.: Availability of letter concluding consultation. Fed Reg 59: 26647–26648

FDA (1994b) Secondary direct food additives permitted in food for human consumption; food additives permitted in feed and drinking water of animals; aminoglycoside 3′-phosphotransferase II. Fed Reg 59: 26700–26711

USDA (1994) Determination of nonregulated status for genetically engineered canola (Docket no. 94-052-2). Fed Reg 59: 55250

USDA APHIS (1992) Interpretive ruling on Calgene, Inc., petition for determination of regulatory status of FLAVR SAVR™ tomato (Docket no. 92-087-2). Fed Reg 57: 47608–47616

USDA APHIS (1994) Availability of determination of nonregulated status of Calgene, Inc., genetically engineered cotton lines (Docket no. 93-116-2). Fed Reg 59: 8452–8453

10 Efficiency of Horizontal Transfer by Natural Genetic Transformation of Plasmids with Homology to the *Pseudomonas stutzeri* Chromosome and the Effect of DNA Restriction

M.G. Lorenz, B. Meyer, and W. Wackernagel

Genetik, Fachbereich Biologie, Carl von Ossietzky Universität, Postfach 2503, 26111 Oldenburg, Germany

Introduction

Natural genetic transformation, like conjugation and transduction, is a mechanism of horizontal gene transfer between bacteria. During transformation DNA is released from a donor cell by excretion or cell lysis and subsequently taken up by a recipient cell which has gained this ability as the result of the development of the physiological state of competence. There are several indirect and direct indications that transformation occurs in the environment and contributes to adaptational processes in bacterial populations (Lorenz and Wackernagel 1993, 1994). A goal of experimental genetic ecology of transformable bacteria is to understand the stimulating and inhibiting factors which in concert tune the rate of transformation in the environment to certain levels. Among other factors DNA restriction has recently been put forward as a possible barrier to transformation (Lorenz and Wackernagel 1994).

Investigations of transformation in the environment should include as many model organisms as possible with well characterised transformation systems. Several features of *P. stutzeri* with regard to its ecology and physiology of transformation make this bacterium a suitable model organism. *P. stutzeri* is a widely occurring denitrifying bacterium living in soils, sediments and water. Competence is regulated by the limitation of nutrients (Lorenz and Wackernagel 1991, 1992), a situation frequently encountered by free-living bacteria. DNA-binding/uptake is selective for homologous DNA, otherwise typical of transformable pathogenic bacteria (Lorenz and Wackernagel 1994).

In this communication the transformation system of *P. stutzeri* is further analysed with regard to the physiology of transformation by plasmids and the influence of DNA restriction on transformation.

E.R. Schmidt/T. Hankeln (Eds.)
Transgenic Organisms and Biosafety

Materials and Methods

For the transformation experiments plasmid pKT210 (Cm^r, Sm^r) was used which is a derivative of the broad host range plasmid RSF1010 (Bagdasarian et al. 1981). To study the effect of homology to the chromosome of *P. stutzeri* on plasmid transformation, plasmid pWI85 was used. This plasmid was isolated from a *P. stutzeri* gene bank which was obtained by ligation of partially *Pst*I-restricted *P. stutzeri* chromosomal DNA to *Pst*I-digested RSF1010. pWI85 carries an 8.5-kb insert of *P. stutzeri* chromosomal DNA complementing the Leu auxotrophy of the *P. stutzeri* recipient. Transformation experiments were done using the plate transformation procedure (Lorenz and Wackernagel 1991).

Results and Discussion

The pKT210 DNA concentration dependence of transformation indicated a two-hit process, i.e. two plasmid molecules were necessary to transform a *P. stutzeri* cell. At a transformation-saturating concentration of DNA (0.5 μg/assay) about 4×10^9 plasmid molecules were needed to obtain a transformant, indicating that transformation of *P. stutzeri* by pKT210 was very inefficient. With pWI85, which carries a DNA fragment of the *P. stutzeri* chromosome complementing the *leu* mutation, about 1.5×10^7 (Sm^r selection) and 6×10^4 (Leu^+ selection) plasmid molecules, respectively, sufficed to generate a transformant. Hence, homology to the chromosome enhanced the efficiency of plasmid transformation about 270 (Sm^r selection) and 66 700-fold (Leu^+ selection). Also, the dependence of transformation on the DNA concentration indicated a one hit process: different from pKT210 only one pWI85 plasmid was required to transform a *P. stutzeri* cell. The one-hit process of transformation with plasmid DNA containing a chromosomal insert resembles the one hit process of transformation with markers of a chromosomal DNA preparation.

In a test for co-transformation of the Sm^r and Leu^+ markers of pWI85 it was found that at Leu^+ selection the Sm^r marker was lost at a frequency of 87%. The few Leu^+ Sm^r transformants contained pWI85, whereas the Leu^+ Sm^s transformants did not contain a plasmid. Probably the region of the plasmid covering the Leu^+ marker was integrated into the recipient's chromosome and the rest of the plasmid was lost. Transformants of a *recA* mutant of *P. stutzeri* all had the non-selected Sm^r marker, and pWI85 was present. Upon Sm^r selection 49% were also Leu^+. They all had pWI85. Sm^r Leu^- clones were plasmid-free, presumably representing spontaneous Sm^r mutants.

These data suggest the following model of plasmid transformation in *P. stutzeri*. During uptake into the cell plasmid DNA (pKT210) is processed in a way that the intracellular generation of an intact replicon depends on the separate entry of two plasmid molecules. Probably plasmid molecules enter the cell in the linear single-stranded form as known to occur in other transformable bacteria (see Lorenz and Wackernagel 1994). Complementary, overlapping regions of two plasmid single strands hybridise, and subsequent repair synthesis generates an intact replicon. When the plasmid (pWI85) contains a region homologous to the chromosome two alternative routes exist for plasmid transformation in *P. stutzeri*. The major route leads to the integration of the plasmid's homologous nucleotide sequences into the chromosome by a replacement process depending on homologous recombination (i.e., no replacement in *recA* mutants). Transformation via the minor route is based on a homology-facilitated regeneration of an intact plasmid replicon. This process might be similar to that described in the model of plasmid transformation of *Bacillus subtilis*. According to this model a double strand break is introduced into the region of the plasmid homologous to the chromosome by a cell surface-associated endonuclease. The entering plasmid single strand synapses to the homologous region in the chromosome, and a double-stranded replicon is generated by DNA repair synthesis (Canosi et al. 1981).

P. stutzeri appears to possess a DNA restriction system. This was concluded from the fact that the frequency of transformation by electroporation with pWI85 isolated from *Escherichia coli* ($pWI85_{Ec}$) was more than two orders of magnitude lower than with pWI85 isolated from *P. stutzeri* ($pWI85_{Ps}$). Natural transformation of *P. stutzeri* by $pWI85_{Ec}$ was 36 times less efficient than natural transformation by $pWI85_{Ps}$ (Leu^+ selection). This indicated that DNA restriction is active during natural transformation of *P. stutzeri*. Co-transformation analysis revealed a substantial reduction of the frequency of the unselected markers when unmodified plasmid ($pWI85_{Ec}$) was used instead of modified ($pWI85_{Ps}$). These results suggest that DNA restriction acts upon both routes of homology-facilitated plasmid transformation, with a stronger effect on the minor route (reconstitution of plasmids; see above).

Conclusions

Natural genetic transformation of the ubiquitous bacterium *P. stutzeri* by a derivative of the broad host range plasmid RSF1010 was found to be rather inefficient. However, the presence on the plasmid of a nucleotide sequence homologous to DNA residing in the recipient cell increased the efficiency of transformation up to 66 700-fold. From an ecological point of view this finding implies that rates of exchange of plasmids by transformation among cells of *P. stutzeri* populations harboring related plasmids could be as high as the exchange of chromosomal genes. Like in the majority of species tested (Lorenz and Wackernagel 1994) DNA restriction in *P. stutzeri* may play a role as a barrier to natural transformation, although not to the extent that interspecies transformation is abolished (for a review, see Lorenz and Wackernagel 1994).

Acknowledgements. This work was supported by the BMFT and the Fonds der Chemischen Industrie.

References

Bagdasarian M, Lurz R, Rückert B, Franklin FCH, Bagdasarian MM, Frey J, Timmis KN (1981) Specific-purpose plasmid cloning vectors. II. Broad host range, high copy number, RSF1010-derived vectors, and a host-vector system for gene cloning in *Pseudomonas*. Gene 16: 237–247

Canosi U, Iglesias A, Trautner TA (1981) Plasmid transformation in *Bacillus subtilis*: effects of insertion of *Bacillus subtilis* DNA into plasmid pC194. Mol Gen Genet 181: 434–440

Lorenz MG, Wackernagel W (1991) High frequency of natural genetic transformation of *Pseudomonas stutzeri* in soil extract supplemented with a carbon/energy and phosphorus source. Appl Environ Microbiol 57: 1246–1251

Lorenz MG, Wackernagel W (1992) Stimulation of natural genetic transformation of *Pseudomonas stutzeri* in extracts of various soils by nitrogen or phosphorus limitation and influence of temperature and pH. Microb Releases 1: 173–176

Lorenz MG, Wackernagel W (1993) Bacterial gene transfer in the environment. In: Wöhrmann W, Tomiuk J (ed) Transgenic organisms – risk assessment of deliberate release. Birkhäuser, Basel, pp 43–64

Lorenz MG, Wackernagel W (1994) Bacterial gene transfer by natural genetic transformation in the environment. Microbiol Rev 58: 563–602

11 Accidental Formation of Replication-Competent Viruses from Gene Transfer by Retroviral Vectors

D. Klehr-Wirth, M. Wirth, R. Grannemann, C. Münk, and H. Hauser

Gesellschaft für Biotechnologische Forschung mbH, Mascheroder Weg 1, 38124 Braunschweig, Germany

Abstract

Retrovirus mediated gene transfer is one of the most powerful tools for the manipulation of a variety of cells. Beside its significance for the transduction of cultured cells for basic research and for biotechnology it has become a predominant method in gene therapy trials. Retroviral vectors provide a number of advantages compared to other transfer systems. However, the release of newly combined viruses with or without autonomous replication competence imposes higher safety conditions in handling such systems and makes them unacceptable in gene therapy. Here, we describe experimental systems which allow to define properties of cells, retroviral helper functions and retroviral vectors leading to the release of newly recombined retroviruses. The results allow us to estimate the safety of particular systems and to construct safer helper cells and vectors.

Introduction

Retroviruses and Retroviral Vectors

Retroviral particles harbour two +-strand RNA genomes encoding the viral core proteins (the gag proteins CA, NC and MA), the enzymatic functions of pol (reverse transcriptase, integrase and protease) and the env proteins (SU and TM) which constitute the viral envelope (for a recent review, see Luciw and Leung 1992). The SU protein determines the host range of the virus: infection by ecotropic viruses is restricted to murine cells only, amphotropic viruses can infect a broad range of cells including murine and human cells and xenotropic SU proteins confer infectibility to certain non-murine cells. After infection, the viral RNA is reversely transcribed into double-stranded DNA by the reverse

E.R. Schmidt/T. Hankeln (Eds.)
Transgenic Organisms and Biosafety

transcriptase. The double-stranded DNA intermediate is then integrated into the host's genome by the integrase. The viral genes are transcribed and translated by the cellular machinery. For virus assembly two full-length transcripts associate with the viral polyproteins gag and gag-pol. The specific packaging of the retroviral RNA is mediated by a structural RNA element, the so-called packaging signal (ψ). During the process of budding from the plasma membrane viruses obtain their outer membrane presenting the SU protein.

For retroviral gene transfer helper cells which carry modified retroviruses are used (for review, see Miller 1992; Miller et al. 1993; Jolly 1994). Most of them are based on the murine Moloney leukemia virus, MLV. Helper cells provide retroviral proteins for the formation of virions but are supposed to release no infectious particles as the retroviral genome has been immobilized by deletion of the packaging signal ψ. The packaging cells are transfected with retroviral vectors in which the retroviral genes are substituted by the genes of interest. Retroviral vectors are composed of non-coding retroviral elements, namely two long terminal repeats (LTRs), a polypurine tract (ppt) and ψ which are needed for reverse transcription, integration and efficient packaging. By means of this physical separation of the viral coding elements required for the formation of virus particles and the viral elements necessary for packaging and reverse transcription of the RNA, the virus mediated gene transfer is intended to be limited to one cycle of infection; as a consequence, only the retroviral vector should be transduced; thus, the infected cell should not be able to release virus particles itself.

Mechanisms for the Accidental Formation of Newly Recombined Viruses

During the last years several observations have been reported which show that retrovirus mediated gene transfer can give rise to newly recombined viruses (Bosselmann et al. 1987; Scarpa et al. 1991; Otto et al. 1994). As a consequence, infected cells obtain undesired properties or even become producer cells themselves, a process which renders the transfer uncontrollable. These findings limit the use of retroviral gene transfer, especially for gene therapy purposes.

There are two predominant mechanisms that are responsible for alterations of the properties of infectants: *recombination* and *retrofection.*

Homologous recombination is a frequent process. The efficiency depends on the degree of homology between the retroviral vector and the helper function. As a result of a recombination event immobilized helper functions regain their capacity to replicate autonomously and thus replication-competent retroviruses (RCRs) spread through the culture.

These RCRs represent the main safety problem accompanying the retrovirus mediated gene transfer; amphotropic RCRs obtained from recombination events in MLV based packaging cells were shown to replicate in non-human primates and to induce lymphoma (Donahue et al. 1992; Vanin et al. 1994).

Apart from the frequent recombination during transfection of cells (extrachromosomal DNA recombination), during cultivation of retroviral helper cells two additional recombination pathways have to be considered. Generally, recombination may occur within transfected packaging cells on the level of DNA or during reverse transcription on the level of RNA. While DNA recombination is a purely cellular process performed by the cellular repair machinery, RNA recombination is coupled with the retroviral life cycle and mediated by a viral enzyme, the reverse transcriptase. The mechanisms of these two pathways are shown in Fig. 1. Up to now the percental contribution of these mechanisms for the formation of RCRs has not been determined.

Prerequisite for RNA based recombination is the copackaging of the vector and the helper RNA into the same virus particle. Although the helper RNA lacks the ψ packaging signal it may be packaged at a low rate. It was shown that the deletion of the ψ packaging signal may reduce the rate of packaging only about 10^3- to 10^4-fold (Mann and Baltimore 1985; Stuhlmann et al. 1990). Therefore, high titer packaging cell lines producing 10^6 virus particles per ml release a significant number of virions carrying the helper RNA. These helper RNAs may recombine and thereby give rise to the formation of RCRs.

In the last few years different helper/vector systems have been developed. In order to increase the level of safety much effort was spent to decrease the risk of recombination in packaging cells. The subdivision of the helper genome in two independent transcription units resulted in helper cells in which more than one recombination event is required for the formation of RCRs. In addition, the extent of homology between vector and helper genes was significantly reduced. In these sophisticated packaging cells (e.g., ΩE cells, Morgenstern and Land 1990) the frequency of RCR formation is supposed to be reduced.

Further investigations showed that even cellular RNAs without any retroviral elements may be packaged to a low extent, a process which was termed retrofection (Linial 1987; Linial and Miller 1990). In MLV based packaging cells the rate of retrofection is reduced 10^6- to 10^8-fold if compared to functional retroviral vectors (Stuhlmann et al. 1990; Klehr-Wirth, unpublished). The risk emanating from the packaging of cellular genes is probably low as their RNAs may not compensate for any viral function and thus could not contribute to the formation of RCRs. However, it cannot be excluded that certain cellular RNAs may form a structure that is similar to the retroviral ψ sequence and are preferentially transmitted. A safety problem arises if these RNAs are able to change the properties of infected cells, as, e.g., oncogenes do.

Another potential problem using the retrovirus mediated gene transfer arises from endogenous retroviral elements. It was shown that particularly rodent cells carry a huge number of endogenous retroviral elements. Usually, they are not able to form infectious viruses themselves, however, they are potential candidates for recombinational events both on DNA and on RNA level and thus may contribute to the formation and release of new, autonomously replicating retroviruses.

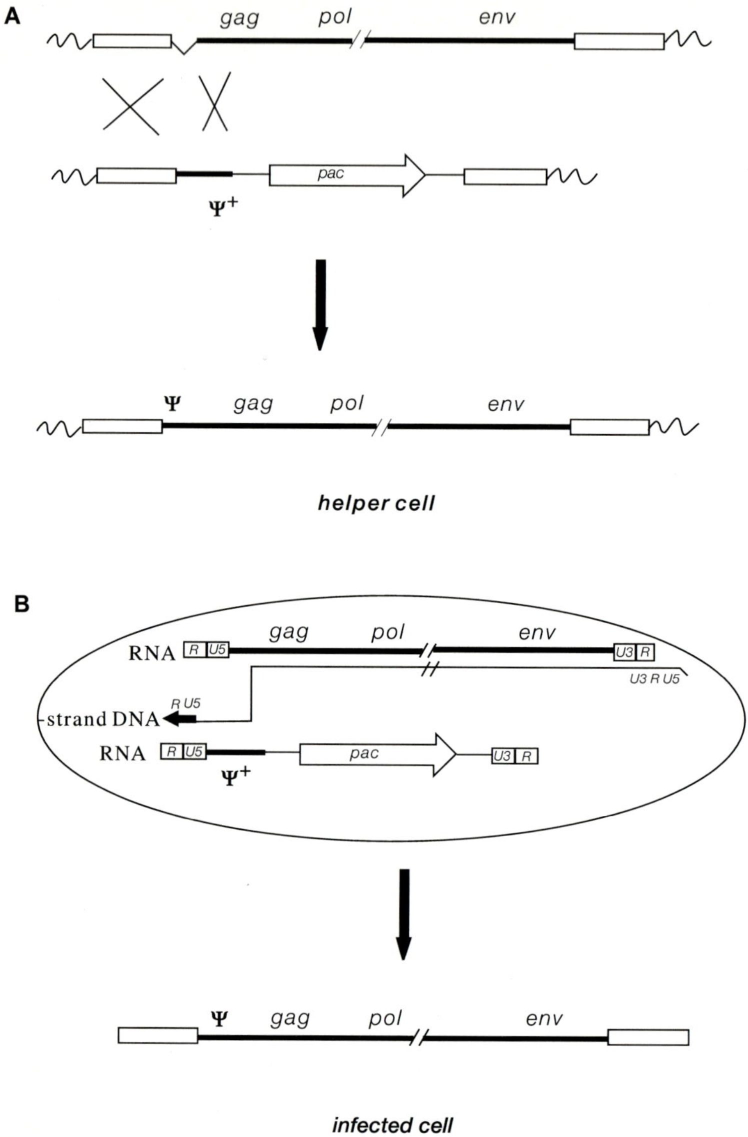

Fig. 1. The formation of replication-competent retroviruses (RCRs) by two different mechanisms, DNA recombination **A** and RNA recombination **B.** *Open boxes,* retroviral long terminal repeats (LTRs); *waved lines,* chromosomal host DNA; *thick lines,* helper genes; *pac,* puromycin acetyl transferase gene. The *oval* represents the viral capsid within the infected cell. See text for further descriptions

Results

A Patch-Repair Assay for the Detection of Functional, Endogenous Elements.

To define the endogenous retroviral elements in cell lines different biochemical assays have been performed. However, these assays (e.g., Southern analysis and PCR) are based on sequence homologies but do not reveal any information about the (partial) functionality of different endogenous elements.

In order to identify functional endogenous elements that have the capacity to complement viral functions we built up a patch-repair assay (Fig. 2). For this assay a virus was used (pMLVdhfr-7; Stuhlmann et al. 1989) which carries a marker gene in the U3 element of the 3′ LTR. This marker (dihydrofolate reductase) confers resistance to methotrexate and infectants may be easily detected by selection. In order to identify endogenous cellular sequences being able to complement a helper function we deleted a 1.8-kb element within the env gene of pMLVdhfr-7. The defective provirus was transfected into different cells. Due to recombination events with functional cellular elements the deficient virus should be repaired and thereby regain its capacity to replicate autonomously. This was proved by testing the supernatant for the presence of virus particles conferring resistance to methotrexate.

Murine NIH3T3 cells, human 293 cells, BHK-21 cells and recombinant NIH3T3 cells carrying transfected copies of promoterless env genes (NIH/env) as controls, were tested for their ability to complement the deletion in the env gene. As depicted in Table 1, autonomously replicating, dhfr transducing retroviruses

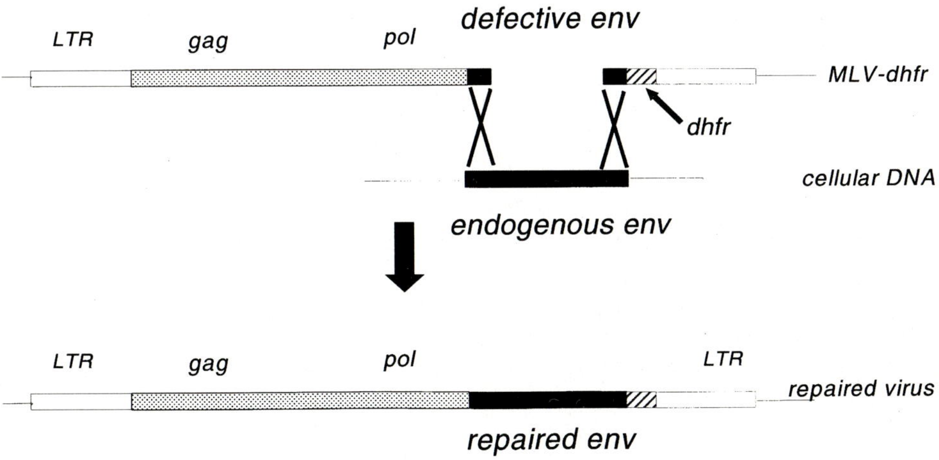

Fig. 2. Patch-repair assay for the detection of functional endogenous env elements. *LTR*, long terminal repeat; *MLV*, Moloney leukemia virus

Table 1. Detection of functional env elements in standard cell lines

Cell line	Reversion[a](patch repair)	PCR	Nested PCR
NIH3T3	3×10^4 (P 14–18)	+++	+++
293	0 (P>50)	–	–
BHK-21	n.d.	–	+
NIH/env	3×10^4 (P, 2–6)	n.d.	n.d.
Ψ_2	n.d.	+++	+++

P, passage number at the time of first virus detection; n.d., not determined; PCR, polymerase chain reaction; +++, gag sequences from murine C-type viruses detected in chromosomal DNA of indicated cells; –, no gag sequences detected by PCR.
[a]Titer of released dhfr-transmitting retroviruses (cfu/10^6 cells per ml).

were released from NIH3T3 cells 14–18 passages after transfection. NIH/env cells harbouring additional, transcriptionally silent copies of env turned positive after two to six passages. In contrast, no recombination event could be detected in human 293 cells, even after extended passages. The results obtained by the patch-repair assay confirmed our concomitant PCR results. However, in contrast to the PCR assay which detects retroviral elements independent of their biological activity the patch-repair assay specifically detects functional elements.

The virus obtained from patch-repair in NIH3T3 cells was cloned and the region of interest was sequenced. Interestingly, the captured env copy reveals a pronounced homology to ecotropic env genes.

These results show that the NIH3T3 cell line provides functional endogenous retroviral elements with the ability to complement a defective retrovirus. Furthermore, the result implies that helper cells based on murine NIH3T3 cells may alter their initially introduced helper function by recombination.

Formation of Replication-Competent Retroviruses During Retrovirus-Mediated Gene Transfer

In order to measure the frequency of recombination in retroviral packaging cells we established a sensitive test system. For this system we used two different ecotropic packaging cell lines, the $\psi 2$ cells developed by Mann et al. (1983) and the ΩE cells constructed by Morgenstern and Land (1990).

$\psi 2$ cells carry a MLV provirus with a deletion of the packaging sequence in order to immobilize the helper function. For the generation of RCRs only one recombination event is sufficient. In ΩE cells the viral genome is subdivided in two independent transcription units. In addition, non-retroviral polyadenylation signals were used and homologies within coding regions were reduced by codon wobbling. In ΩE cells at least two recombination events would be needed for the generation of RCRs.

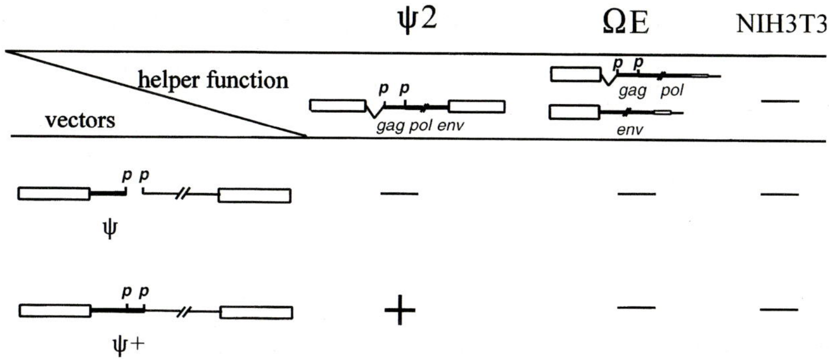

Fig. 3. Homologous recombination in retroviral packaging cells. +, recombination-competent retroviruses (RCRs) detectable as described in the text

These cells were transfected with retroviral vectors carrying a resistance marker (puromycin acetyl transferase, pac) and propagated in selective medium. The vectors differ in the length of the packaging sequence: both the minimal (ψ, nt 212–563, according to the MLV sequence published by Shinnick et al. 1981) and an extended encapsidation signal (ψ^+, nt 212–743) was used (Fig. 3). The ψ^+ sequence is part of the 5′ gag coding region and thereby provides 180 bp of homology to the helper function. Since extension of the minimal packaging sequence significantly increases the virus titer, retroviral vectors frequently carry packaging elements (nt 212–1038) which provide even 475 bp of homology (Armentano et al. 1987; Bender et al. 1987; Morgenstern und Land 1990).

The release of RCRs was detected by a marker rescue assay. For this assay, the primary infectants are tested for their ability to release pac transducing viruses. Infected cells can efficiently release virus particles themselves if they have received a helper function as a consequence of a recombination event between a copackaged vector and helper RNA. This replication-competent retrovirus is able to spread through the culture of infected cells and mobilizes retroviral vectors.

As shown in Fig. 3, RCRs were observed after transfection of the ψ^+ vector into $\psi 2$ packaging cells. These RCRs were detected shortly after selection of the helper cells, indicating that the recombination event occurred at an early time point. No RCR formation was observed using a retroviral vector without any internal homology region even after prolonged culture (16 passages after transfection). This result reflects that the rate of recombination significantly depends on the presence of internal homologous elements in vector and helper functions.

ΩE cells did not give rise to RCR formation with any of the vectors during extended passaging. This is due to the fact that in ΩE cells a single recombination event is insufficient to form RCRs.

Mechanisms of Replication-Competent Retrovirus Formation

For the safe use of the retroviral gene transfer it is predominantly necessary to rule out recombination events between helper functions and retroviral vectors. Reduction of homologies and the subdivision of helper functions are helpful but insufficient precautions. In order to generally rule out the possibility of recombination profound knowledge of the mechanism of recombination is required. In particular, it is required to define the relevance of the two different pathways, namely RNA and DNA recombination.

If DNA recombination occured only during transfection, a helper cell line producing recombinant retroviruses, once established, could be characterized and defined to be 'safe'. In contrast, if RNA recombination has to be considered and/or DNA recombination also occurs after establishment of retroviral vectors, a producer cell cannot be entitled 'safe'.

In order to distinguish between the two different recombination pathways we constructed cells carrying transcriptionally active and inactive helper functions (Fig. 4). For this approach, NIH3T3 cells were stably transfected with the indicated helper functions. To eliminate cells expressing the promoterless construct due to insertion into a transcriptionally active site, cells were selected for the absence of env expression by FACS sorting. In contrast, NIH3T3 cells transfected with helper functions driven by the SV40 early promoter were sorted according to a high expression of env. These 'packaging' cells were then stably transfected with the ψ^+ vector. The formation of RCRs was monitored in mixtures of transfected 'packaging' cells. As shown in Fig. 4, all cell populations encoding the transcribed helper function released RCRs. However, only one of

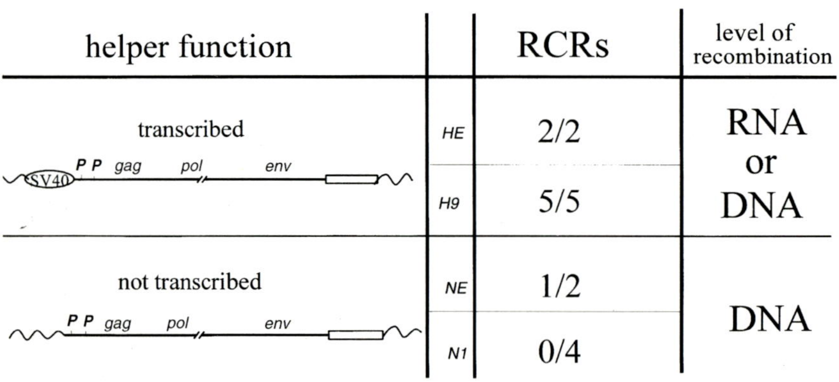

Fig. 4. Formation of recombination-competent retroviruses (RCRs) depends on transcriptionally active helper functions. The number of cultures that turned out positive is indicated. *HE*, pools of transfected NIH3T3 cells expressing the helper genes; H9, selected high producer cell clone; *NE*, pools of transfected NIH3T3 cells containing silent helper genes; N1, cell clone with silent helper genes

six pools encoding the transcriptionally 'silent' function gave rise to RCR formation.

The low rate of recombination in cells carrying the non-transcribed helper function indicates that DNA recombination with transcriptionally inactive genes is a rare event. At present we do not know if the chromatin status of the transcriptionally active helper function might influence the rate of DNA recombination.

Discussion

In order to increase biological safety in using the retroviral gene transfer, undesired events and their frequencies have to be uncovered. This should allow us to improve the retroviral vector and helper cell lines with respect to biological safety. We have established experimental systems for the evaluation of existing retroviral transfer systems. The results from these systems should enable the definition of the predominant molecular level which leads to the formation of unwanted retroviruses.

The patch-repair assay was performed to identify endogenous retroviral elements that have the capacity of modifying helper functions. It was shown that endogenous elements of NIH3T3 cells can be involved in the creation of newly recombined autonomously replicating viruses. This result demonstrates that the chromosomal background of helper cells directly influences the safety of the retroviral gene transfer, since endogenous elements supply recombination targets for helper functions. This holds true for all MLV based packaging cells based on the NIH3T3 cell, even the benefit of highly sophisticated packaging cells like ΩE cells is limited.

For improvement of the retroviral gene transfer our results suggest the utilization of cell lines that do not harbour functional endogenous retroviral elements. In this respect, non-rodent cell lines represent a favourable basis for the development of helper cells. In recent developments problems concerning the cellular background are taken into account. Construction and characterization of helper cells based on human 293 or dog D17 and CF2 cells are in progress (M. Wirth, unpublished; Jolly 1994). While our results show that 293 cells can be considered to be suitable for helper cell development, corresponding characterization of the dog cells is still missing.

Recombination analysis shows that retroviral vectors providing a high retrovirus titer (ψ^+ vectors) carry an increased risk for recombination due to the extended packaging sequence. With these vectors, RCRs are frequently generated and easily detectable if only one recombination event is required to generate RCRs, e.g., as in ψ2 cells. Therefore, the use of these high titer vectors in simple packaging cells is not advisable.

Although these vectors do not give rise to RCRs in the safer ΩE packaging cells after short term cultivation, the occurrence of RCR formation cannot be generally excluded during extended cultivation. Since in these cells at least two recombination events are needed for RCR formation, RCRs will hardly be measurable in laboratory scale. However, even in these cells RCR formation could occur assuming subsequent recombination events with endogenous elements. As a conclusion, the use of the ψ^+ vectors even in ΩE cells cannot be considered to guarantee a safe retrovirus mediated gene transfer.

Alternative packaging signals with reduced homology to the helper genes could help to minimize the risk of recombination but guarantee a high rate of packaging. The rat VL30 packaging element (Torrent et al. 1992, 1994) is a possible candidate. Since rodent cells carry hundreds of endogenous VL30 elements again a non-rodent background for helper cells would be preferred.

In order to control the frequency of recombination we need to understand the contribution of the two recombination pathways for the formation of RCRs. We could show that both transcriptionally active and inactive helper functions may recombine. Therefore, it has to be assumed that both recombination pathways contribute to RCR formation.

To eliminate the risk of DNA-based recombination, recombination-deficient cells are required. The use of cells with a low recombinogenic potential as a basis for packaging cells should significantly reduce the level of chromosomal recombination and thus guarantee a stable, non-rearranged helper and vector persistance. Apart from the benefit of such a cell line with respect to safety aspects, it would be advantageous for biotechnological purposes since these cells will maintain introduced genes in a stable, non-rearranged manner.

On the other hand, regulation of transcription could specifically reduce RNA recombination. An inducible helper gene expression would limit the time span in which recombination is possible. In future a specific modification of the reverse transcriptase, in order to reduce its tendency to template switches, might be of interest.

There are some reports indicating that transcription may also enhance the rate of DNA recombination (Nickoloff and Reynolds 1990; Nickoloff 1992), indicating that an open chromatin structure as provided in transcriptionally active genes is required for the action of the cellular recombination machinery mediating DNA recombination. In this respect, even DNA recombination might be obviated by a strict regulation of helper gene transcription.

References

Armentano D, Yu S-Y, Kantoff PW et al. (1987) Effect of internal viral sequences on the utility of retroviral vectors. Virology 183: 611–619

Bender MA, Palmer TD, Gelinas RE, Miller AD (1987) Evidence that the packaging signal of moloney murine leukemia virus extends into the gag region. J Virol 61: 1639–1646

Bosselmann RA, Hsu R-Y, Bruszewski J, Hu S, Martin F, Nicoson M (1987) Replication-defective chimeric helper proviruses and factors affecting generation of competent virus: expression of moloney murine leukemia virus structural genes via the metallothionein promoter. Mol Cell Biol 7: 1797–1806

Donahue RE, Kessler SW, Bodine D, McDonagh K, Dunbar C, Goodman S, Agricola B, Byrne E, Raffeld M, Moen R, Bacher J, Zsebo KM, Nienhuis AW (1992) Helper virus induced T cell lymphoma in nonhuman primates after retroviral mediated gene transfer. J Exp Med 176: 1125–1135

Jolly D (1994) Viral vectors systems for gene therapy. Cancer Gene Therapy 1: 51–64

Linial ML, (1987) Creation of a processed pseudogene by retroviral infection. Cell 49: 93–102

Linial ML, Miller AD (1990) Retroviral RNA packaging: sequence requirements and implications, In: Swanstrom R and Vogt PK (eds.), Retroviruses: strategies of replication, Springer, Berlin p. 125–152

Luciw AP, Leung NJ (1992) Mechanisms of retrovirus replication. In: Levy JA (ed) The retroviridae. Plenum Press, New York London p159–298

Mann R, Baltimore D (1985) Varying the position of a retrovirus packaging sequence results in the encapsidation of both unspliced and spliced RNAs. J Virol 54: 401–407

Mann R, Mulligan RC, Baltimore D (1983) Construction of a retrovirus packaging mutant and its use to produce helper-free defective retrovirus. Cell 33: 153–159

Miller AD (1992) Retroviral vectors. In: Muzyczka N (ed.) Viral expression vectors. Springer, Berlin Heidelberg New York, p 1–24

Miller AD, Miller DG, Garcia JV, Lynch CM, (1993) Use of retroviral vectors for gene transfer and expression. Meth Enzymol 217: 581–599

Morgenstern JP, Land H (1990) Advanced mammalian gene transfer: high titre retroviral vectors with multiple drug selection markers and a complementary helper-free packaging cell line. Nucleic Acids Res 18: 3587–3596

Nickoloff JA, Reynolds RJ (1990) Transcription stimulates homologous recombination in mammalian cells. Mol Cell Biol 10: 4837–4845

Nickoloff JA (1992) Transcription enhancers intrachromosomal homologous recombination in mammalian cells. Mol Cell Biol 12: 5311–5318

Otto E, Jones-Trower A, Vanin EF, Stambaugh K, Mueller SN, Anderson WF, McGarrity GJ (1994) Characterization of a replication-competent retrovirus resulting from recombination of packaging and vector sequences. Hum Gene Ther 5: 567–575

Scarpa M, Cournoyer D, Muzny DM, Moore KA, Belmont JW, Caskey CT (1991) Characterization of recombinant retroviruses from moloney-based vectors in ecotropic and amphotropic packaging cell lines. Virology 180: 849–852

Stuhlmann H, Jaenisch R, Mulligan RC (1989) Construction and properties of replication-competent murine retroviral vectors encoding methotrexate resistance. Mol Cell Biol 9: 100–108

Stuhlmann H, Dieckmann M, Berg P (1990) Transduction of cellular neo mRNA by retrovirus-mediated recombination. J Virol 64: 5783–5796

Torrent C, Wang P, Darlix J-L (1992) A murine leukemia virus derived retroviral vector with a rat VL30 packaging psi sequence. Bone Marrow Transplant 9: 143–147

Torrent C, Gabus C, Darlix J-L (1994) A small and efficient dimerization/packaging signal of rat VL30 RNA and its use in murine leukemia virus-VL30-derived vectors for gene transfer. J Virol 68: 661–667

Shinnick TM, Lerner RA, Sutcliffe JG (1981) Nucleotide sequence of Moloney murine leukaemia virus. Nature (London) 293: 543–548

Vanin EF, Kaloss M, Broscius C, Nienhuis AW (1994) Characterization of replication-competent retroviruses from nonhuman primates with virus-induced T-cell lymphomas and observations regarding the mechanism of oncogenesis. J Virol 68: 4241–4250

12 Transgenic Expression of a Toxin-Coding Killer Virus of the Yeast *Zygosaccharomyces bailii* in *Saccharomyces cerevisiae*: Genetic Evidence for a Possible Function of "Cryptic" Mycoviruses in the Evolution of Their Hosts

F. Neuhausen and M.J. Schmitt

Institut für Mikrobiologie und Weinforschung, Johannes Gutenberg-Universität Mainz, 55099 Mainz, Germany

Abstract

The killer toxin-secreting yeast *Zygosaccharomyces bailii* 412 contains two cytoplasmically inherited double-stranded RNA (dsRNA) viruses (ZbV-L, ZbV-M) responsible for the expression of a killer phenotype in its infected host. While ZbV-L functions as a classical helpervirus by providing capsid (cap) and RNA polymerase functions (cap/pol) necessary for packaging and replication of both viruses, M-dsRNA-containing killer viruses (ZbV-M) are satellites of ZbV-L that contain the genetic information for toxin production only. Both viruses were shown to be sufficient to confer the *Z. bailii* killer phenotype upon transfected spheroplasts of a *S. cerevisiae* non-killer strain, resulting in toxin-secreting transfectants. Genetic analysis of transfected *S. cerevisiae* strains indicated that the heterologous expression of a *Z. bailii* killer phenotype depends on the expression of a single chromosomal gene: *KEX3*. Thus, *kex3* mutants of *S. cerevisiae* are phenotypically non-killers showing normal, i.e. wild-type levels of the toxin-coding M-dsRNA. In the genetic background of a *kex3* mutation, the stably maintained ZbV-M killer viruses would be regarded as "cryptic", because of the lack of any dsRNA-associated phenotype. Cytoductive transfer of ZbV-M from "cryptic" non-killers into *KEX3* non-killer recipient strains resulted in heterozygous (*kex3/KEX3*) diploids that expressed a strong killer phenotype. However, meiotic segregants derived from dissection of *kex3/KEX3* diploids showed (a) a loss in cell viability and (b) a significant selection of a certain genotype among the surviving haploid progenies. We describe a model in which Zb-M killer viruses lead to a suicidal phenotype of haploid segregants under conditions in which Kex3p is fully expressed and an additional mutation within either of two chromosomally encoded toxin-sensitivity genes (*TOS1* and/or *TOS2*) is present.

E.R. Schmidt/T. Hankeln (Eds.)
Transgenic Organisms and Biosafety

Survival of these strains is only possible in the presence of a *kex3* mutation, leading to stably maintained ZbV-M viruses that, because of the lack of any detectable phenotype, would be regarded as "cryptic". In this respect we discuss a more general model postulating a remarkable effect of such "cryptic" mycoviruses on the evolution of the virus-infected host.

Transfection of *S. cerevisiae* with the ZbV-Killer Virus of *Z. bailii*

The wild-type killer yeast *Z. bailii* 412 harbors three distinct, cytoplasmically inherited dsRNA viruses (ZbV-L, ZbV-Z, ZbV-M) which in the case of ZbV-L and ZbV-M are responsible for the expression of a stable killer phenotype in this strain (Table 1). While the 4.7-kB LA-dsRNA of ZbV-L encodes the major capsid protein (cap) and the RNA-dependent RNA polymerase (cap/pol), the M-dsRNA genome (1.8 kB) of ZbV-M was shown to encode the secreted 10-kDa killer toxin KT412 responsible for the receptor-mediated killing of sensitive strains (Radler et al. 1993; Schmitt and Neuhausen 1994). Both ZbV-L and ZbV-M resemble icosahedral viral particles 40 nm in diameter consisting of the LA-encoded 85-kDa major capsid protein and of a minor 180-kDa cap/pol-fusion protein (Schmitt and Neuhausen 1994).

In order to identify a possible transgenic effect of the toxin-coding ZbV-M mycovirus naturally persisting in the wild-type killer yeast *Z. bailii* 412 after heterologous expression in *S. cerevisiae*, we used purified ZbV-L and ZbV-M viruses to infect spheroplasts of the non-killer strain *S. cerevisiae* GG100-14D; a strain which was previously shown to be suitable for successful uptake and homologous expression of ScV-M killer viruses derived from K1, K2 and K28 killer strains of the same yeast species (Schmitt and Tipper 1990; Schmitt and Neuhausen 1994). The resulting transfectants (strains MS412a and MS412b)

Table 1. dsRNA-viruses of the killer yeast *Zygosaccharomyces bailii* 412

Virus	dsRNA/size (kB)	Encoded gene(s) and function(s)[a]
ZbV-L	LA (4.7)	(cap), major capsid protein (85 kDa) (cap/pol), minor capsid/polymerase fusion protein (necessary for viral packaging and replication)
ZbV-M	M (1.8)	(pptox), killer toxin precursor (only expressing toxicity, no immunity)
ZbV-Z	Z (2.8)	(cap), major capsid protein (35 kDa) (cap/pol), capsid/polymerase fusion protein (autonomously replicating dsRNA-virus not involved in the killer phenomenon of its host)

ZbV, *Z. bailii* virus.
[a]For references, see Schmitt and Neuhausen (1994) and Wickner (1992).

showed a characteristic *Z. bailii* killer phenotype and a stable inheritance of the toxin-coding M-dsRNA genome. Heat-curing of the transfectant *S. cerevisiae* MS412a resulted in the non-killer derivative MS412c which was shown to have lost its killing ability by a simultanous loss of its M-dsRNA.

Effects of Intergeneric Transfer of ZbV-M in *S. cerevisiae*

In order to test maintenance and expression of the toxin-coding M-dsRNA in different genetic backgrounds of *S. cerevisiae*, the ZbV-M killer virus of the transfectant strain MS412 was transferred to various non-killer strains by the means of virus transfection, cytoduction and/or mating (Schmitt and Tipper 1992; Schmitt and Neuhausen 1994). As summarized in Table 2, tetrad analysis of meiotic segregants derived from crosses of the ZbV-M donor MS412 to the toxin-sensitive non-killer strain 1938 revealed a significant and dramatic loss in cell viability which was only in the range of 15%–27% (Table 2, cross B3). Since the same decrease in viability was observed in haploid progenies derived from the cross (B4) of MS412 to the resistant strain 1938X4 (a *sen1* mutant which is resistant because of a lack of primary toxin receptors within the yeast cell wall), we argued that the toxin-coding M-dsRNA genome is responsible for this dramatic effect. Furthermore, ZbV-M viruses are not encoding any immunity de-

Table 2. Viability and phenotype of meiotic segregants derived from *S. cerevisiae* diploids harboring the ZbV-M killer virus of the yeast *Z. bailii*

Cross	Phenotype of parental strains	Meiotic segregants after tetrad analysis		
		ZbV-M	Survival (%)	Phenotype of survivors
A	MS412[a] × 2285 (K+) (K–)	+	100	K+
B1	MS412c[b] × 1938 (K–) (K–)	–	75	K–
B2	GG100 – 14D[c] × 1938 (K–) (K–)	–	77	K–
B3	MS412 × 1938 (K+) (K–)	+	27	K+
B4	MS412 × 1938X4[d] (K+) (K–)	+	15	K+

K+, killer phenotype; K–, non-killer.
[a]ZbV-M transfectant of *S. cerevisiae* GG100-14D.
[b]Heat-cured non-killer derivative of *S. cerevisiae* MS412 lacking ZbV-M.
[c]*S. cerevisiae* non-killer host used for transfection with ZbV-M.
[d]Toxin resistant *sen1* mutant of *S. cerevisiae* 1938 defective in killer toxin binding to the yeast cell wall.

Table 3. Toxin sensitivity of ZbV-M-transfected and non-transfected strains of *S. cerevisiae*

Strain[a]	ZbV-M	Toxin sensitivity[b] (mm)	Survival rate (%)[c]	Sensitivity class
GG100-14D	–	4	1.5	I
MS412	+	4	1.5	I
MS412c	–	4	1.5	I
2285	–	4	3.0	I
2285–21	+	4	n.d.	I
1938	–	12	0.15	II
1938X4	–	4	1.5	I

[a]Strain 2285–21 corresponds to strain 2285 after cytoduction to the ZbV-M transfectant MS412; strain 1938X4 is a *sen1* mutant of strain 1938 having a decreased capacity of binding exogenously applied killer toxin to the cell wall. For the other strains, see Table 2.
[b]Toxin sensitivity, determined by the well test (agar diffusion) method, is expressed in the size of growth inhibition after application of 7×10^4 units killer toxin KT412.
[c]Survival rate, determined in liquid media in the presence of 7×10^5 units killer toxin KT412, is expressed as percentage colony-forming units (cfu) after cultivation at 20°C for 6 h.

terminant which could potentially protect the killer cell against the action of its own toxin. This lack of immunity in *S. cerevisiae* strains harboring ZbV-M was confirmed when transgenic *S. cerevisiae* strains expressing ZbV-M were assayed for sensitivity against exogenously applied killer toxin KT412 (Table 3). Independent of the presence or absence of ZbV-M, all *S. cerevisiae* strains tested showed a type-I sensitivity which was clearly distinguishable from the type-II sensitivity of strain 1938, which was designated to be hypersensitive (Table 3). In fact, all *S. cerevisiae* strains that have been tested so far belong to either of the two sensitivity groups, while resistant strains can only be identified among strains belonging to the genus Zygosaccharomyces (data not shown). Thus, unlike in *S. cerevisiae* killer strains, the natural M species of ZbV-M is either not involved in or not necessary for the expression of toxin immunity in *Z. bailii*.

Transgenic Expression of ZbV-M in *S. cerevisiae* Requires *KEX3* Function

Since the non-killer strain *S. cerevisiae* 2285 (which is phenotypically type-I sensitive) was the only strain that did not show a decrease in cell viability after crossing to the killer strain MS412 (see Table 2, cross A), we used this strain (2285) as recipient for ZbV-M from strain MS412 by the means of cytoduction. Cytoduction is a genetic technique, by which the cytoplasmic genome of killer strain MS412 was transferred to strain 2285 without diploidization by transient heterokaryon formation, since strain 2285 is a *kar1* mutant defective in nuclear fusion (Schmitt and Tipper 1992; Schmitt and Neuhausen 1994). Although the resulting cytoductant (strain 2285–21) expressed a weak *Z. bailii* killer phenotype, it still remained type-I sensitive against exogenous killer toxin KT412. Crossing MS412 to 2285 resulted in diploids which, after sporulation and tetrad dissection, showed a meiotic 2:2 segregation for strong and weak killers. Since all haploid segregants showed the same level of the toxin-coding M-dsRNA, we proposed the presence of one chromosomal gene in strain MS412 which is responsible and sufficient for the expression of a strong killer phenotype. In analogy to the *KEX1* and *KEX2* genes (whose products are responsible for the expression of a killer phenotype in *S. cerevisiae*) we designated this gene *KEX3*. Since diploids, heterozygous for this gene (*KEX3/kex3*) always expressed a strong killer phenotype, it can be concluded that the *kex3* mutation in strain 2285 is recessive.

Viability of ZbV-M-Infected *S. cerevisiae* Strains Depends on a Certain Set of Chromosomal Host Genes

By studying the Kex3p-dependent expression of a *Z. bailii* killer virus in the heterologous *S. cerevisiae* system, another set of chromosomal host genes has to be considered; genes, that are involved in or responsible for the establishment of two different sensitivity classes in *S. cerevisiae* (type I and type II, see Table 3). As soon as the Kex3$^+$ killer transfectant *S. cerevisiae* MS412 (which expresses a type-I sensitivity against exogenous killer toxin KT412) is crossed to a type-II sensitive parental strain (such as 1938), a dramatic decrease in cell viability among the haploid meiosegregants was observed. Since the same was true in segregants derived from the cross of MS412 to 1938X4 (which is a *sen1* mutant expressing a type-I sensitivity because of a defect in toxin binding to the cell wall), it can be hypothesized that KT412 might kill sensitive cells by either acting in the cytoplasm, in or at the plasma membrane, or shortly after being secreted into the periplasm. Toxin-binding studies as well as sensitivity tests on the isogenic yeast strains 1938 and its resistant derivative 1938X4 indicated that a toxin resistance at the cell wall level (dependend on a mutation in the chromosomal *SEN1* gene) must be distinguished and differentiated from a toxin-sensitivity caused by intracellular expression of the toxin-coding M-dsRNA genome. Based on the extremely low amounts of killer-active survivors derived after sporulation and meiosis either of the diploids MS412/1938 or MS412/1938X4 (see Table 2), we postulate the existance of at least two chromosomally encoded toxin-sensitivity genes (*TOS1,TOS2*) that, when mutated, can independently lead to the type-II hypersensitivity seen in strain 1938. The corresponding gene products Tos1p and Tos2p could either represent interacellular toxin receptors necessary for killing sensitive yeast cells or represent the main target of the viral killer toxin. Although the signal transduction of this toxin is still unknown, our data indicate the existence of different chromosomal host

Table 4. Influence of the chromosomal genes *KEX3*, *TOS1*, and *TOS2* on the viability of meiotic segregants of *S. cerevisiae* harboring a ZbV-M killer virus

Genotype[a]	Presence of ZbV-M[b]	Phenotypic characteristics		
		Killer	Sensitivity[a] class	Viability[a]
Parental strains				
MS412 *KEX3 TOS1 TOS2*	+	K	I	+
1938 *kex3 tos1 tos2*	–	NK	II	+
Possible progenies from the cross MS412 × 1938				
1*KEX3 TOS1 TOS2*	+	K	I	+
2 *KEX3 TOS1 tos2*	+	K	II	Lethal
3 *KEX3 tos1 TOS2*	+	K	II	Lethal
4 *KEX3 tos1 tos2*	+	K	II	Lethal
5 *kex3 TOS1 TOS2*	+	NK	I	+
6 *kex3 TOS1 tos2*	+	NK	II	+
7 *kex3 tos1 TOS2*	+	NK	II	+
8 *kex3 tos1 tos2*	+	NK	II	+

K, killer phenotype; NK, non-killer phenotype.
[a]Genotype designation: *KEX3*, chromosomal gene responsible for killer expression; *TOS1*, *TOS2*, chromosomal toxin-sensitivity genes; mutations within either *TOS1* or *TOS2* lead to a toxin hypersensitivity defined as type-II sensitivity (see Table 3) which is lethal in the presence of *KEX3*; mutations in *KEX3* cause a non-killer phenotype.
[b]Stable cytoplasmic inheritance of the killer toxin-coding dsRNA-virus ZbV-M.

genes that are involved in the expression of an "internal" sensitivity against this toxin. Our current model (Table 4) summarizes the postulated genotypes of the parental yeast strains MS412 and 1938, and shows both, genotypes and phenotypes of all possible meiotic segregants derived from that cross (see Table 2, cross B3). Remarkably, only 27% of all meiotic segregants were viable which – in the great majority of all cases (93%–97%) – expressed a non-killer phenotype although they showed normal, i.e. wild-type levels of the cytoplasmically inherited ZbV-M killer virus. Among these non-killer survivors, both sensitivity classes (type-I and type-II of the parental strains) as well as intermediate phenotypes were found, while all viable killer toxin-secreting segregants always portrayed the type-I sensitivity phenotype of the parental strain MS412. This characteristic distribution of Kex3$^+$ killers expressing a type-I-sensitivity indicates that a type-II-hypersensitivity can be realized and manifested by just a single mutation within one of the two *TOS* genes. Thus, Kex3$^+$ derivatives are only viable when both wild-type alleles of *TOS1* and *TOS2* are present.

Possible Consequences of "Cryptic" Mycoviruses on the Evolution of Their Hosts

Our genetic data suggest that - under certain circumstances - a toxin-coding dsRNA mycovirus can have a dramatic effect on the quality and quantity of generative progenies derived from sporulation of its infected host. Such an effect was only observable in an artificial and newly created system, in which a natural killer virus of the yeast *Z. bailii* was transferred and heterologously expressed in a non-killer strain of the yeast *S. cerevisiae*. The stable maintenance of this killer virus in *S. cerevisiae* was only possible because both, the two yeast genera as well as the dsRNA-viruses of both yeasts, are very similar, probably reflecting the close phylogenetic relationship between both systems (Schmitt and Neuhausen 1994; Hendriks et al. 1992). Already Buck (1986) speculated that similarities between different mycoviruses might reflect a common evolutionary ancestor for the different hosts. Because of the transfer of a *Z. bailii* killer virus into a "non-adapted" *S. cerevisiae* host, an artificial system was created that bypassed the natural long-term effect of evolutionary selection and adaptation. Therefore, this newly created virus/host system led to a dramatic decrease in cell viability of haploid segregants accompanied by a strong selection for a certain genotype among the surviving yeast population. The primary effect of this selection resulted in a Kex3-mediated, phenotypic inactivation of the ZbV-M killer virus such that more than 93% of all haploid progenies expressed a "cryptic" non-killer phenotype showing "normal" amounts of the toxin-coding M-dsRNA, indistinguishable from the dsRNA copy numbers in $Kex3^+$-killer derivatives. The strong selection for a *TOS1/TOS2* genotype observed in all killer-active ($Kex3^+$) segregants is indicative for an early "adaptation", leading to survivors whose vitality (after further crossings) would be significantly more pronounced than the viability of the F1 generation. In this case, the ZbV-M killer virus would cause a forced selection of a certain combination of chromosomal host genes allowing survival in the presence of a toxin-coding mycovirus. Survivors haboring ZbV-M would either be non-killers (because of a chromosomal *kex3* mutation) or $Kex3^+$ killers expressing a type-I sensitivity (because of having both dominant allels of the postulated *TOS* genes). In the case of $Kex3^-$ non-killers (which represent the great majority of all haploid progenies), ZbV-M would be regarded as "cryptic" since no dsRNA- associated phenotype would be detectable. The potential 'power and danger' of these virus-harboring yeasts becomes only visible after sexual reproductions, leading to a strong decrease in cell viability and a selection of certain genotypes among the surviving progenies. The immediately arising question whether this selection for a given genotype would be of any advantage for the virus-infected host can only be addressed once the biological functions of the corresponding gene products of *KEX3* and *TOS1/TOS2* have been identified. However, the biological significance of "cryptic" mycoviruses becomes even more striking by the fact that more than 30% of all fungi that have been tested so far, carry cytoplasmically inherited dsRNA- or dsDNA-mycoviruses without expressing any detectable phenotype (Buck 1986). Only in some cases it could be shown that these mycoviruses can cause certain viral diseases (like in *Agaricus bisporus*) leading to a dramatic decline in cell viability of an infected population (Buck 1986). The role or possible function of the great majority of "cryptic" mycoviruses still remains unclear, although our data on the transgenic expression of a *Z. bailii* killer virus in the yeast *S. cerevisiae* suggest that these viruses might have had (and still may have) a remarkable effect on the evolution of their infected hosts.

Acknowledgment. This work was kindly supported by a grant (Schm 541/3-2) from the Deutsche Forschungsgemeinschaft.

References

Buck KW (1986) Fungal virology. In: Buck KW (ed) Fungal virology, CRC Press, Boca Raton, pp 1–84

Hendriks AJ, Goris A, Van de Peer Y, Neefs JM, Vancanneyt M, Kersters K, Berny JF, Hennebert GL, De Wachter R (1992) Phylogenetic relationships among ascomycetes and ascomycetes-like yeasts as deduced from small subunit RNA sequences. System Appl Microbiol 15: 98–104

Radler F, Herzberger S, Schönig I, Schwarz P (1993) Investigation of a killer strain of Zygosaccharomyces bailii. J Gen Microbiol 139: 495–500

Schmitt MJ, Tipper DJ (1990) K28, a unique double-stranded RNA killer virus of Saccharomyces cerevisiae. Mol Cell Biol 10: 4807–4815

Schmitt MJ, Tipper DJ (1992) Genetic analysis of maintenance and expression of L and M double-stranded RNAs from yeast killer virus K28. Yeast 8: 373–384

Schmitt MJ, Neuhausen F (1994) Killer toxin-secreting double-stranded RNA mycoviruses in the yeasts Hanseniaspora uvarum and Zygosaccharomyces bailii. J Virol 69: 1765–1772

Wickner RB (1992) Double-stranded and single-stranded RNA viruses of Saccharomyces cerevisiae. Annu Rev Microbiol 46: 347–376

13 Foreign DNA in Mammalian Cells and Organisms

W. Doerfler[1], K. Fechteler[1,2], H. Heller[1], C. Kämmer[1], M. Knoblauch[1,3], P. Koetsier[1,4], G. Orend[1,5], J. Schröer[1], and R. Schubbert[1]

[1]Institut für Genetik, Cologne University, Weyertal 121, 50931 Cologne, Germany
Current addresses: [2]Boehringer-Ingelheim, 55216 Ingelheim, Germany
[3]Max-Delbrück-Centrum, 13122 Berlin-Buch, Germany
[4]Qiagen, 40724 Hilden, Germany
[5]La Jolla Cancer Research Foundation, Cancer Research Center, 10901 North Torrey Pines Road, La Jolla, CA 92037, USA

Introduction

Mammalian cells have the capacity for taking up foreign DNA and to integrate this DNA into their own genomes under appropriate conditions. This integration may not be stable in all cells upon their continued replication. The functional consequences of foreign DNA integration depend on a number of factors, like chromosomal site of integration, extent of organizational alterations in the recipient genome at locations close to and remote from the insertion site, on the de novo methylation and transcriptional inactivation of the inserted foreign genes, and on factors not yet understood at the present time. The acquisition of foreign DNA can be advantageous or deleterious for the recipient cell. It may lead to insertional mutagenesis, cell transformation, cell death, or may bestow upon the cell new functional facilities. Since we have shown that foreign DNA ingested with the daily food supply survives in part the digestive regimen in the gastrointestinal tract of mice, reaches the bloodstream of these animals and can be traced to the nuclei of intestinal epithelia, peripheral leukocytes, spleen and liver cells, it will be important to study the fate of foreign DNA also in mammalian organisms. The long-term sequelae of this life-long influx of foreign DNA into the mammalian organism are not known.

Over the last three decades, my laboratory has studied problems of foreign DNA uptake and integration, as well as some of the consequences of integration in mammalian cell systems. Under different experimental conditions, adenovirus type 12 or type 2 DNA, plasmid DNA, or the DNA of bacteriophage lambda or M13 was used in these studies. Some of this work has been performed with adenovirus-transformed cells or with adenovirus type 12 (Ad12)-induced tumor cells in culture. We are aware of the possibility that the conclusions adduced from these studies can be influenced by the character of the experimental system under investigation.

E.R. Schmidt/T. Hankeln (Eds.)
Transgenic Organisms and Biosafety

In this chapter, we shall summarize recent data from our laboratory. In this survey, the following topics will be covered (details have been described in the references cited in each section):

1. Chromosomal association of Ad12 DNA in human or hamster cells after Ad12 infection or after the addition of Ad12 DNA to the culture medium (Schröer and Doerfler 1996).
2. The structure of the insertion site in an Ad12-transformed hamster cell line and in an Ad12-induced tumor (Knoblauch et al. 1996).
3. Studies on the mechanism of Ad12 DNA integration into cellular DNA in a cell-free model system: purification of recombination-active proteins (Fechteler et al. 1995).
4. Initiation and spreading of de novo DNA methylation in integrated Ad12 genomes (Orend et al. 1995a).
5. Are incompletely methylated foreign DNA segments less stably integrated (Orend et al. 1995b)?
6. Stability of transgene methylation patterns in mice: position effects, strain specificity and cellular mosaicism (Koetsier et al. 1996).
7. Alterations in DNA methylation patterns in cellular DNA segments located far remote from the sites of foreign (Ad12, plasmid, or bacteriophage lambda) DNA integration (Heller et al. 1995).
8. Food-ingested foreign DNA is taken up via the gastrointestinal wall epithelia and Peyer's patches into peripheral leukocytes, spleen and liver (Schubbert and Doerfler 1996).

Chromosomal Association of Ad12 DNA in Human or Hamster Cells After Ad12 Infection or After Addition of Ad12 DNA to the Culture Medium (Schröer and Doerfler 1996)

Human adenovirus type 12 (Ad12) infects human cells productively and leads to viral replication, whereas the infection of hamster cells remains abortive with total blocks in viral DNA replication and late viral gene transcription. The intranuclear fate of Ad12 DNA has been followed in productively infected human and in abortively infected hamster cells by using the fluorescent in situ hybridization (FISH) technique. Cells have also been investigated by the same method after the addition of foreign (Ad12) DNA to the culture medium. Human HeLa cells or primary human umbilical cord fibroblasts, hamster BHK21 cells or the Ad12-transformed T637 hamster cell line have been studied. As early as 2 h after the infection of these cells, extensive association of Ad12 DNA with metaphase chromosomes can be demonstrated by FISH. This association continues until late (24–28 h) after infection. In productively infected cells, large nuclear areas are taken over by viral DNA replication late in the infection cycle, as detected by FISH in interphase nuclei. Chromosomal association is frequently limited to one chromatid only, but signals in adjacent positions on both chromatids are occasionally observed. Upon simply adding Ad12 DNA to the medium of cells growing in culture, Ad12 DNA becomes also associated with the chromosomes, frequently with both chromatids. The efficiency of this association is markedly increased when Ad12 DNA with the terminal protein still covalently attached is used in these experiments. Upon long-term cultivation and passage of abortively infected BHK21 cells for 96 h after infection, a gradual decrease of viral DNA association with chromosomes has been observed. Integration of Ad12 DNA has previously been documented in hamster cells early after infection, and recombination between viral and cellular DNA was also shown in human cells. The FISH data on extensive chromosomal association of foreign DNA suggest a means to study the pathway of Ad12 DNA from early steps in viral infection via chromosomal interactions to integration events which may be stably achieved in only a small number of cells.

Structure of the Insertion Site in an Ad12-Transformed Hamster Cell Line and in an Ad12-Induced Tumor
(Knoblauch et al. 1996)

Foreign DNA can integrate into the genomes of mammalian cells, and this process plays major roles in viral oncogenesis, in the generation of transgenic organisms and will be important in evolving regimens for human somatic gene therapy. In previous work, we have studied mechanisms of foreign DNA integration by using hamster cells transformed by human adenovirus type 12 (Ad12), Ad12-induced hamster tumor cells or a cell-free system derived from highly purified fractions of nuclear extracts from BHK21 hamster cells grown in culture. The insertion sites of Ad12 DNA genomes have been analyzed in detail in the Ad12-transformed hamster cell line T637, its revertants, which have lost most of the >20 Ad12 genome equivalents integrated chromosomally in cell line T637, and in the Ad12-induced tumor T191. Some of these junction sites have been molecularly cloned, and the nucleotide sequences at the sites of transition between viral and cellular DNAs have been determined. These sites of linkage between the hamster cellular and the foreign (viral) DNA are characterized by the frequent occurence of patch homologies between the recombination partners. The cellular junction sites investigated here are not transcriptionally active. One of the cellular DNA sequences abutting the right Ad12 DNA terminus in cell line T637 (os2) is represented only once in the hamster genome and has a strikingly low abundance of 5′-CG-3′ dinucleotide sequences. One 5′-GCGC-3′ sequence close to the Ad12 DNA integrate is heavily methylated in normal, in Ad12-

transformed, and in Ad12-induced tumor cells. The second such sequence is located more remote from the junction site, is partly methylated in BHK21 hamster cells, and shows differences in methylation in different Ad12-transformed cell lines. This site is unmethylated in liver DNA. The cellular DNA sequences at the site of Ad12 linkage in the tumor T191 exhibit homologies to highly repetitive sequences of the Alu family and to an origin of hamster DNA replication with an Alu element. A number of here and previously investigated junction sites between Ad12 DNA and hamster or mouse DNA in Ad12-transformed cell lines or Ad12-induced tumor cell lines are characterized by striking stem-loop structures encompassing the junction sites.

Studies on the Mechanism of Ad12 DNA Integration into Cellular DNA in a Cell-Free Model System: Purification of Recombination-Active Proteins
(Fechteler et al. 1995)

Integrative recombination between the nucleotide 20 885 to 24 053 fragment of Ad12 DNA and the hamster preinsertion DNA sequence p7, which has served as viral DNA integration site in the Ad12-induced tumor CLAC1 (Stabel et al. 1980), has been studied in a cell-free system. Nuclear extracts initially from uninfected BHK21 hamster cells have been fractionated by a series of chromatographic procedures. The purified protein fraction VI contains a limited number of protein bands still capable of catalyzing the cell-free recombination reaction. The recombination-active proteins range in size between approximately 40 and 90/91 kDa and presumably form a multi-protein complex, named IRAC (integrative recombination active complex). Cell-free generated recombinants have been identified and characterized by an *Escherichia coli* transfection assay (Jessberger et al. 1989) or by the polymerase chain reaction (PCR; Tatzelt et al. 1993). The integrative recombination reaction has been more realistically imitated by using fractionated nuclear extracts prepared from BHK21 hamster cells 36 h after the infection with Ad12. In this way, the contribution of Ad12-encoded proteins in integrative recombination can be assessed. Upon gel filtration over a Sephacryl S-300 column, fraction II in the purification scheme of extracts from Ad12-infected cells exhibits a marked stimulation of cell-free recombination when compared to fractions from uninfected BHK21 cells. The recombinationally active proteins from Ad12-infected cell extracts have been purified by the same regimen. Upon polyacrylamide gel electrophoresis followed by silver staining, these proteins present a distribution similar to that from uninfected cells. Purified fractions from extracts of uninfected or of Ad12-infected BHK21 cells form protein DNA-complexes with both single-stranded and double-stranded DNA segments from the preferred cellular target site for Ad12 DNA integrative recombination. The protein-DNA complexes with single-stranded DNA are sequence-, possibly structure-specific. The cellular recombination system apparently suffices to catalyze cell-free recombination between Ad12 and hamster cell DNA. Ad12-specific proteins may modify or enhance the reaction directly or via the modification of cellular factor(s).

Initiation and Spreading of De Novo DNA Methylation in Integrated Ad12 Genomes (Orend et al. 1995a)

The de novo methylation of foreign DNA integrated into the mammalian genome is a fundamental process whose mechanism has not yet been elucidated. We have studied de novo methylation in Ad12 genomes inserted into the genomes of Ad12-induced hamster tumor cells. De novo methylation of Ad12 DNA, which is not methylated in the virion, is initiated in two paracentrally located regions and spreads from there across the integrated Ad12 genomes.

After extensive cultivation of cloned Ad12-induced hamster tumor cell lines, the same segments in integrated Ad12 DNA become methylated or remain unmethylated, depending on their positions in the viral genome.

When Ad12 DNA or Ad12 DNA fragments are transfected into hamster cells and permanent cell lines are established by selection for the cotransfected neomycin phosphotransferase gene, patterns of de novo methylation in terminally or internally located segments of Ad12 DNA are different from those in Ad12-induced tumor cell lines.

A detailed study on the topology of the integrated viral genomes in the Ad12-transformed hamster cell lines T637 and A2497-3 and in the Ad12-induced hamster tumors T191, T1111(1) and T181 has been performed. Some of the integrated viral genomes are inserted into the cellular genome in an orientation colinear with the virion genome, others have been rearranged. An originally internally located Ad12 DNA segment has become juxtaposed to the left terminal sequences of the viral genome in several cell lines and tumors. In the complete Ad12 genomes, the internally located PstI-D fragment becomes extensively methylated at the 5′-CCGG-3′ and 5′-GCGC-3′ sequences. When this DNA segment has been juxtaposed to the left terminal, hypomethylated fragment of Ad12 DNA in rearranged genomes, the PstI-D fragment remains unmethylated. We, therefore, reason that the initiation of de novo methylation in integrated Ad12 DNA cannot be exclusively directed by nucleotide sequence. Other parameters, such as site of integration, conformation of integrates, mode of cell selection, or chromatin structure related to transcriptional activity may play decisive roles.

Are Incompletely Methylated Foreign DNA Segments Less Stably Integrated (Orend et al. 1995b)?

We have studied the stability of integrated Ad12 DNA sequences and the relation of foreign DNA persistence to the state of methylation of this DNA. In the Ad12-transformed hamster cell line T637, multiple copies of Ad12 DNA are chromosomally integrated. Some of these integrated viral genomes are rearranged in that internal parts of the viral DNA have become juxtaposed to its left terminus. FISH hybridization analyses demonstrate that the Ad12 DNA in cell line T637 and in some of its revertants is located at one site on one of the hamster chromosomes. Major portions of the integrated viral genomes in cell line T637 have become extensively de novo methylated in specific patterns. Most of the rearranged Ad12 DNA sequences in the T637 genome are un- or hypomethylated. In the morphological revertants of the Ad12-transformed hamster cell line T637, the majority of the integrated Ad12 genomes have been lost. Surprisingly, we have found that in the revertants the un- or hypomethylated rearranged viral sequences have been selectively lost, in contrast to some of the methylated sequences that have been stably retained.

Stability of Transgene Methylation Patterns in Mice: Position Effects, Strain Specificity and Cellular Mosaicism (Koetsier et al. 1996)

The methylation status of a transgene that carries the adenovirus type 2 E2A late promoter linked to the chloramphenicol acetyltransferase gene has been studied in the DNA of three transgenic mouse lines (5–8, 7–1, 8–1). These lines have been analyzed over a large number of offspring generations beyond the founder animal. In mating experiments, the influence of the parent of origin and strain specific background on the transgene methylation patterns has been assessed and found to have no effect on the preestablished methylation patterns in mouse lines 5–8 and 8–1.

The founder animal 7–1 carried two groups of a total of ten transgenes which were located on two different chromosomes. These arrays of transgenes could be segregated into separate mouse lines 7–1A and 7–1B. The transgenes of 7–1A animals exhibit cellular mosaic methylation patterns that are demethylated in approximately 10% of the offspring in a mixed genetic background.

Upon further transmission of these transgenes in a mixed genetic background, the grandparental methylation patterns have been reestablished in most progeny. Mating to inbred DBA/2 mice results in the maintenance of the demethylated pattern or in further demethylation of the transgenes in approximately 50% of the offspring. In contrast, an equal number of transgenic siblings from matings to C57BL/6 mice show a return to the original methylation pattern. The mosaic methylation status of this locus is apparently controlled by mouse strain specific factors. The methylation patterns of the 7–1B transgenes are not cellular mosaic and have remained stable in all offspring, as has been found for the lines 5–8 and 8–1. Hence, the strain-dependent and cellular mosaic transgene methylation patterns of 7–1A animals have probably been a consequence of the chromosomal integration site of the transgenes (position effect).

Alterations in DNA Methylation Patterns in Cellular DNA Segments Located Far Remote from the Sites of Foreign (Ad12, Plasmid, or Bacteriophage Lambda) DNA Integration

(Heller et al. 1995)

The insertion of foreign DNA into an established mammalian genome can lead to extensive alterations in the patterns of cellular DNA methylation. Ad12-transformed hamster cells, Ad12-induced hamster tumor cells, or hamster cells carrying integrated plasmid DNA or DNA of bacteriophage lambda have been used as model systems. DNA methylation levels have been examined by cleavage of cellular DNA with HpaII, MspI, or HhaI, followed by Southern blot hybridization using ^{32}P-labeled, randomly selected cellular DNA probes. For several, but not all, cellular DNA segments investigated, extensive increases in DNA methylation have been found in comparison to the methylation patterns in BHK21 or primary Syrian hamster cells. In eight different Ad12-induced hamster tumors moderate increases in DNA methylation have been observed. Increased methylation of cellular genes has also been documented in two hamster cell lines with integrated Ad12 DNA without the Ad12-transformed phenotype, in one cloned BHK21 cell line with integrated plasmid DNA, and in at least three cloned BHK21 cell lines with integrated lambda DNA.

By FISH, the cellular hybridization probes have been localized to different hamster chromosomes. The endogenous intracisternal A particle (IAP) genomes have shown a striking distribution on many of the hamster chromosomes, frequently on their short (p) arms. When BHK21 hamster cells have been abortively infected with Ad12, increases in cellular DNA methylation have not been seen. Thus, early Ad12 gene products are not directly involved in increasing cellular DNA methylation. We attribute the alterations in cellular DNA methylation at least in part to the insertion of foreign DNA. Can alterations in the methylation profiles of hamster cellular DNA contribute to the generation of the oncogenic phenotype?

The finding that cellular DNA methylation patterns are very significantly altered in virus-transformed cells and in Ad12-induced tumors raises questions as to the role that such changes could play in the generation of the oncogenic phenotype. The question of whether these alterations are cause or consequence of viral transformation can obviously not be answered. Increases in DNA methylation might be associated with the shut-off of cellular functions, decreases in DNA methylation with activation of previously silent genes or DNA segments. The IAP sequences do not necessarily follow this inverse correlation. Of course, alterations in cellular transcription patterns will have to be assessed gene by gene. We have only begun such studies and found that among 40 randomly selected hamster genes, five show altered expression patterns, which are different in individual transformed cell lines (Rosahl and Doerfler 1992). We have observed increases in the levels of transcription of the IAPI DNA sequences in the Ad12-transformed cell lines T637, TR3, and A2497-3 as compared to uninfected or Ad12-infected BHK21 cells or primary hamster cells.

Food-Ingested Foreign DNA Is Taken Up Via the Gastrointestinal Wall Epithelia and Peyer's Patches into Peripheral Leukocytes, Spleen and Liver (Schubbert and Doerfler 1996)

Food ingested foreign DNA is not completely degraded in the gastrointestinal (GI) tract of mice. Minute amounts ($\leq 0.1\%$) of this DNA can be recovered from peripheral leukocytes, from spleen and liver. Phage M13mp18 DNA has been pipette-fed or added to the food supply. Its fate in the animals has been followed by dot blot or Southern blot hybridization, by FISH, or by PCR and by the resequencing of PCR products. In 84 animals, fragments of the 7250 bp M13mp18 DNA have been detected between 1 and 7 h after feeding in the contents of the small intestine, cecum (until 12 h), the large intestine, or in the feces. In 227 animals, M13mp18 DNA fragments of up to 976 bp have been found in blood 2–8 h after feeding. In control animals, DNA from gut contents and blood remains devoid of any detectable M13mp18 DNA homologies. M13mp18 DNA fragments have been demonstrated by PCR and FISH in about one out of 1000 peripheral leukocytes between 2 and 8 h after feeding and in spleen or liver cells up to 18 h after feeding. In histological sections, M13mp18 DNA can be traced by FISH in the columnar epithelial cells and in leukocytes in Peyer's patches of the cecum wall, as well as in liver cells and in leukocytes of the spleen. These findings suggest transport of foreign DNA through the intestinal wall and Peyer's patches to peripheral blood leukocytes and into organs. The medical implications of this transport and uptake may be considerable.

Acknowledgments. This research was supported by the Deutsche Forschungsgemeinschaft through SFB274-A1 and by project Do-165/1, by the Minister of Research and Technology through Genzentrum Köln, TP 2.03, by the

Wilhelm Sander Stiftung, München, and the Boehringer Ingelheim Fonds, Stuttgart (fellowship to K.F.).

References

Fechteler K, Tafzelf J, Huppertz S, Wilgenbus P, Doerfler W (1995) The mechanism of adenovirus DNA integration: Studies in a cell-free system. Current Topics Microbiol Immunol 199/II: 109–137

Heller H, Kämmer C, Wilgenbus P, Doerfler W (1995) Chromosomal insertion of foreign (adenovirus type 12, plasmid or bacteriophage lambda) DNA is associated with enhanced methylation of cellular DNA segments. Proc Natl Acad Sci USA 92: 5515–5519

Jessberger R, Heuss D, Doerfler W (1989) Recombination in hamster cell nuclear extracts between adenovirus type 12 DNA and two hamster preinsertion sequences. EMBO J 8: 869–878

Knoblauch M, Schröer J, Schmitz B, Doerfler W (1996) The structure of adenovirus type 12 DNA integration sites in the hamster cell genome J Virol 70:

Koetsier P, Mangel L, Schmitz B, Doerfler W (1996) Stability of transgene methylation patterns in mice: position effects, strain specificity and cellular mosaicism. Transgenic Res 5: 1–10

Orend G, Knoblauch M, Kämmer C, Tjia ST, Schmitz B, Linkwitz A, Meyer zu Altenschildesche G, Maas J, Doerfler W (1995a) The initiation of de novo methylation of foreign DNA integrated into a mammalian genome is not exclusively targeted by nucleotide sequence. J Virol 69: 1226–1242

Orend G, Knoblauch M, Doerfler W (1995b) Selective loss of unmethylated segments of integrated Ad12 genomes in revertants of the adenovirus type 12-transformed cell line T637. Virus Res 38: 261–267

Rosahl T, Doerfler W (1992) Alterations in the levels of expression of specific cellular genes in adenovirus-infected and -transformed cells. Virus Res 26: 71–90

Schröer J, Doerfler W (1996) Adenovirus type 12 DNA associates with mammalian chromosomes early after virus infection or after DNA transfer by addition to the cell culture medium (submitted)

Schubbert R, Doerfler W (1996) Foreign DNA ingested by mice reaches peripheral leukocytes, spleen and liver via the intestinal wall mucosa (submitted)

Stabel S, Doerfler W, Friis RR (1980) Integration sites of adenovirus type 12 DNA in transformed hamster cells and hamster tumor cells. J Virol 36: 22–40

Tatzelt J, Fechteler K, Langenbach P, Doerfler W (1993) Fractionated nuclear extracts from hamster cells catalyze cell-free recombination at selective sequences between adenovirus DNA and a hamster preinsertion site. Proc Natl Acad Sci USA 90: 7356–7360

II Stability of DNA

14 Persistence of DNA in the Environment and Its Potential for Bacterial Genetic Transformation

W. Wackernagel

Genetik, Fachbereich Biologie, Carl von Ossietzky Universität Oldenburg, Postfach 2503, 26111 Oldenburg, Germany

Introduction

Previously it was assumed that naked DNA molecules appearing in bacterial habitats such as in soil, sediment or freshwater and sea water would almost instantaneously be degraded by nucleolytic enzymes. Such enzymes are known to be released from microorganisms including bacteria, yeasts and fungi which live in the terrestrial and aquatic environment. The low molecular weight degradation products of biological macromolecules provided by the action of extracellular enzymes can serve as nutrients for the microorganisms. More recently, experimental data suggest that extracellular DNA in terrestrial environments can persist for considerable time periods. The persistence appears to depend on additional parameters besides the presence and activity of extracellular DNases. Some of these findings will be summarized in the following. Further, the phenomenon of natural genetic transformation in bacteria has been documented for an increasing number and a wide variety of bacterial species (for references, see Lorenz and Wackernagel 1994). Natural transformation is the ability of bacteria to take up DNA actively, to inherit such DNA and to express its genetic information (Lorenz and Wackernagel 1994).

Thus, based on the presence of free DNA in the environment and the DNA uptake ability of bacteria, a gene flow in microbial habitats can be proposed in which the extracellular DNA constitutes a bacterial gene pool outside of cells. To this gene pool DNA bacteria continuously add DNA, whereas DNA is removed from the pool by degradation and by bacterial DNA uptake. For safety considerations regarding the use of organisms genetically modified by recombinant DNA technology (GMO), it must be taken into account that the recombinant DNA of these GMOs can also enter the extracellular gene pool of bacteria populations and communities. Interspecies DNA transfer by natural transformation has been identified in gram-positive bacteria, the α, β, and γ subgroups of proteobacteria and cyanobacteria (for references, see Lorenz and Wackernagel 1994).

E.R. Schmidt/T. Hankeln (Eds.)
Transgenic Organisms and Biosafety

Bacterial DNA Uptake Ability Is an Evolutionarily Old Trait

Active DNA uptake by bacteria occurs when the physiological status of competence is reached. Competence requires the expression of many chromosomal genes. As an example, in *Bacillus subtilis* many genes organized in several *com* loci and necessary for transformation have been identified (Dubnau 1991a; Hahn et al. 1993; Londono-Vallejo and Dubnau 1993). Further genes are involved in regulatory processes and the steps required for integration of ingested DNA into the chromosome by homologous recombination (Dubnau 1991b). Interestingly, several of the genes involved in the uptake of DNA have nucleotide sequence similarity to genes effecting other transmembrane translocation processes of DNA (as active in conjugative DNA transfer) and of other macromolecules (as in secretion of proteins and assembly of pili) as recently summarized (Dreiseikelmann 1994).

Up till now natural transformation has been detected in more than 40 bacterial species and archaea (Lorenz and Wackernagel 1994). Of these, 12 are human pathogens or clinical isolates. Among the others there are members of various trophic and taxonomic groups including gram-positive bacteria, proteobacteria and cyanobacteria. They live in habitats of the environment including soil, sediment and water. This suggests that the genetic determinants for the DNA uptake machinery have evolved early in bacterial evolution. It also suggests that DNA uptake ability is beneficial for cells to an extent which provides an evolutionary potential strong enough to conserve the DNA uptake capability among many (perhaps even most) bacteria until present.

Interestingly, there is some evidence that bacteria excrete DNA during specific growth phases besides the release during cell death coupled with cell lysis. In *B. subtilis* there appears to be a temporal coordination between DNA release and competence, because cells at the end of the logarithmic growth develop competence almost coincidently with the excretion of chromosomal and plasmid DNA (Lorenz et al. 1991). It would be interesting to examine whether the pulse of DNA-release is regulated by the same genetic elements which regulate competence development. If this were the case, one could argue that specific growth situations in *B. subtilis* serve the exchange of genetic material within a population.

Association of DNA with Mineral Surfaces

In extensive studies to elucidate the fate of DNA in soils and sediments, the interaction of DNA with various purified mineral constituents and under a variety of conditions was examined. It was found that in almost all cases binding of DNA to mineral material was observed. The forces that allow the anionic polymer DNA to associate with the negatively charged mineral surfaces are probably dependent on the presence of cations (to reduce repulsive forces) and are assumed to be nonionic interactions (van der Waals forces; Lorenz and Wackernagel 1987). Also, divalent cation-bridging as a binding mechanism has been proposed (Lorenz and Wackernagel 1987). The fact that Mg^{2+} and Ca^{2+} have approximately the same effect on DNA binding as 100-fold higher concentrations of monovalent cations is in accordance with this (Lorenz and Wackernagel 1987; Romanowski et al. 1991). Adsorption isothermes of DNA (linear duplex DNA and supercoiled plasmid) with quartz sand (Lorenz and Wackernagel 1987; Romanowski et al. 1991) and various clay minerals (Lorenz and Wackernagel 1992a) indicated high binding capacity for sand and even higher capacities for clay minerals, which bound more than 45 mg DNA per gram of clay. Kinetic studies revealed that the association of DNA with quartz sand, clays and mineral material sampled from a groundwater aquifer was rapid, giving half maximum adsorption at 23 °C within seconds or minutes. The efficient DNA binding to "dirty" groundwater aquifer mineral particles (sediment) suggested that the covering of minerals with inorganic precipitates and organic materials does not prevent or reduce DNA adsorption (Romanowski et al. 1993a).

In conclusion, minerals including quartz, feldspar, heavy minerals and various clays (kaolinite, bentonite, montmorillonite) have the ability to adsorb DNA. The ionic conditions for this are fully met by solvents like natural groundwater or soil solutions. After the DNA-mineral complex has formed, it is relatively stable. In a flow through microcosm filled with groundwater aquifer minerals and loaded with DNA, the elution with artificial groundwater (sterile) for about 190 h removed less than 10% of the mineral-associated DNA (Romanowski et al. 1993a).

According to a model of polyelectrolyte adsorption to charged surfaces (Hesselink 1983) the DNA would form a complex with the mineral surface in which adsorbed (trains) and nonadsorbed (loops) segments of DNA would al-

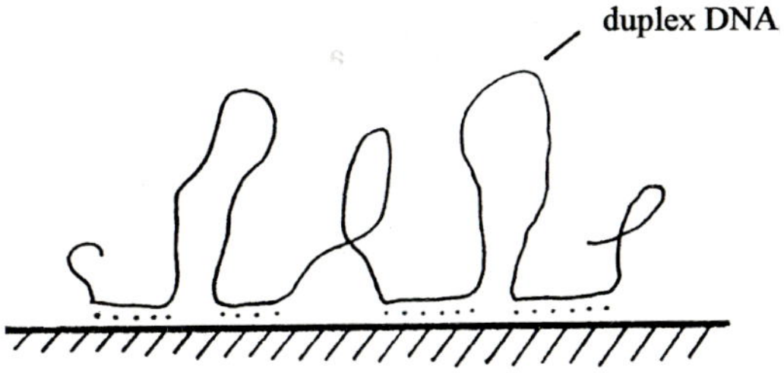

Fig. 1. Adsorption of a duplex DNA molecule to a solid surface of a mineral particle. The scheme is based on the polyelectrolyte adsorption model of Hesselink (1983). Bound segments are termed "trains," nonadsorbed segments "loops." There is a dynamic equilibrium between trains and loops. Repulsion forces between the negatively charged surface and DNA are overcome by monovalent cations and divalent cations. The *dots* symbolize divalent cations involved in bridging between the negatively charged phosphate groups of DNA and the silicate anions of the mineral

ternate between loosely dangling ends (Fig. 1). The association would result from a dynamic equilibrium between trains and loops.

Binding of DNA to Minerals Provides Protection Against DNases

The discovery that the binding of linear duplex DNA to sea sand protected the DNA against degradation by pancreatic DNaseI (Lorenz and Wackernagel 1987) prompted extensive investigation of this phenomenon. Also supercoiled plasmid DNA molecules were protected on sea sand (Romanowski et al. 1991). Not only quartz, but also feldspar and heavy minerals (Lorenz and Wackernagel 1987) as well as various clay minerals gave protection against DNaseI (Lorenz and Wackernagel 1992a; Khanna and Stotzky 1992; Paget et al. 1992). The strongest protection was seen on groundwater aquifer material in the presence of groundwater, where more than 1000-fold higher amounts of DNaseI were required to degrade the bound DNA to the same extent as free DNA in groundwater (Romanowski et al. 1993a). These experiments suggested that free DNA appearing in bacterial habitats could escape degradation by extracellular microbial DNases by binding to mineral surfaces. In fact, Ahrenholtz et al. (1994) demonstrated that bacterial DNA with a potential to transform the soil bacterium *Acinetobacter calcoaceticus* genetically was highly protected against the extracellular nuclease of the ubiquitious bacterium *Serratia marcescens* when the DNA was bound on groundwater aquifer minerals in the presence of groundwater. This finding was important because it showed that the protective effect of mineral-binding of DNA is not a phenomenon restricted to the eukaryotic DNaseI but holds true also for a typical microbial extracellular DNase. It also showed that the type of DNA-mineral association that would occur in natural solutes present in aqueous habitats or soils without added extra association-promoting ions suffices to produce protection of DNA against microbial DNases.

Persistence of DNA in Natural Waters, Sediments and Soils

The half-life periods of DNA molecules in bacterial habitats have been determined by a variety of techniques. In waste water, the half-life of free plasmid DNA measured by the decrease of supercoiled molecules was only a few minutes (Phillips et al. 1989). In freshwaters and marine waters half-life measured as loss of acid-precipitable material was in the range of several hours (Paul et al. 1989). In soils, degradation of DNA was measured by several methods following inoculation of nonsterile soil samples with plasmid DNA and extraction of DNA after various time periods (Romanowski et al. 1992). The highly sensitive

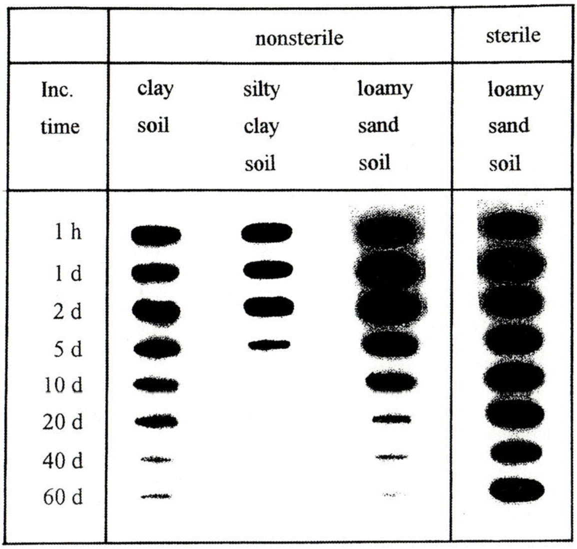

Fig. 2. Dot blot hybridization used to monitor the persistence of naked plasmid DNA introduced into three different nonsterile soils. Total DNA was isolated from soil samples (kept wet at 23 °C) at various times after inoculation with plasmid DNA using a hot sodium dodecylsulfate extraction procedure (Romanowski et al. 1992). A ^{32}P-labeled probe was employed for hybridization; an autoradiogram is shown

monitoring procedures included production of acid-soluble material, loss of hybridizable DNA in dot blot (Fig. 2) and Southern transfer procedures and decay of the transforming activity of the DNA. The studies revealed that the half-life of the transforming activity of plasmids depended on the soil type and ranged between several hours and about one day when determined during the first days after inoculation (Romanowski et al. 1992). The biphasic inactivation curves of the transforming potential suggested lower DNA degradation rates at later times. Using further improved techniques with even higher monitoring-sensitivity (electroporation for transformation and quantitative PCR for determination of the number of intact plasmid genes) it was found that 2 months after inoculation with naked plasmid DNA transforming plasmid molecules were still detectable in nonsterile soil (0.001%–0.01%) and gene sequences were identified at a residual frequency of 0.01%–1% (Romanowski et al. 1993b). This long persistence of naked DNA in nonsterile soils was impressive (Fig. 3). It was speculated that adsorption of DNA to mineral material in the various soil types protected DNA from soil-borne DNases and thereby contributed to DNA persistence (Romanowski et al. 1993b). Very recently detailed analyses revealed that the fraction of extracellular DNA in natural soil which is bound to particulate material is only very slowly degraded. In contrast, the DNA that remains in the soil solution is completely broken down to acid-soluble material within about 24 h at 23 °C (Blum, Lorenz and Wackernagel, in preparation; see also Chap. 15).

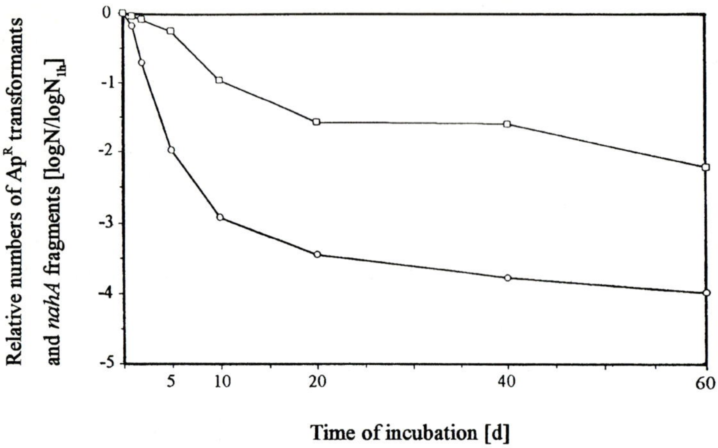

Fig. 3. Inactivation of the transforming potential of plasmid DNA molecules (pUC8-ISP; *circles*) and of the *nahA* gene (*squares*) introduced into nonsterile loamy sand soil during incubation at 23 °C. In total DNA extracted, the transforming activity was monitored by electroporation of *Escherichia coli* SF8, and the *nahA* gene number was measured by quantitative polymerase chain reaction (PCR) (Romanowski et al. 1993b). The data are plotted relative to the numbers of plasmids and of genes recovered 1 h after inoculation (N_{1h})

Mineral-Associated DNA Is Available for Uptake by Bacteria

As detailed above DNA bound on mineral particles resisted desorption during extended elution. Thus, at millimolar concentrations of divalent cations (Ca^{2+}, Mg^{2+}) as present in natural groundwater (Romanowski et al. 1993a) or extracts of nonsterile marsh soil (Lorenz and Wackernagel 1991) or podsoil, brown earth or fluvial marsh soil (Lorenz and Wackernagel 1992b), the equilibrium between bound and free DNA is far on the side of the mineral-DNA complex. The question arises whether such DNA is available for uptake by competent bacteria. Studies with microcosms containing sand loaded with transforming DNA indicated that competent cells of the soil bacteria *B. subtilis* (Lorenz et al. 1988), *Pseudomonas stutzeri* (Lorenz and Wackernagel 1990) and *Acinetobacter calcoaceticus* (Chamier et al. 1993) are effectively transformed by the DNA-sand complex. Control experiments established that not the DNA desorbing from the minerals during incubation with bacteria provided the major transforming activity. Rather, DNA appeared to be ingested by bacteria directly from the mineral surface. Also supercoiled plasmid DNA bound on "dirty" minerals sampled from a groundwater aquifer were taken up by competent *B. subtilis* (Romanowski et al. 1993b) and *A. calcoaceticus* (Chamier et al. 1993). Other experiments showed that natural transformation also occurs with DNA bound on clay particles (Lorenz and Wackernagel 1992a, 1993; Khanna and Stotzky 1992). We assume that the dynamic system of loop-train transitions in surface-associated DNA molecules (Fig. 1) allows uptake of DNA into a cell without extra energy for the removal of the polymer from its adsorbent. Once a cell has bound to an end or a loop of adsorbed DNA, internalization can proceed as the train following the end or the loop turns into a loop itself. It was observed that competent *B. subtilis* cells bind more effectively to sand grains when these were complexed with DNA (Lorenz et al. 1988).

Another important question is whether the chemical environment in aquatic and terrestrial habitats allows naturally competent cells to bind DNA and to take it up. Studies addressing this point have provided convincing evidence that transformation occurs in sea water (Paul et al. 1991, 1992), in groundwater (Lorenz et al. 1992; Chamier et al. 1993; Romanowski et al. 1993b) and in soil extract medium (Lorenz et al. 1992; Lorenz and Wackernagel 1991, 1992b). Thus, chemical constituents of environmental media seem not to pose a barrier to DNA-uptake during natural transformation.

It is not yet clear to what extent bacteria living in natural habitats develop the state of competence. Since competence is a regulated property in most naturally transformable bacteria (for references, see Lorenz and Wackernagel 1994), it was interesting to note that in the soil bacterium *Pseudomonas stutzeri* transformability is strongly increased upon nutrient limitation (e.g., limitation of the phosphate, nitrogen or organic carbon source; Lorenz and Wackernagel 1991, 1992b). The regulatory principles and the biological significance of this observation need to be elucidated.

Bacterial Sex in the Natural Habitat

Besides conjugation and transduction, the natural genetic transformation is a means exchanging genetic material among bacteria. Since transformation depends on chromosomally located genes (and not on "parasitic genetic elements" like viruses or plasmids) it can be considered to be the genuine bacterial mechanism of horizontal gene transfer. At present we do not know how efficient natural transformation occurs in bacterial habitats. Recently, a comparison was presented of transformation frequencies obtained under optimized liquid medium conditions and in microcosms simulating marine, sediment or soil habitats (Lorenz and Wackernagel 1994). Although the microcosm frequencies were mostly smaller, the data suggested considerable transformation efficiencies in complex nonsterile systems.

That horizontal gene transfer occurs among bacteria in natural habitats is evident by the highly sexual (panmictic) population structure of several species examined, among these pathogenic, commensal and soil bacteria (for ref. see Lorenz and Wackernagel 1994). It is noteworthy that among the species with panmictic population structure most are naturally transformable, among them the soil bacteria *Rhizobium meliloti* (Maynard Smith et al. 1993), *B. subtilis*

(Istock et al. 1992) and *Bacillus licheniformis* (Duncan et al. 1995). These observations together with the identification of mosaic genes in bacteria composed of nucleotide sequences of different evolutionary history (for ref. see Lorenz and Wackernagel 1994) suggest that horizontal gene transfer processes are active in microbial habitats. A recent compilation of experimentally documented interspecies natural transformation events involving chromosomal genes and plasmids identified about 70 cases (Lorenz and Wackernagel 1994) suggesting that exchanges may encompass the genetic material of perhaps all prokaryotic and even of eukaryotic sources.

Conclusions

Experimental data from a wide variety of studies suggest that a gene flow between bacteria occurs under natural conditions. The release of DNA from bacteria and the uptake by other bacterial cells in the course of genetic transformation provide the basis for a continual exchange of genetic material. Free DNA present in aquatic habitats, in soils and sediments represents an extracellular gene pool of bacteria living in the environment. Recombinant DNA will enter this gene pool and will underly the same genetic translocation principles as natural nucleotide sequences. In nature, gene flow occurs even from bacteria to eukaryotes in certain instances, like from *Agrobacterium tumefaciens* to plants. A possible gene flow from an eukaryote (yeast) to bacteria involving recombinant DNA is suggested by observations of Graupner and Wackernagel (see Chap. 18, this volume). They show the spontaneous and induced production of plasmid-like DNA structures from a yeast chromosome. These structures can transform bacteria. The gene flow by naked DNA in the environment is probably facilitated by the protection of DNA against DNases as a consequence of the association with solid surfaces, a mechanism that would increase DNA persistence in aquatic and terrestric habitats of bacteria and thereby improve the chance for DNA uptake by bacteria. Although there is ample evidence that in addition to gene transfer routes within species also routes between different species are opened by natural transformation, there may be barriers. These could be posed by uptake discrimination between homologous and heterologous DNA (as seen in several species including *P. stutzeri*; for references, see Lorenz and Wackernagel 1994), limited nucleotide sequence identity blocking recombinational integration, and DNA restriction (Lorenz and Wackernagel 1994; see also Chap. 10, this volume).

In summary, natural bacterial transformation is a fundamental process of bacterial sex and is assumed to contribute to the adaptation of natural bacterial populations to changing environments and thereby drives bacterial speciation and evolution.

Acknowledgments. The work in the laboratory of the author was supported by the Bundesminister für Forschung und Technologie, by the Bayer AG (Wuppertal,

Germany) within the "Verbundprojekt Sicherheitsforschung Gentechnik" and by the Fonds der Chemischen Industrie.

References

Ahrenholtz I, Lorenz MG, Wackernagel W (1994) A conditional suicide system in *Escherichia coli* based on the intracellular degradation of DNA. Appl Environ Microbiol 60: 3746–3751

Chamier B, Lorenz MG, Wackernagel W (1993) Natural transformation of *Acinetobacter calcoaceticus* by plasmid DNA adsorbed on sand and groundwater aquifer material. Appl Environ Microbiol 59: 1662–1667

Dreiseikelmann B (1994) Translocation of DNA across bacterial membranes. Microbiol Rev 58: 293–316

Dubnau D (1991a) Genetic competence in *Bacillus subtilis*. Microbiol Rev 55: 395–424

Dubnau D (1991b) The regulation of genetic competence in *Bacillus subtilis*. Mol Microbiol 5: 11–18

Duncan KE, Ferguson N, Kimura K, Zhou X, Istock CA (1995) Fine scale genetic and phenotypic structure in natural populations of *Bacillus subtilis* and *Bacillus licheniformis*: implications for bacterial evolution and speciation. Evolution (in press)

Hahn J, Inamine G, Kozlov Y, Dubnau D (1993) Characterization of *comE*, a late competence operon of *Bacillus subtilis* required for the binding and uptake of transforming DNA. Mol Microbiol 10: 99–111

Hesselink FT (1983) Adsorption of polyelectrolytes from dilute solution. In: Parfitt GD, Rochester CH (eds) Adsorption from solution at the solid/liquid interface. Academic, London, pp 317–412

Istock CA, Duncan KE, Ferguson E, Zhou X (1992) Sexuality in a natural population of bacteria – *Bacillus subtilis* challenges the clonal paradigm. Mol Ecol 1: 95–103

Khanna M, Stotzky G (1992) Transformation of *Bacillus subtilis* by DNA bound on montmorillonite and effect of DNase on the transforming ability of bound DNA. Appl Environ Microbiol 58: 1930–1939

Londono-Vallejo JA, Dubnau D (1993) *comF*, a *Bacillus subtilis* late competence locus, encodes a protein similar to ATP-dependent RNA/DNA helicases. Mol Microbiol 9: 119–131

Lorenz MG, Wackernagel W (1987) Adsorption of DNA to sand and variable degradation rates of adsorbed DNA. Appl Environ Microbiol 53: 2948–2952

Lorenz MG, Wackernagel W (1990) Natural genetic transformation of *Pseudomonas stutzeri* by sand-adsorbed DNA. Arch Microbiol 154: 380–385

Lorenz MG, Wackernagel W (1991) High frequency of natural genetic transformation of *Pseudomonas stutzeri* in soil extract supplied with a carbon/energy and a phosphorus source. Appl Environ Microbiol 57: 1246–1251

Lorenz MG, Wackernagel W (1992a) DNA binding to various clay minerals and retarded enzymatic degradation of DNA in a sand/clay microcosm. In: Gene Transfers and Environment. MJ Gauthier (ed) Springer (Berlin, Heidelberg, New York) 115–126

Lorenz MG, Wackernagel W (1992b) Stimulation of natural genetic transformation of *Pseudomonas stutzeri* in extracts from various soils by nitrogen or phosphorus limitation and influence of temperature and pH. Microbial Releases 1: 173–176

Lorenz MG, Wackernagel W (1993) Transformation as a mechanism for bacterial gene transfer in soil and sediment – studies with a sand/clay microcosm and the cyanobacterium *Synechocystis* OL50. In: Trends in Microbial Ecology. R Guerrero and C Pedrós-Alió (eds.) Spanish Society for Microbiolgy, 325–330

Lorenz MG, Wackernagel W (1994) Bacterial gene transfer by natural genetic transformation in the environment. Microbiol Rev 58: 563–602

Lorenz MG, Aardema BW, Wackernagel W (1988) Highly efficient genetic transformation of *Bacillus subtilis* attached to sand grains. J Gen Microbiol 134: 107–112

Lorenz MG, Gerjets D, Wackernagel W (1991) Release of transforming plasmid and chromosomal DNA from two cultured soil bacteria. Arch Microbiol 156: 319–326

Lorenz MG, Reipschläger K, Wackernagel W (1992) Plasmid transformation of naturally competent *Acinetobacter calcoaceticus* in non-sterile soil extract and groundwater. Arch Microbiol 157: 355–360

Maynard Smith J, Smith NH, O'Rourke M, Spratt BG (1993) How clonal are bacteria? Proc Natl Acad Sci USA 90: 4384–4388

Paget E, Simonet LJ, Monrozier P (1992) Adsorption of DNA on clay minerals: protection against DNaseI and influence on gene transfer. FEMS Microbiol Lett 97: 31–40

Paul JH, Frischer ME, Thurmond JM (1991) Gene transfer in marine water column and sediment microcosms by natural plasmid transformation. Appl Environ Microbiol 57: 1509–1515

Paul JH, Jeffrey WH, David AW, DeFlaun MF, Cazares LH (1989) Turnover of extracellular DNA in eutrophic and oligotrophic freshwater environments of southwest Florida. Appl Environ Microbiol 55: 1823–1828

Paul JH, Thurmond JM, Frischer ME, Cannon JP (1992) Intergeneric natural plasmid transformation between *E. coli* and a marine *Vibrio* species. Mol Ecol 1: 37–46

Phillips SJ, Dalgarn DS, Young SK (1989) Recombinant DNA in waste water: pBR322 degradation kinetics. J Water Pollut Control Fed 61: 1588–1595

Romanowski G, Lorenz MG, Wackernagel W (1991) Adsorption of plasmid DNA to mineral surfaces and protection against DNaseI. Appl Environ Microbiol 57: 1057–1061

Romanowski G, Lorenz MG, Wackernagel W (1992) Persistence of free plasmid DNA in soil monitored by various methods, including a transformation assay. Appl Environ Microbiol 58: 3012–3019

Romanowski G, Lorenz MG, Wackernagel W (1993a) Plasmid DNA in a groundwater aquifer microcosm: adsorption, DNase resistance and natural genetic transformation of *Bacillus subtilis.* Molec Ecol 2: 171–181

Romanowski G, Lorenz MG, Wackernagel W (1993b) Use of polymerase chain reaction and electroporation of *Escherichia coli* to monitor the persistence of extracellular plasmid DNA introduced into natural soils. Appl Environ Microbiol 59: 3438–3446

15 Persistence of DNA in Natural Soils: Adsorption to Particulate Material Provides Protection Against Nucleolytic Degradation

S.A.E. Blum, M.G. Lorenz, and W. Wackernagel

Genetik, Fachbereich Biologie, Carl von Ossietzky Universität Oldenburg, 26111 Oldenburg, Germany

Introduction

The release of genetically engineered organisms to the environment bears the risk of a transfer of recombinant DNA to organisms in natural habitats. In particular, the presence of in vitro recombined nucleotide sequences in bacteria of the environment could lead to their increased fitness or other new phenotypes resulting in disturbance of the microbial community. Among bacteria, genes can be horizontally transmitted by conjugation, transduction, and transformation. In our laboratory, the various aspects of genetic transformation of soil bacteria is studied which is the cell-to-cell transfer of naked DNA. In bacterial habitats, such as soil and sediment, DNA-degrading enzymes produced by bacteria are ubiquitously present. This raises the question of whether DNA may persist long enough to interact with transformable soil bacteria and to be finally taken up by them. Here we present data showing that DNA can persist in soil for several days if the DNA adsorbs to soil particulate material.

Materials and Methods

DNA

[^{3}H]Thymidine-labelled DNA of phage P22 and of plasmid pKT210 (Bagdasarian et al. 1981) isolated according to Hermanns and Wackernagel (1977) and Chamier et al. (1993), respectively, were used.

E.R. Schmidt/T. Hankeln (Eds.)
Transgenic Organisms and Biosafety

Soils

The soils employed were a brown earth, a loamy soil, and a podsoil collected from field plots near Braunschweig and Langen, Germany. The soils were air-dried, sieved (2 mm mesh size) and used non-sterilized in 0.7 g portions for the glass column microcosm experiments.

Preparation of Soil Extracts

Soil extracts were prepared as described previously (Lorenz and Wackernagel 1991).

Adsorption and Elution of DNA

In a flow-through glass column system (Lorenz and Wackernagel 1987) the soils were rehydrated to water holding capacity with sterile destilled water containing 300 ng of DNA and incubated at 23°C for various periods of time. In some experiments the soils were rehydrated with sterile destilled water and incubated at 23°C for 8 days. Then the DNA in an aqueous extract of the corresponding soil was added and the incubation continued. Adsorption of DNA to the column system itself was below 0.5% of applied DNA. At the end of incubation time the non-adsorbed DNA was obtained by washing with 3×1 ml of the corresponding soil extract (washing fluid). Subsequently, adsorbed DNA was eluted with 3×1 ml of a 0.12 *M* sodium phosphate buffer (Ogram et al. 1987) or with a 50 m*M* Tris-HCl, pH 8.8, 500 m*M* NaCl buffer (TN588; eluate). The volume of 3 ml was sufficient both to remove the non-adsorbed DNA in the washing step and to obtain the elutable DNA in the elution step. In washing fluid and eluate the radioactivity was determined by liquid scintillation counting (Kontron β-Matic 300, Kontron, Munich, Germany). The amount of adsorbed DNA was calculated by subtracting the radioactivity obtained in the washing fluid from the total radioactivity applied to the column.

Degradation of DNA

The extent of degradation of both non-adsorbed and adsorbed DNA was determined by the quantification of acid-soluble degradation products in washing fluid and eluate as described previously (Romanowski et al. 1991).

Results and Discussion

DNA Adsorption

In the first set of experiments the DNA was applied to air dried soils in the course of rehydration. Adsorption of DNA to soil particulate material was rapid. The maximum amount of DNA adsorbed within 20 min and remained constant for at least 5 h (Table 1). After 24 h the amount of adsorbed DNA had decreased in brown earth and loamy soil and almost doubled in podsoil (Table 1). Similar results were obtained previously for adsorption of DNA to purified soil components for which the half maximum adsorption was reached within 15 min at 23°C (Lorenz and Wackernagel 1987, 1992; Khanna and Stotzky 1992; Paget et al. 1992; Greaves and Wilson 1969).

Of the adsorbed DNA on the average 27% were elutable with a 0.12 *M* sodium phosphate buffer (Table 1) according to Ogram et al. (1987). The velocity of adsorption and the amount of elutable DNA were similar for linear duplex and supercoiled DNA (Table 2).

It was shown previously that concentrations as low as 0.5 m*M* Mg^{2+} or Ca^{2+} were sufficient for extensive DNA-binding to mineral surfaces (several µg of DNA per g sand; Lorenz and Wackernagel 1987; Romanowski et al. 1991). Similarly, divalent cations (mainly Mg^{2+} or Ca^{2+}) that were present in groundwater at concentrations of 1.2 m*M* effectively mediated DNA adsorption to material of a groundwater aquifer (Romanowski et al. 1993). The rapid and almost complete adsorption of DNA to brown earth and loamy soil in the experiments shown here indicates the presence of sufficiently high concentrations of divalent cations in the soils. The relatively low DNA binding capacity of podsoil compared to that of

Table 1. Adsorption of linear DNA applied to air-dried soils and elutable fraction of adsorbed DNA

Time (h)	DNA adsorbed as percentage of DNA applied			DNA eluted as percentage DNA adsorbed		
	Brown earth	Loamy soil	Podsoil	Brown earth	Loamy soil	Podsoil
0.3	84	93	24	16	17	21
1	87	92	21	16	22	33
2	80	91	21	19	26	33
3	90	90	27	16	27	30
4	94	90	18	17	26	33
5	93	87	24	24	24	33
24	48	47	46	21	43	72

Non-adsorbed DNA was removed by washing with the corresponding soil extract. Adsorbed DNA was eluted with 0.12 *M* sodium phosphate buffer, pH 8.0 (Ogram et al. 1987)

Table 2. Adsorption of supercoiled DNA applied to air-dried soils and elutable fraction of adsorbed DNA

Time (h)	DNA adsorbed as percentage of DNA applied			DNA eluted as percentage of DNA adsorbed		
	Brown earth	Loamy soil	Podsoil	Brown earth	Loamy soil	Podsoil
1	89	86	35	34	27	29
3	85	88	35	36	22	29
5	92	86	39	33	24	26
24	41	50	57	27	32	56

Non-adsorbed DNA was removed by washing with the corresponding soil extract. Adsorbed DNA was eluted with 0.12 *M* sodium phosphate buffer, pH 8.0 (Ogram et al. 1987)

the other soils (Table 1) could have resulted from the almost complete absence of clay minerals in podsoil. Previously clay was shown to have an at least 500-fold higher DNA-binding capacity than sand (Lorenz and Wackernagel 1994).

DNA Recovery

It was attempted to improve the recovery of adsorbed DNA obtained by elution with the sodium phosphate buffer. In general, a buffer for elution of DNA from soil should have two main properties. The buffer should have a high DNA elution efficiency and the coelution of humic substances should be as low as possible because these substances interfere with the subsequent analysis of the DNA (including quantification, PCR and hybridization). Among the various buffers tested the best results were obtained with 50 m*M* Tris, pH 8.8, 500 m*M* NaCl

Table 3. Elution of adsorbed linear DNA with either 0.12 *M* sodium phosphate buffer (NaP) or 50 m*M* Tris-HCl, pH 8.8, 500 m*M* NaCl (TN588)

Soil type	DNA eluted as percentage of DNA adsorbed	
	0.12 *M* NaP	TN588
Brown earth	11	62
Loamy soil	26	62
Podsoil	21	59

The soils were loaded with 300 ng DNA during rehydration and incubated for 2 h at 23 °C. Prior to elution, the non-adsorbed DNA was removed by washing with the corresponding soil extract

(TN588). The elution efficiency of TN588 on the average was three times the efficiency of the sodium phosphate buffer (Table 3). Humic substances (brownish colour) were hardly eluted by TN588 as indicated by the nearly colourless eluates. The improved elution buffer was employed in the following experiments.

Degradation of DNA

The extent of DNA degradation was determined separately for adsorbed and non-adsorbed DNA. For this purpose, the amount of acid-soluble degradation products was quantified in washing fluids and eluates 5 h and 24 h after addition of DNA to the soils. After 5 h degradation products in both the washing fluid and the eluate were not detectable (limit of detection 0.2%). After 24 h the degradation of non-adsorbed DNA was complete (brown earth and loamy soil) or almost complete (podsoil; 76%) whereas the adsorbed DNA obtained in the eluate was only 13%–26% acid-soluble. Supercoiled and linear DNA introduced into the soils were degraded to the same extent suggesting that endonucleases were involved in the degradation of the DNA in the three soils.

The results are in accord with previous data which showed that plasmid DNA adsorbed to chemically pure sand (Romanowski et al. 1991) or to material of a groundwater aquifer (Romanowski et al. 1993) is 100–1000 times more resistant against degradation by DNase I than DNA free in solution. Employing the extracellular nuclease of the ubiquitous bacterium *Serratia marcescens* the degradation of DNA adsorbed to material of a groundwater aquifer required about 80 times higher amounts of the nuclease than were needed for the degradation of dissolved DNA (Ahrenholtz et al. 1994). The data reported here suggest that protection from degradation by soil-borne DNases is mediated by the adsorption of DNA to particulate material of natural soils.

Adsorption of DNA to Rehydrated Soils

In these experiments DNA was introduced into wet soil. The soils were rehydrated with sterile demineralized water and incubated at 23°C for 8 days. Then the DNA dissolved in the corresponding soil extract was added and the incubation continued for up to 5 days.

Similar to what was observed with dry soils, the adsorption of DNA to particulate material of the rehydrated soils was rapid. The amount of DNA that was adsorbed within 1 h hardly changed in podsoil during 5 days of further incubation and slightly decreased in loamy soil and brown earth (Table 4). Adsorbed DNA was recovered by elution with the TN588 buffer. The elution efficiency was high (51% of the adsorbed DNA on the average) during the first 24 h. However, the amount of adsorbed DNA elutable with TN588 decreased

Table 4. Adsorption of linear DNA applied to rehydrated (8 days, 23 °C) soils and elutable fraction of adsorbed DNA

Time (h)	DNA adsorbed as percentage of DNA applied			DNA eluted as percentage of DNA adsorbed		
	Brown earth	Loamy soil	Podsoil	Brown earth	Loamy soil	Podsoil
1	78	88	36	56	59	50
3	82	70	42	56	59	43
5	77	64	39	52	59	49
24	68	64	42	49	53	29
72	59	65	30	36	49	20
120	58	56	49	29	23	8

The soils were loaded with 300 ng DNA and incubated for indicated times. Non-adsorbed DNA was removed by washing with the corresponding soil extract. Adsorbed DNA was eluted with 50 m*M* Tris-HCl, pH 8.8, 500 m*M* NaCl (TN588)

until the end of incubation (Table 4). Perhaps smaller molecules produced by continual fragmentation of DNA translocated to sites from were they could no longer be recovered by the buffers employed. It is also possible that DNA or its degradation products were ingested by soil microorganisms which did not detach during elution.

It is concluded that the condition of the soils (air-dried or rehydrated to water holding capacity) at the moment of DNA application was of minor influence on the velocity and extent of DNA adsorption.

Degradation of DNA in Rehydrated Soils

The degradation of DNA to acid-soluble products in the soils was quantified separately for adsorbed and non-adsorbed DNA. The non-adsorbed DNA was degraded almost completely within 24 h. In contrast, the adsorbed DNA contained only between 10% and 28% acid-soluble material after 1 d. Even after 5 days only between 17% and 36% of the adsorbed DNA eluted with TN588 were acid-soluble. Figure 1, as an example, shows degradation of non-adsorbed and adsorbed DNA in loamy soil.

As in the first set of experiments where the DNA was applied during the rehydration of the air-dried soils, at most one third of the adsorbed DNA was degraded whereas all of the non-adsorbed DNA was degraded.

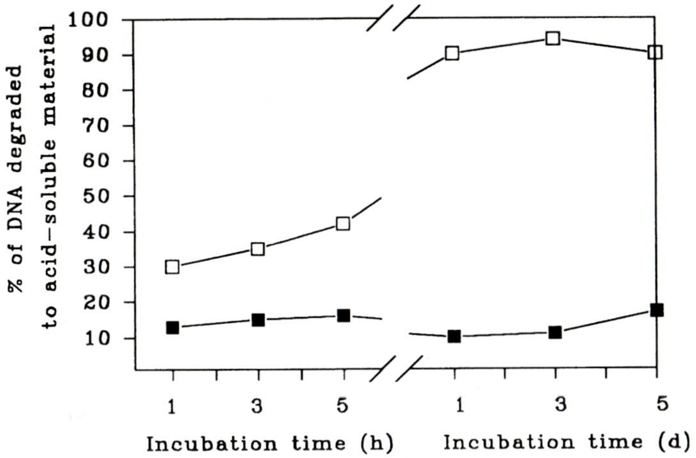

Fig. 1. Degradation of linear adsorbed (*closed squares*) and non-adsorbed (*open squares*) DNA in loamy soil. DNA was applied to rehydrated (8 days, 23 °C) soil. Non-adsorbed DNA was obtained by washing with soil extract. Adsorbed DNA was eluted with TN588

Conclusions

In the microcosm experiments with three non-sterile soils extracellular DNA persisted as high molecular weight material for at least 5 days. The distinct analysis of adsorbed and non-adsorbed DNA revealed that most of the DNA that adsorbed to soil particles escaped the nucleolytic degradation whereas the non-adsorbed DNA in the soil liquid phase was effectively degraded by soil-borne nucleases. The rapid transition of DNA from the soil liquid phase where nucleolytic degradation occurs at a high rate to the surfaces of particulates therefore is suggested to be the mechanism by which extracellular DNA can persist in soil.

References

Ahrenholtz I, Lorenz MG, Wackernagel W (1994) The extracellular nuclease of *Serratia marcescens*: studies on the activity in vitro and effect on transforming DNA in a groundwater aquifer microcosm. Arch Microbiol 161: 176–183

Bagdasarian M, Lurz R, Rückert B, Franklin FCH, Bagdasarian MM, Frey J, Timmis KN (1981) Specific-purpose plasmid cloning vectors. II. Broad host range, high copy number, RSF1010-derived vectors, and a host-vector system for gene cloning in *Pseudomonas*. Gene 1: 6237–247

Chamier B, Lorenz MG, Wackernagel W (1993) Natural transformation of *Acinetobacter calcoaceticus* by plasmid DNA adsorbed on sand and groundwater aquifer material. Appl Environ Microbiol 59: 1662–1667

Greaves MP, Wilson MJ (1969) The adsorption of nucleic acids by montmorillonite. Soil Biol Biochem 1: 317–323

Hermanns U, Wackernagel W (1977) The recBC enzyme of *Escherichia coli* K12: premature cessation of catalytic activities in vitro and reactivation by potassium ions. Eur J Biochem 76: 425–432

Khanna M, Stotzky G (1992) Transformation of *Bacillus subtilis* by DNA bound on montmorillonite and effect of DNase on the transforming ability of bound DNA. Appl Environ Microbiol 58: 1930–1939

Lorenz MG, Wackernagel W (1987) Adsorption of DNA to sand and variable degradation rates of adsorbed DNA. Appl Environ Microbiol 53: 2948–2952

Lorenz MG, Wackernagel W (1991) High frequency of natural genetic transformation of *Pseudomonas stutzeri* in soil extract supplemented with a carbon/energy and phosphorus source. Appl Environ Microbiol 57: 1246–1251

Lorenz MG, Wackernagel W (1992) DNA binding to various clay minerals and retarded enzymatic degradation of DNA in a sand/clay microcosm. In: Gauthier MJ (ed) Gene transfers and environment. Springer, Berlin, Heidelberg, New York, pp 103–113

Lorenz MG, Wackernagel W (1994) Bacterial gene transfer by natural genetic transformation in the environment. Microbiol Rev 58: 563–602

Ogram A, Sayler GS, Barkay T (1987) The extraction and purification of microbial DNA from sediments. J Microbiol Meth 7: 57–66

Paget E, Simonet LJ, Monrozier P (1992) Adsorption of DNA on clay minerals: protection against DNaseI and influence on gene transfer. FEMS Microbiol Lett 97: 31–40

Romanowski G, Lorenz MG, Wackernagel W (1991) Adsorption of plasmid DNA to mineral surfaces and protection against DNaseI. Appl Environ Microbiol 57: 1057–1061

Romanowski G, Lorenz MG, Wackernagel W (1993) Plasmid DNA in a groundwater aquifer microcosm – adsorption, DNAase resistance and natural genetic transformation of *Bacillus subtilis*. Molec Ecol 2: 171–181

16 Degradation of Transforming DNA in a Groundwater Aquifer Microcosm by the Extracellular Nuclease of *Serratia marcescens* SM6 and Application of the Nuclease in a New Conditional Suicide System for *Escherichia coli*

I. Ahrenholtz, M.G. Lorenz, and W. Wackernagel

Genetik, Fachbereich Biologie, Carl von Ossietzky Universität Oldenburg, Postfach 2503, 26111 Oldenburg, Germany

Abstract

We have examined whether association of high molecular weight DNA with mineral material sampled from the environment (groundwater aquifer) would provide protection of the DNA against a typical extracellular microbial nuclease. Such enzymes are assumed to destroy DNA in the environment. It was found that in a microcosm consisting of groundwater and groundwater aquifer material mineral-adsorbed transforming DNA was about 80 times more resistant to degradation by the extracellular nuclease of the ubiquitous bacterium *Serratia marcescens* than the same DNA dissolved in groundwater. It was concluded that mineral association provides protection of DNA against nucleases in natural environments. The persistence of DNA increases the chance of natural genetic transformation of bacteria in their habitats.

Because of the potential persistence of DNA in the environment which would include also recombinant DNA from genetically engineered microbes we have developed a new bacterial suicide system which combines the controlled killing of cells with the destruction of their genetic material. For the construction of the suicide cassette the cloned gene of the extracellular nuclease of *S. marcescens* was deleted for the leader-coding sequence (normally required for transmembrane transport of the nuclease) and the truncated gene was put under the control of the λ P_L promoter in *Escherichia coli.* After thermal inactivation of the λ cI857 repressor the survival of the cells declined rapidly and more than 80% of the intracellular DNA was converted to acid-soluble material. The RecBCD enzyme of *E. coli* (exonuclease V) contributed to the intracellular DNA degradation. The new suicide system kills cells effectively and at the same time destroys the DNA

E.R. Schmidt/T. Hankeln (Eds.)
Transgenic Organisms and Biosafety

of the cells that otherwise would be potentially available for horizontal gene transfer processes including genetic transformation.

Introduction

Potential risks of intentional or unintentional releases of genetically engineered microorganisms are seen in the uncontrolled dissemination of the microorganisms and their recombinant DNA in the environment. It has been shown that free DNA can persist in the environment and is available for uptake by bacterial cells that are naturally transformable. The main course of nucleic acid degradation in the environment presumably is due to the activity of microbial nucleases which have been found in a variety of bacterial habitats (for a review, see Lorenz and Wackernagel 1994).

The extracellular nuclease of the ubiquitous enterobacterium *Serratia marcescens* (in the following termed *Serratia* nuclease) cleaves RNA and DNA to acid-soluble material (Nestle and Roberts 1969). We found that the enzyme introduces single- and double-strand cuts into duplex DNA. Our data further suggest that the *Serratia* nuclease may contribute to DNA turnover in the environment (Ahrenholtz et al. 1994a).

The present study had two aims. The first was to examine whether association of high molecular weight DNA with mineral material sampled from the environment would provide protection of the DNA against the *Serratia* nuclease. The second was the development of a new bacterial suicide system which combines the controlled killing of cells with the destruction of their genetic material. Previously published suicide systems (for a review, see Molin et al. 1993) used killing functions which led to the destruction of the cell membrane or cell wall and would eventually lead to the release of cellular DNA into the bacterial habitat.

Materials and Methods

Strains and Plasmids. The bacterial strains and plasmids used in this study are described in Table 1.

Microcosms. A flowthrough glass column system was used as microcosm (Lorenz and Wackernagel 1987; Lorenz et al. 1988). Columns filled with 0.7 g mineral material (the interstitial volume of the bed was 0.2 ml) were equilibrated for 15 min with groundwater (GW). The chemical and physical characteristics of the GW and the groundwater aquifer material (GWA) have been described previously (Romanowski et al. 1993). Before use, GWA material was sterilized by dry heat (180 °C, 5 h) and GW was sterilized by autoclaving for 20 min at 121 °C. The

Table 1. Bacterial strains and plasmids

Strain or plasmid	Description	Source or reference
E. *coli* TGE900	F', *sup*$^+$, *ilv*, *bio*(λcI857Δ*Bam*ΔHI)	Courtney et al. 1984
E. coli AH2	TGE900 *recB*21 *argA*:: Tn10, Tc^R	This work
A. calcoaceticus BD4	Prototrophic	DSM no. 586
A. calcoaceticus BD413	*trpE*27	DSM no. 588
pSF1E	Ap^R, 4149 bp; pSF1 was digested with *Eco*RI for deletion of the *ssb* gene and religated	Ahrenholtz et al. 1994b
pNuc4	Ap^R, 4128 bp; gene for the extracellular nuclease of *S. marcescens* SM6 (*nuc*) in vector pUC18	Ball et al. 1987
pAH10	Ap^R, 3660 bp; leader-free nuclease gene of *S. marcescens* (*nuc*ΔL) under control of the T7 ϕ10 promoter in vector pET81F$^+$	Ahrenholtz et al. 1994b
pAH12	Ap^R, 5227 bp; leader-free nuclease gene of *S. marcescens* (*nuc*ΔL) under control of the lambda P_L promoter in vector pSF1E	Ahrenholtz et al. 1994b

columns were loaded with 2 µg chromosomal DNA in 0.2 ml GW as previously described (Lorenz and Wackernagel 1987; Lorenz et al. 1988) and incubated for 2 h at 23 °C. Nonadsorbed chromosomal DNA was removed by washing with GW for 25 min at a flow rate of 12 ml h^{-1}.

Treatment of Transforming Bacterial DNA with the Serratia Nuclease in Groundwater. The experiments were performed as previously described (Ahrenholtz et al. 1994a). Chromosomal DNA for transformation of *Acinetobacter calcoaceticus* BD413 (*trpE*27; Table 1) was isolated from *A. calcoaceticus* BD4 (prototrophic; Table 1). The transformation frequency is defined as the number of *trp*$^+$ transformants per viable cell. The relative transformation efficiency is defined as the transformation frequency with nuclease-treated DNA divided by the transformation frequency obtained without nuclease treatment. The reversion frequency of competent *A. calcoaceticus* BD413 cells to prototrophy (*trp*$^+$) was $<6\times10^{-8}$.

Treatment of Transforming Bacterial DNA with the Serratia Nuclease in the Groundwater Aquifer Material Microcosm. The experiments were performed as previously described (Ahrenholtz et al. 1994a). The transformation frequency was determined as described above.

Determination of Survival After Thermoinduction of Killing. The survival of cells after thermoinduction of the suicide system was determined as previously described (Ahrenholtz et al. 1994b). The survival (N/N_0) is the viable count of the culture after the indicated times (N) divided by the viable count of the culture before induction (N_0).

Intracellular DNA Degradation. The cellular DNA was labeled with [methyl-^{3}H]-thymidine and the fraction of trichloroacetic acid-insoluble radioactivity was determined following various incubation periods at 28 °C, 42 °C (thermo-induction of the suicide system) and at 42 °C in the presence of chloramphenicol as previously described (Ahrenholtz et al. 1994b).

Results and Discussion

Degradation of Transforming DNA

The degradation of bacterial chromosomal DNA dissolved in groundwater or adsorbed on groundwater aquifer material was monitored by the loss of its

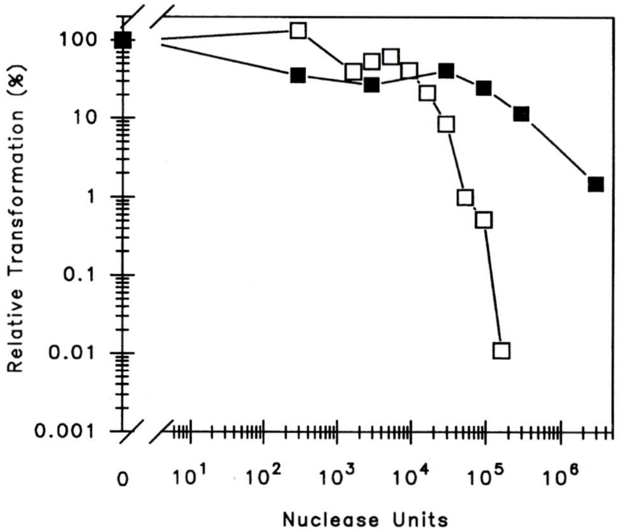

Fig. 1. Inactivation of the transforming activity of bacterial chromosomal DNA dissolved in groundwater (GW) (*open squares*) or adsorbed on groundwater aquifer material (GWA) in a GWA/GW microcosm (*closed squares*) by various concentrations of purified *Serratia* nuclease (Ahrenholtz et al. 1994a). One unit of DNase activity is defined as the amount of enzyme which causes the relaxation or linearization of 10^9 supercoiled plasmids in 30 min at 37 °C in reaction buffer (50 m*M* Tris-HCl, pH 8.0, 10 m*M* $MgCl_2$, 0.5 mg ml^{-1} acetylated bovine serum albumin) The relative transformation is the transformation frequency obtained with nuclease-treated DNA divided by the transformation frequency obtained without nuclease treatment. A relative transformation of 100% refers to a transformation frequency of 7×10^{-4} in the experiments in GW and to a transformation frequency of 8.5×10^{-5} in the experiments in the GWA/GW microcosms. The data are mean values of two independent experiments

transforming activity for competent cells of *Acinetobacter calcoaceticus*. To reduce the transforming activity of DNA dissolved in groundwater to 1%, 5.2×10^4 enzyme units of purified *Serratia* nuclease were necessary, whereas in a GWA/GW microcosm for the same reduction 4.3×10^6 enzyme units were necessary (Fig. 1). Thus, in a groundwater aquifer microcosm mineral-adsorbed transforming DNA was about 80 times more resistant to nucleolytic degradation than was dissolved DNA (Ahrenholtz et al. 1994a). We conclude that mineral-association provides a partial protection of DNA against nucleases in natural environments and thereby increases the persistence of DNA in bacterial habitats. In previous studies on the increased resistance of DNA against nucleolytic degradation resulting from adsorption of DNA on various purified minerals (Lorenz and Wackernagel 1987, 1992; Romanowski et al. 1991; Paget et al. 1992; Khanna and Stotzky 1992) or on "dirty" mineral mixes sampled from the environment (Romanowski et al. 1993) always bovine pancreatic DNase I was used, which is normally not present in soil, sediment and water. Our results are the first demonstration that adsorption to natural mineral material makes transforming DNA refractory to the activity of a DNase, which is released by a ubiquitous bacterial species into its environment.

Conditional Suicide System

A conditional suicide system is presented in which cell death is caused by the controlled derepression of a nuclease gene resulting in the degradation of intracellular DNA to acid-soluble material.

Construction of the Suicide Cassette

The gene of the extracellular nuclease of *S. marcescens* (Ball et al. 1987) was deleted for the leader-coding sequence and the truncated gene (*nuc*ΔL) was put under the control of the λ P_L promoter in *E. coli* (Fig. 2; Ahrenholtz et al. 1994b). In *E. coli* TGE900 which carries the gene for the thermosensitive λ cI857 repressor in the chromosome the expression of the *nuc*ΔL gene is repressed at 28 °C and can be thermally induced.

Thermoinduction of Killing

After thermal inactivation of the λ cI857 repressor the survival of the cells declined rapidly and reached its minimum after 30 min at 42 °C (2×10^{-5}; Fig. 3, Table 2). The cell killing resulted specifically from the derepression of the nuclease gene on plasmid pAH12 in the cells, whereas at 28 °C the nuclease expression from pAH12 is very low or absent (Ahrenholtz et al. 1994b).

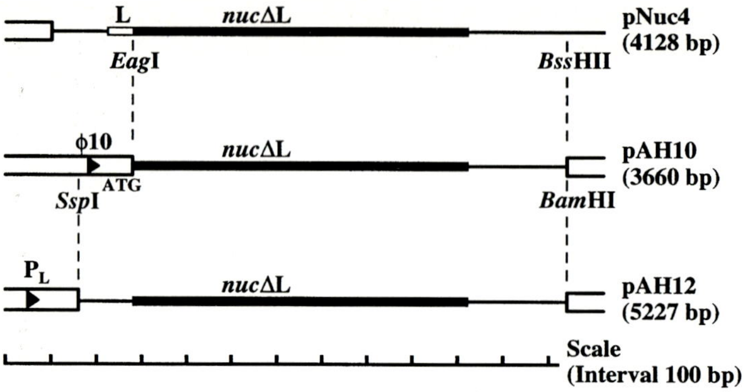

Fig. 2. Construction of the containment plasmid pAH12 (Ahrenholtz et al. 1994b) *L*, nucleotide sequence of the leader peptide; *nucΔL*, coding sequence for the mature *Serratia* nuclease; ϕ10, T7 ϕ10 promoter, P_L, λ P_L promoter; *boxes*, vector DNA; *line*, insert DNA; *ATG*, start codon. *Arrowheads* indicate location and direction of the promoters

Additionally, the survival and the intracellular DNA degradation (see below) after thermoinduction of killing in a *recB* derivative of TGE900 pAH12 was examined. In this derivative the RecBCD enzyme of *E. coli* is inactive. This enzyme has ATP-dependent exonuclease activity, degrades single- and double-stranded DNA and is part of the *recA*-dependent recombinational repair system in *E. coli* (for review, see Taylor 1988). In the *recB* strain AH2 pAH12 (Table 1) the thermoinduced killing was reduced (2×10^{-4} after 30 min at 42 °C; Fig. 3,

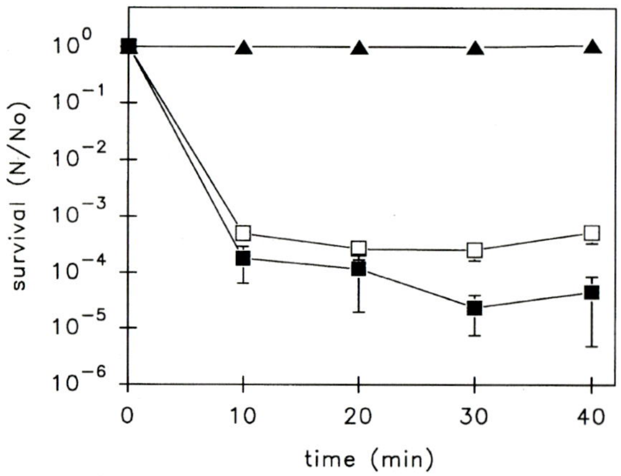

Fig. 3. Survival of *Escherichia coli* TGE900 with the suicide plasmid pAH12 (*closed squares*) or the vector pSF1E (*closed triangles*) and of the *recB* derivative AH2 with pAH12 (*open squares*) after thermoinduction by incubation at 42 °C (0–40 min). The survival (N/N_0) is the viable count of the culture at the indicated times (N) divided by the viable count of the culture before induction (N_0). The data are mean values of two independent experiments

Table 2. Survival of *Escherichia coli* strains grown at 28 °C without and with thermoinduction of the λ P_L promoter (42 °C)

Strain and description	Survival (N/N_0)[a]	
	42 °C culture	28 °C culture
TGE900 pAH12 (*nuc*ΔL)	2.3×10^{-5} ($\pm$ $1.6{\times}10^{-5}$)	1.5
TGE900 pSF1E (vector)	1.2	1.5
AH2 pAH12 (*nuc*ΔL, *recB*)	1.7×10^{-4} ($\pm$ 9.310^{-5})	1.1
AH2 pSF1E (vector, *recB*)	1.0	1.0

[a]The survival (N/N_0) is the viable count of the culture after 30 min at 42 °C or 28 °C (N) divided by the viable count of the culture at time 0 (N_0) The data are mean values of two independent experiments; the numbers in parenthesis are deviations from the mean

Table 2) compared with the *recB*$^+$ counterpart. The data suggest that the RecBCD enzyme enhances the killing efficiency of the *Serratia* nuclease after thermoinduction of the suicide system.

Examination of Surviving Clones

Clones which survived the thermoinduced killing after 30 min at 42 °C were sensitive to a second thermotreatment. Among 25 clones examined the majority (84%) was as sensitive as the original strain, the others showed a somewhat slower killing kinetics or a slightly higher final level of survivors (data not shown). This suggests that the system presented is not very prone to mutational escape events (Ahrenholtz et al. 1994b).

Intracellular DNA Degradation

Thermoinduction of the suicide system in *E. coli* TGE900 pAH12 and in the *recB* strain AH2 pAH12 led to the intracellular degradation of cellular DNA (Fig. 4). More than 80% of the intracellular DNA was converted to acid-soluble material within 2.5 h in the strain TGE900 pAH12. The RecBCD enzyme of *E. coli* slightly contributed to the intracellular DNA degradation, probably by its exonucleolytic activity for duplex DNA (Fig. 4). The data suggest that the cellular DNA first is cleaved by the *Serratia* nuclease (duplex DNA endonuclease) and that afterwards the RecBCD enzyme initiates exonucleolytic degradation of the linear DNA fragments.

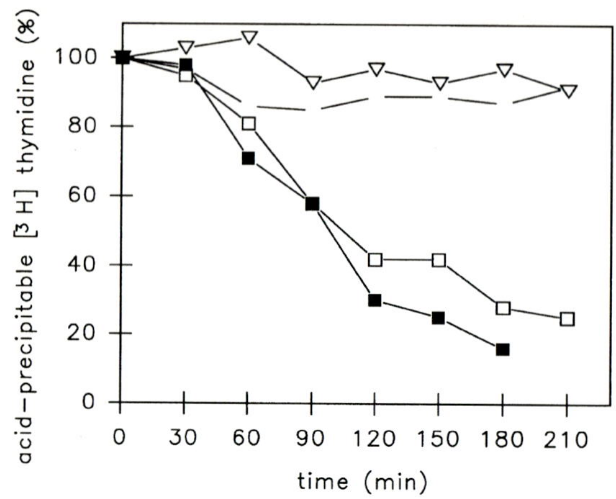

Fig. 4. Intracellular DNA degradation in *Escherichia coli* TGE900 pAH12 and in the *recB* derivative AH2 pAH12 following thermoinduction. The cellular DNA was labeled by the growth of the cells at 28 °C in the presence of [^{3}H]thymidine. Thermoinduction occurred by shifting cultures to 42 °C from 0 to 30 min, followed by 28 °C from 30 min to 210 min. At the indicated times, the trichloroacetic acid-insoluble radioactive material was determined in the thermoinduced cultures of TGE900 pAH12 (*closed squares*) and the *recB* derivative (*open squares*), in the cultures thermoinduced in the presence of Cm (*open triangles*) and in the cultures kept at 28 °C throughout (*dashed line*)

In summary, the described suicide system combines the regulated killing of bacterial cells with the destruction of the intracellular DNA that otherwise would be potentially available for horizontal gene transfer processes including conjugation, transduction and particularly transformation. More strain design and experimental work is required to make this basic bacterial suicide system usable for various applications in biotechnology where survival of recombinant bacteria and persistence of their DNA is not wanted after cells have done their job.

Acknowledgement. We thank Dr. M.J. Benedik for *E. coli* JM101 with the plasmid pNuc4, Dr. C. Urbanke for *E. coli* TGE900 and the plasmid pSF1, Dr. S.M. Tanhauser for the plasmid pET81F$^+$ and Brigitte Thoms for help during the construction of *E. coli* strain AH2. This work was supported by the Fonds der Chemischen Industrie.

References

Ahrenholtz I, Lorenz MG, Wackernagel W (1994a) The extracellular nuclease of *Serratia marcescens*: studies on the activity in vitro and effect on transforming DNA in a groundwater aquifer microcosm. Arch Microbiol 161: 176–183

Ahrenholtz I, Lorenz MG, Wackernagel W (1994b) A conditional suicide system in *Escherichia coli* based on the intracellular degradation of DNA. Appl Environ Microbiol 60: 3746–3751

Ball TK, Saurugger PN, Benedik MJ (1987) The extracellular nuclease gene of *Serratia marcescens* and its secretion from *Escherichia coli*. Gene 57: 183–192

Courtney M, Buchwalder A, Tessier L-H, Jaye M, Benavente A, Balland A, Kohli V, Lathe R, Tolstoshev P, Lecocq J-P (1984) High-level production of biologically active human α_1-antitrypsin in *Escherichia coli*. Proc Natl Acad Sci USA 81: 669–673

Khanna M, Stotzky G (1992) Transformation of *Bacillus subtilis* by DNA bound on montmorillonite and effect of DNase on the transforming ability of bound DNA. Appl Environ Microbiol 58: 1930–1939

Lorenz MG, Aardema BW, Wackernagel W (1988) Highly efficient genetic transformation of *Bacillus subtilis* attached to sand grains. J Gen Microbiol 134: 107–112

Lorenz MG, Wackernagel W (1987) Adsorption of DNA to sand and variable degradation rates of adsorbed DNA. Appl Environ Microbiol 53: 2948–2952

Lorenz MG, Wackernagel W (1992) DNA binding to various clay minerals and retarded enzymatic degradation of DNA in a sand/clay microcosm. In: Gauthier MJ (ed) Gene transfers and environment. Springer, Berlin Heidelberg New York, pp 103–113

Lorenz MG, Wackernagel W (1994) Bacterial gene transfer by natural genetic transformation in the environment. Microbiol Rev 58: 426–465

Molin S, Boe L, Jensen LB, Kristensen CS, Givskov M (1993) Suicidal genetic elements and their use in biological containment of bacteria. Annu Rev Microbiol 47: 139–166

Nestle M, Roberts WK (1969) An extracellular nuclease from *Serratia marcescens*. II. Specificity of the enzyme. J Biol Chem 244: 5219–5225

Paget E, Monrozier LJ, Simonet P (1992) Adsorption of DNA on clay minerals: protection against DNase I and influence on gene transfer. FEMS Microbiol Lett 97: 31–40

Romanowski G, Lorenz MG, Wackernagel W (1991) Adsorption of plasmid DNA to mineral surfaces and protection against DNase I. Appl Environ Microbiol 57: 1057–1061

Romanowski G, Lorenz MG, Wackernagel W (1993) Plasmid DNA in a groundwater aquifer microcosm – adsorption, DNase resistance and natural genetic transformation of *Bacillus subtilis*. Mol Ecol 2: 171–181

Taylor A (1988) RecBCD enzyme of *Escherichia coli*. In: Smith GR, Kucherlapati SK (ed) Genetic recombination. American Society for Microbiology, Washington DC, pp 230–263

17 A Conditional Suicide System for Genetically Engineered Bacteria Based on the Endonuclease I Gene of *Escherichia coli*

M. Jekel, M.G. Lorenz, and W. Wackernagel

Genetik, Fachbereich Biologie, Universität Oldenburg, Postfach 2503, 26111 Oldenburg, Germany

Introduction

The use of genetically engineered microorganisms (GEMs) in science and industrial production bears the risk of their unintended release into the environment. Moreover, several applications (e.g., in agriculture or waste treatment) rely directly on the intended release of large quantities of GEMs. Such releases may interfere with the balanced microbial ecosystems by the survival and reproduction of GEMs and/or by horizontal transfer of their recombinant DNA to indigenous bacteria via conjugation, transduction or transformation (see Lorenz and Wackernagel 1993). Such considerations have led to the development of biological containment systems which cause a controlled cell death by the regulated expression of genes with cell-killing functions. Mostly, these functions cause cell death by membrane destabilization or cell lysis (for a review, see Molin et al. 1993). As a consequence, genetic material may be released from cells into microbial habitats after induction of the killing genes. In other studies it was shown that free DNA can persist in soils for weeks and months (Romanowski et al. 1992, 1993) and may be taken up by cells in the course of natural genetic transformation (for a review, see Lorenz and Wackernagel 1994).

Because of the high potential that released recombinant DNA persists and is taken up by microorganisms in their natural habitat, we constructed a conditional suicide system which (a) directs a controlled killing of the cells and (b) causes the intracellular degradation of DNA before it can enter the environment. For this purpose, we chose the periplasmic endonuclease I of *Escherichia coli*. This enzyme is a sequence-unspecific endonuclease which degrades double-stranded DNA to oligodeoxyribonucleotides (Lehman et al. 1962).

E.R. Schmidt/T. Hankeln (Eds.)
Transgenic Organisms and Biosafety

Results and Discussion

Cloning and Sequencing of the endA Gene from Escherichia coli

The *endA* gene encoding endonuclease I of *E. coli* was not available when we started the project. Therefore, we identified the *endA* gene in a lambda library of the *E. coli* K12 chromosome and cloned the gene. It is located in the 63.6 min of the genetic map of *E. coli* and at position 3104 of the physical map (Jekel and Wackernagel 1994), respectively. The nucleotide sequence (see Fig. 1) revealed that the open reading frame (ORF) of *endA* codes for a precursor protein of 235 amino acids (aa) with a signal peptide and a predictable signal peptide cleavage site. The presence of a signal peptide is in agreement with the periplasmic location of the mature endonuclease I in *E. coli* (Cordonnier and Bernardi 1965; Obinata and Mizuno 1968). The predicted aa sequence of endonuclease I turned out to have a high degree of identity to the aa sequences of the extracellular DNases from *Vibrio cholerae* (60%) and *Aeromonas hydrophila* (54%). It was shown that upstream of the ORF is a region which exhibits low promoter activity (Jekel and Wackernagel 1995). We confirmed the identity of the presumed signal peptide cleavage site by purifying endonuclease I from the periplasm of an overproducing strain and by sequencing the aminoterminus of the mature protein. The data indicate that a 22-aa signal peptide is cleaved during the export of endonuclease I into the periplasmic space resulting in a mature protein of 213 aa. The predicted relative molecular mass of the mature endonuclease I (M_r, 24454) is in accordance with that estimated from a sodium dodecyl sulfate (SDS) polyacrylamide gel of the purified endonuclease I (data not shown).

Construction of the Suicide System

To produce intracellularly active endonuclease I we deleted the signal peptide-coding nucleotide sequence by cloning the *Eae*I/*Tfi*I fragment of the *endA* gene (see Fig. 1) into the expression cassette of pET81 (Tanhauser et al. 1992). This put the nucleotide sequence of the mature endonuclease I (*endA*ΔL) with its first codon adjacent to the startcodon and also joined the T7Φ10 promoter, a translation initiation site, and a T7 terminator to *endA*ΔL. To control the expression of *endA*ΔL by thermoinduction we cloned the expression cassette including the T7Φ10 promoter and the *endA*ΔL into pSF1E (for details, see Ahrenholtz et al. 1994) resulting in pMJ20. This put the *endA*ΔL with the T7Φ10 promoter and the translation initiation site downstream of the lambda p_L promoter. In the *E. coli* strain TGE900 which carries the gene for the thermosensitive lambda *c*I857 repressor in the chromosome, the *endA*ΔL gene is repressed at 28°C and can be induced by shifting the temperature to 42 °C.

```
AAGCTTTTATTGAAGAAGTGGTACCGCACGAACTGGCACATTTGCTGGTATGGAAACATTTCGGCCGCGTAGCGCACATGGCAAAGAGTGGAAGTGGATG  100
                   KpnI

ATGGAAAACGTGCTGGGTGTTCCCGCCCGTCGTACGCATCAGTTCGAACTGCAATCCGTGCGTCGCAACACCTTCCCCTACCGCTGCAAGTGCCAGGAGC  200

                                                                                   TTGACA
ATCAGCTTACCGTACGCCGCCATAATCGCGTAGTTCGTGGCGAGGCCGTCTATCGCTGTGTTCACTGCGGTGAACAGCTGGTTGCGAAATAACCATCTGA  300
                                                                             PvuII

  TATAAT                      TGGACA                     TATAAT                     RBS      M  Y  R  Y  L    5
ACTATCAGGAACTTTCCTGATCTGGCTGATTGCATACCAAAACAGCTTTCGCTACGTTGCTGGCTCGTTTTAACACGGAGTAAGTGATGTACCGTTATTT  400
                                                   SCS
 S  I  A  A  V  V  L  S  A  A  F  S  G  P  A  L  A ↓[E  G  I  N  S  F  S  Q  A  K  A  A  A  V  K] V   38
GTCTATTGCTGCGGTGGTACTGAGCGCAGCATTTTCCGGCCCGGCGTTGGCCGAAGGTATCAATAGTTTTTCTCAGGCGAAAGCCGCGGCGGTAAAAGTC  500
                                               EaeI

 H  A  D  A  P  G  T  F  Y  C  G  C  K  I  N  W  Q  G  K  K  G  V  V  D  L  Q  S  C  G  Y  Q  V  R    71
CACGCTGACGCGCCCGGTACGTTTTATTGCGGATGTAAAATTAACTGGCAGGGCAAAAAAGGCGTTGTTGATCTGCAATCGTGCGGCTATCAGGTGCGCA  600

K  N  E  N  R  A  S  R  V  E  W  E  H  V  V  P  A  W  Q  F  G  H  Q  R  Q  C  W  Q  D  G  G  R  K  N  105
AAAATGAAAACCGCGCCAGCCGCGTAGAGTGGGAACATGTCGTTCCCGCCTGGCAGTTCGGTCACCAGCGCCAGTGCTGGCAGGACGGTGGACGTAAAAA  700

 C  A  K  D  P  V  Y  R  K  M  E  S  D  M  H  N  L  Q  P  S  V  G  E  V  N  G  D  R  G  N  F  M  Y   138
CTGCGCTAAAGATCCGGTCTATCGCAAGATGGAAAGCGATATGCATAACCTGCAGCCGTCAGTCGGTGAGGTGAATGGCGATCGCGGCAACTTTATGTAC  800
                                                 PstI

 S  Q  W  N  G  G  E  G  Q  Y  G  Q  C  A  M  K  V  D  F  K  E  K  A  A  E  P  P  A  R  A  R  G  A    171
AGCCAGTGGAATGGCGGTGAAGGCCAGTACGGTCAATGCGCCATGAAGGTCGATTTCAAAGAAAAAGCTGCCGAACCACCAGCGCGTGCACGCGGTGCCA  900

I  A  R  T  Y  F  Y  M  R  D  Q  Y  N  L  T  L  S  R  Q  Q  T  Q  L  F  N  A  W  N  K  M  Y  P  V  T  205
TTGCGCGCACCTACTTCTATATGCGCGACCAATACAACCTGACACTCTCTCGCCAGCAAACGCAGCTGTTCAACGCATGGAACAAGATGTATCCGGTTAC 1000
                                                               PvuII

 D  W  E  C  E  R  D  E  R  I  A  K  V  Q  G  N  H  N  P  Y  V  Q  R  A  C  Q  A  R  K  S  *          235
CGACTGGGAGTGCGAGCGCGATGAACGCATCGCGAAGGTGCAGGGCAATCATAACCCGTATGTGCAACGCGCTTGCCAGGCGCGAAAGAGCTAACCTACA 1100

                                                   T2  \          /  T2
CTAGCGGGATTCTTTTTGTTAACCCCTACCCCACGCGTACAACCGCGTGGGGAGACGACGCGGATTTTTAACTATGCGTATCCCCCGCATTTATCATCCT 1200
        TfiI                 T1      /       \      T1

GAACCACTGACCAGCCATTCTCACATCGCGCTTTGCGAAGATGCCGCCAACCATATCGGGCGCGTACTGCGCATGGGGCCGGGGCAGGCGTTGCAATTGT 1300

TTGACGGTAGCAACCAGGTCTTTGACGCCGAAATTACCAGCGCCAGCAAAAAAAGCGTGGAAGTGAAGGTGCTGGAAGGCCAGATCGACGATCGCGAATC 1400

TCCGCTGCATATTCACCTCGGTCAGGTGATGTCGCGTGGTGAAAAAATGGAATTTACTATCCAGAAATCGATCGAACTCGGTGTAAGCCTCATTACGCCA 1500

CTTTTTTCTGAGCGCTGCGGCGTTAAACTGGATAGTGAACGTCTGAACAAGAAGCTT                                            1557
```

Fig. 1. The nucleotide (nt) sequence of the cloned *Hin*dIII fragment containing the *endA* gene from *Escherichia coli* (Jekel and Wackernagel 1995). The –10 and –35 regions of putative promoters are *underlined*, and the consensus sequences of *E. coli* σ^{70} promoters are shown for comparison. The nt sequence of the putative ribosome binding site (RBS) and restriction sites are *underlined*. The *converging arrows* indicate inverted repeats of possible transcription terminators (*T1*, *T2*). The deduced amino acid (aa) sequence is typed above the nt sequence. The stop codon is indicated by an *asterisk*. The determined aa sequence of the mature periplasmic endonuclease I is *framed*. The signal peptide cleavage site (*SCS*) is indicated by a *downward arrow*. The nt sequence has been assigned GenBank/EMBL accession number X65169

Induction of Killing

The killing of cells was determined following temperature induction of the suicide system. Briefly, log phase-cells of TGE900 (pMJ20) were grown at 28 °C in LB with ampicillin (100 µg/ml). One half of the culture was shifted to 42 °C, whereas the other half was further incubated at 28 °C. After different times the

Table 1. Survival of *Escherichia coli* TGE900 with the suicide plasmid pMJ20 following temperature shift from 28 °C to 42 °C and without thermal treatment

Time after temperature shift (min)	Surviving cells[a] at 28 °C (%)	Surviving cells[a] at 42 °C (%)
0	100	100
10	n.d.	0.011
20	n.d.	0.0096
30	160	0.0036
60	230	0.00099
90	n.d.	0.00083
120	440	0.00073

n.d., not determined.
[a]Viable counts and the survival were determined as described in "Induction of Killing". The data represent the means of three independent experiments.

viable counts of the cultures were determined by plating dilutions of the cultures on LB-Agar containing ampicillin (100 μg/ml) and incubating for 24 h at 28 °C. Survival of cells is the viable counts at indicated times (n) divided by the viable counts of the culture before induction (n_0) of the suicide system.

The survival of TGE900 (pMJ20) decreased rapidly after thermoinduction and reached its minimum of 10^{-5} after 60 min at 42 °C. Longer incubation at 42 °C did not substantially increase the killing of cells (see Table 1). Cell killing was not observed when TGE900 (pMJ20) was maintained at 28 °C or when cells of TGE900 harbouring a construct similar to pMJ20, but with the *ssb* gene of *E. coli* inserted instead of *endA*ΔL, or harbouring only the vector pSF1E were shifted to 42 °C (Ahrenholtz et al. 1994). This suggests that the killing of TGE900 (pMJ20) resulted specifically from thermal derepression of *endA*ΔL and was not caused by induction of any other plasmid-encoded activity or by protein overproduction per se. In all thermoinduction experiments lysis of cells did not occur (determined microscopically) during the incubation at 42 °C. This suggests that the induction of the suicide system is not accompanied by the release of the cellular DNA (and also recombinant DNA) into the environment.

Conclusions

We established a new conditional suicide system based on the endonuclease I gene of *E. coli*. After induction of the suicide system a rapid and efficient cell killing was observed, which is caused specifically by the induced overproduction of a toxic intracellular substance, the endonuclease I. In further studies it will be examined to what extent the DNA of the cells is degraded following induction of the suicide system. Other experiments will aim at increasing the killing efficiency and at putting the system under the control of other regulatory elements responsive to environmental factors besides thermoshock. A recently established suicide system of a similar concept successfully used the induced production of the signal peptide-devoid extracellular nuclease (DNase and RNase) of *Serratia marcescens* for cell killing (Ahrenholtz et al. 1994). In this system cell killing was accompanied with the conversion of over 80% of cellular DNA to acid soluble products. An even more effective suicide system could consist of a combination of the *Serratia* nuclease with endonuclease I into one suicide cassette. Now that the basic systems have been established, in which cell killing is accompanied by the intracellular degradation of DNA, reducing the chance that the DNA can participate in horizontal gene transfer processes, further steps can be taken. These are specific adaptations of the suicide system for use in industrial or environmental tasks. The suicide system aids various safe applications of genetically engineered microorganisms in biotechnology.

References

Ahrenholtz I, Lorenz MG, Wackernagel W (1994) A conditional suicide system in *Escherichia coli* based on the intracellular degradation of DNA. Appl Environ Microbiol 60: 3746–3751

Cordonnier C, Bernardi G (1965) Localization of *E. coli* endonuclease I. Biochem Biophys Res Comm 20: 555–559

Jekel M, Wackernagel W (1994) Location of the *endA* gene coding for endonuclease I on the physical map of the *Escherichia coli* K-12 chromosome. J Bacteriol 176: 1550–1551

Jekel M, Wackernagel W (1995) The periplasmic endonuclease I of *Escherichia coli* has amino acid sequence homology to the extracellular DNases of *Vibrio cholerae* and *Aeromonas hydrophila*. Gene (in press)

Lehman IR, Roussos GG, Pratt EA (1962) The deoxyribonucleases of *Escherichia coli*. II. Purification and properties of a ribonucleic acid-inhibitable endonuclease. J Biol Chem 237: 819–828

Lorenz MG, Wackernagel W (1993) Bacterial gene transfer in the environment. In: Wöhrmann W, Tomiuk J (ed) Transgenic organisms. Risk assessment of deliberate releases. Birkhäuser, Basel, pp 43–64

Lorenz MG, Wackernagel W (1994) Bacterial gene transfer by natural genetic transformation in the environment. Microbiol Rev 58: 563–602

Molin S, Boe L, Jensen LB, Kristensen CS, Giskov M (1993) Suicidal genetic elements and their use in biological containment of bacteria. Annu Rev Microbiol 47: 139–166

Obinata M, Mizuno D (1968) Intracellular localization of deoxyribonucleases in *Escherichia coli*. Biochim Biophys Acta 155: 98–106

Romanowski G, Lorenz MG, Sayler G, Wackernagel W (1992) Persistence of free plasmid DNA in soil monitored by various methods, including a transformation assay. Appl Environ Microbiol 58: 3012–3019

Romanowski G, Lorenz MG, Wackernagel W (1993) Use of polymerase chain reaction and electroporation of *Escherichia coli* to monitor the persistence of extracellular plasmid DNA introduced into natural soils. Appl Environ Microbiol 59: 3438–3446

Tanhauser SM, Jewell DA, Tu CK, Silverman DN, Laipis PJ (1992) A T7 expression vector optimized for site-directed mutagenesis using oligodeoxyribonucleotide cassettes. Gene 117: 113–117

18 Identification and Characterization of Extrachromosomal Circular DNA Released from a Genetically Modified Chromosome of *Hansenula polymorpha*

S. Graupner and W. Wackernagel

Genetik, Fachbereich Biologie, Universität Oldenburg, 26111 Oldenburg, Germany

Introduction

Recombinant nucleotide sequences in industrial microorganisms are often located on plasmids. It is known that plasmids can be transferred among bacteria in their habitats by a variety of mechanisms including conjugation, mobilization, transduction and natural transformation (see Lorenz and Wackernagel 1993). Therefore, to reduce spread of recombinant DNA by horizontal gene transfer processes, the locatation of recombinant nucleotide sequences in the chromosome of microorganisms has been suggested . This also reduces loss of recombinant sequences by segregation.

The methylotrophic yeast *Hansenula polymorpha* has become important in biotechnology for the production of heterologous proteins (Gellissen et al. 1992). We have studied the genetic stability of a recombinant chromosomal insert in this species. In particular, we have examined whether the *H. polymorpha* LR9 strain with a chromosomal insertion of a recombinant plasmid (*H. polymorpha-Escherichia coli* shuttle vector) would release in vivo from the chromosome DNA structures that are replicated and propagated in *E. coli* when transferred to this organism. Two *H. polymorpha* LR9 strains were employed which had a single insertion or a tandem quadruple insertion of plasmid pFPMT. This plasmid of 9.2 kb contains the colE1 origin of replication and an ampicillin resistance gene, the *LEU2* and *URA3* genes plus an autonomously replicating sequence from *H. polymorpha* (HARS), and the structural gene for aprotinin (eukaryotic protease inhibitor).

E.R. Schmidt/T. Hankeln (Eds.)
Transgenic Organisms and Biosafety

Material and Methods

Strains and Plasmids. E. coli NM522 (Woodcock et al. 1989), *E. coli* DH5α (Hanahan 1983), *E. coli* SF8 (Romanowski et al. 1992) and *E. coli* SF8*recA* (a *recA56* derivative of *E. coli* SF8, obtained by P1 transduction) were grown in LB medium. If needed, it contained 40 µg/ml ampicillin. The *H. polymorpha* strains were HM1-39 (*ura3 leu2* obtained from C.P. Hollenberg), LR9 (*ura3;* Roggenkamp et al. 1986) and two LR9 derivatives with a single (LR9-1) or a tandem quadruple (LR9-4) insertion of pFPMT (obtained from G. Gellissen). The *H. polymorpha* strains were grown in standard SD (minimal medium) and YPD (broth) media (Sherman et al. 1986) at 37 °C. The plasmid pFPMT is a pFMD13025 derivative (Gellissen et al. 1991) which additionally contains the aprotinin gene and a leader sequence for its secretion.

DNA Preparations and Transformation Procedures. Plasmid DNA was prepared from *E. coli* according to Birnboim and Doly (1979). Plasmid DNA from *H. polymorpha* was isolated as described by Devenish and Newlon (1982). Total DNA from *H. polymorpha* spheroblasts was prepared as described by Devenish and Newlon (1982) and purified as described by Marmur (1961). Electrotransformation of *E. coli* was carried out according to Romanowski et al. (1993). *H. polymorpha* was transformed according to Dohmen et al. (1991) or by an improved method (Graupner and Wackernagel, in preparation).

Treatment of DNA with Restriction Endonucleases and Polymerase Chain Reaction. Restriction enzymes were used as recommended by the manufacturer. PCR amplification was carried out in a thermal cycler (Perkin Elmer Cetus) with Taq DNA polymerase (Appligene, Heidelberg) by the method of Romanowski et al. (1993). Electrophoresis of the amplified products (395 bp) was carried out in a 1% (w/v) agarose gel (3 h, 120 V).

Results and Discussion

E. coli Transformants Obtained with DNA from *H. polymorpha*

Electrotransformation of *E. coli* SF8 with total DNA extracted from both *H. polymorpha* LR9-1 and LR9-4 strains resulted in the formation of ampicillin-resistent clones with a frequency of about 10^2/µg DNA. As a result of the *recB21 recC22* mutations, *E. coli* SF8 has no active exonuclease V (RecBCD-enzyme; recombination enzyme which effectively degrades linear duplex DNA in the cell; Taylor 1988) so that linear DNA fragments can exist for a longer period in the cytoplasm of the bacterial cells. Despite the lack of the RecBCD-enzyme,

Table 1. Effect of degradation by exonuclease V of duplex DNA from *H. polymorpha* on the frequency of Ap^R transformants after electrotransformation of *E. coli* SF8*recA* with the DNA

	Number of Ap^R transformants/4 × 10^7 extracted *H. polymorpha* cells	
	– Exonuclease V	+ Exonuclease V
LR9-4	222	280
	317	333
	231	229
LR9-1	53	66
	56	58
	49	57

The spontaneous frequency of Ap^R clones was lower than 1.0×10^{-11}. The efficiency of electroporation was 5×10^7 transformants μg^{-1} of pFPMT DNA

this strain is capable of homologous recombination due to *sbcB sbcC* mutations. Further, this strain is defective for restriction of non-modified DNA. It is conceivable that in *E. coli* SF8 recombinational circularization of linear duplex DNA fragments taken up can occur. In fact, all Ap^R transformants of *E. coli* SF8 examined contained plasmid molecules (see below) termed aberrant plasmids. To find out whether the plasmids were formed in the bacteria or whether they were already present in the DNA isolated from the *H. polymorpha* cells, a recombination-deficient derivative, *E. coli* SF8 *recA56*, was treated with total DNA from both *H. polymorpha* strains. The same frequencies of Ap^R transformants were obtained as in a *rec*$^+$ strain suggesting that circular DNA structures were not formed in *E. coli* by homologous recombination. It was hypothesized that circular molecules were already present in the DNA isolated from the *H. polymorpha* cells. This presumption was supported by the observation that transformants with plasmids were also obtained with the strains *E. coli* NM522 and DH5α containing exonuclease V. Such strains of *E. coli* degrade linear DNA fragments very effectively. The following experiments provided direct evidence that the aberrant plasmids were already present in the DNA extracted from *H. polymorpha*. The total DNA isolated from both *H. polymorpha* strains was treated with purified exonuclease V which in the presence of ATP degrades linear duplex DNA but leaves circular duplex DNA intact. When more than 95% of total DNA was degraded by exonuclease V (as followed by agarose gelelectrophoresis), the number of *E. coli* transformants containing plasmids remained as high as before exonuclease V treatment (Table 1). This showed that the transforming DNA molecules in the total DNA were not available for exonuclease V, most likely because they were already in a circular conformation at this time.

Characterization of Plasmids

Plasmids from 103 transformed *E. coli* clones were isolated (Table 2). The plasmid's size ranged from 2.3 to 36 kb. An extensive analysis of these aberrant plasmids included (a) restriction with XbaI to test for the single XbaI site located in the HARS region, (b) test for the presence of the aprotinin gene by PCR amplification and (c) test for the presence of the *URA3* and *LEU2* genes by transformation of an *ura3 leu2* mutant of *H. polymorpha* and selection of ura^+ and leu^+ prototrophs. The results are summarized in Table 2 and indicate that the aberrant plasmids were derivatives of pFPMT with highly different genetic structures. About 30% of the aberrant plasmids had no XbaI site indicating that the HARS sequence was deleted. The aprotinin gene was identified in 57% of aberrant plasmids. The *URA3* gene was present in 49%, the *LEU2* gene in 80% of aberrant plasmids. With the exception of plasmids no. 85 and 86 the aberrant plasmids were not identical with pFPMT in terms of size and genetic and molecular markers. This suggested that they were not formed by homologous recombination in the *H. polymorpha* cells (these cells are stably haploid cells in which homologous recombination is repressed). In plasmids no. 85 and 86 about 1 kb of the non-coding region between the colE1 origin and the *URA3* gene was deleted. DNA of about the same size was inserted. Probably, illegitimate recombination events were responsible for plasmid formation in *H. polymorpha.*

Table 2. Characteristics of aberrant plasmids from *H. polymorpha* LR9-1 (one insert of pFPMT; nos. 1–33) and LR9-4 (four tandem inserts of pFPMT; nos. 34–103) obtained in *Escherichia coli*

	Size (kb)	Presence of			
		XbaI sites	Aprotinin gene	*URA3* gene	*LEU2* gene
1	2.3	0	+	–	+
2	2.3	0	–	–	+
3	2.7	0	–	–	+
4	2.7	0	+	–	–
5	2.9	0	–	–	–
6	2.9	1	+	–	+
7	2.9	1	+	–	+
8	3.0	1	+	–	+
9	3.0	0	–	–	+
10	3.0	0	–	–	+
11	3.5	0	–	–	+
12	3.5	0	–	–	+
13	3.9	0	+	–	+
14	4.7	1	–	–	+
15	5.8	0	–	–	–
16	5.8	0	+	–	–
17	7.3	0	–	+	+
18	7.3	0	–	+	+
19	7.3	0	–	+	+

Table 2. (*Contd.*)

	Size (kb)	Presence of			
		XbaI sites	Aprotinin gene	*URA3* gene	*LEU2* gene
20	7.6	0	+	–	+
21	7.6	0	+	–	–
22	7.8	0	+	–	+
23	8.6	0	+	–	+
24	8.6	0	+	+	+
25	8.8	1	+	–	+
26	9.8	1	+	+	+
27	11.5	0	+	+	–
28	14.6	4	+	+	–
29	21.5	6	+	–	+
30	24	4	+	+	+
31	27	9	+	+	+
32	29	1	+	+	+
33	31	6	+	+	+
34	3.1	0	–	–	+
35	3.4	1	–	–	+
36	3.5	1	–	–	+
37	4.0	1	–	–	+
38	4.0	1	–	–	+
39	4.0	1	–	–	+
40	4.0	1	–	–	–
41	5.2	0	–	–	+
42	5.2	0	–	–	+
43	5.4	0	–	–	+
44	5.9	0	–	–	–
45	6.2	0	–	–	–
46	6.2	1	–	–	+
47	6.2	1	–	–	+
48	6.2	1	–	–	+
49	6.6	1	–	–	+
50	7.2	0	–	+	+
51	7.6	1	+	+	+
52	7.6	1	–	–	–
53	7.8	1	–	–	–
54	8.0	1	–	+	+
55	8.2	1	+	+	+
56	8.2	1	+	+	+
57	8.2	1	+	+	+
58	8.2	1	+	+	+
59	8.2	1	+	+	–
60	8.2	1	+	–	+
61	8.2	1	+	+	+
62	8.2	1	+	–	+
63	8.2	1	+	+	–
64	8.4	1	+	+	+
65	8.4	1	+	+	+
66	8.4	1	+	+	+
67	8.4	1	+	+	+

Table 2. (*Contd.*)

	Size (kb)	Presence of			
		XbaI sites	Aprotinin gene	*URA3* gene	*LEU2* gene
68	8.4	1	–	–	+
69	8.4	1	–	+	+
70	8.4	1	–	–	+
71	8.5	1	–	+	–
72	8.6	1	+	+	+
73	8.6	1	+	+	–
74	8.6	1	–	+	–
75	8.6	1	–	–	–
76	8.6	1	+	–	–
77	8.6	1	+	–	–
78	8.8	1	+	+	+
79	8.8	0	-	+	+
80	8.8	1	+	+	+
81	8.9	1	+	+	+
82	8.9	2	+	+	+
83	9.0	1	+	–	+
84	9.2	0	–	–	+
85	9.2	1	+	+	+
86	9.2	1	+	+	+
87	9.2	2	+	+	+
88	9.4	1	+	+	+
89	9.6	1	+	+	+
90	9.8	1	–	–	+
91	10.1	2	+	–	+
92	10.2	2	+	+	+
93	10.3	1	+	+	+
94	10.3	2	+	+	+
95	10.5	1	+	–	+
96	10.5	2	+	+	+
97	10.6	2	+	–	+
98	10.7	2	–	+	+
99	11	2	+	–	+
100	13	0	–	+	+
101	16	1	–	+	+
102	17.5	3	+	+	+
103	36	6	+	–	+

[a]The lack of an *Xba*I site indicates that the HARS sequence is deleted. The presence of one or more *Xba*I sites does not indicate the presence or absence of HARS, since *Xba*I sites might also be present in an insertion of *H. polymorpha* chromosomal DNA
[b]A plus indicates the presence, a minus the absence of the aprotinin gene determined by polymerase chain reaction (PCR)
[c]For the *URA3* and *LEU2* genes, a plus represents the presence of the corresponding gene (prototrophic *H. polymorpha* transformants were obtained), whereas a minus indicates that by transformation with the corresponding aberrant plasmid, no prototrophic *H. polymorpha* transformants were detected. The transformation efficiency was mostly low (one to five transformants/μg plasmid DNA) so that, though there is a minus the corresponding gene might be present. The spontaneous frequency of LEU^+ or URA^+ clones was lower than 2.0×10^{-9}

Autonomous replication of aberrant plasmids containing HARS apparently did not occur when they were transformed into *H. polymorpha*. This was concluded from the observation that DNA isolated from *H. polymorpha* transformants obtained with plasmids no. 13, 26, 28, 31, 56, 66, 91 and 102 gave rise to only rare and varying aberrant plasmids after transformation of *E. coli*. Presumably, aberrant plasmids integrate into the *H. polymorpha* genome and are released at low frequency as new aberrant plasmids just as the original plasmid pFPMT itself.

Effect of the Age of the Culture and Ultraviolet Irradiation

Conditions which might stimulate the formation of aberrant plasmids in recombinant *H. polymorpha* were examined.

Age of the Culture. *H. polymorpha* LR9-4 cells were grown for 1 to 5 days in SD medium respectively in YPD medium (Sherman et al. 1986) at 37 °C on a shaker. After different incubation times the DNA was extracted from cells and used for transformation of *E. coli* SF8*recA*. The frequencies of transformants (Table 3) show that over a period of 5 days, the number of aberrant plasmids apparently increased with the age of the culture. Perhaps chromosomal instability is enhanced during aging of the culture.

Ultraviolet Irradiation. Figure 1 shows that UV irradiation stimulated the formation of aberrant plasmids. A UV dose of 36 Jm^{-2} increased the production of aberrant plasmids in *H. polymorpha* LR9-4 about 20- to 90-fold (Fig. 1). At this UV dose the number of viable yeast cells was 78%. Possibly, DNA damages produced by UV irradiation or the DNA-repair processes active after UV irradiation stimulate illegitimate recombination.

Table 3. Effect of the age of *H. polymorpha* LR9-4 cells on the frequency of Ap^R transformants (*E. coli* SF8*recA*) with the DNA of the cells

	No. of Ap^R transformants/5 × 10^7 extracted *H. polymorpha* cells	
Growth phase (age)	SD	YPD
Log (1 × 10^8 cells ml^{-1})	1	5
Stat (1 day)	1	5
Stat (2 days)	4	10
Stat (3 days)	10	8
Stat (4 days)	14	9
Stat (5 days)	11	11

Cells were grown in SD (minimal) or in YPD (broth) media

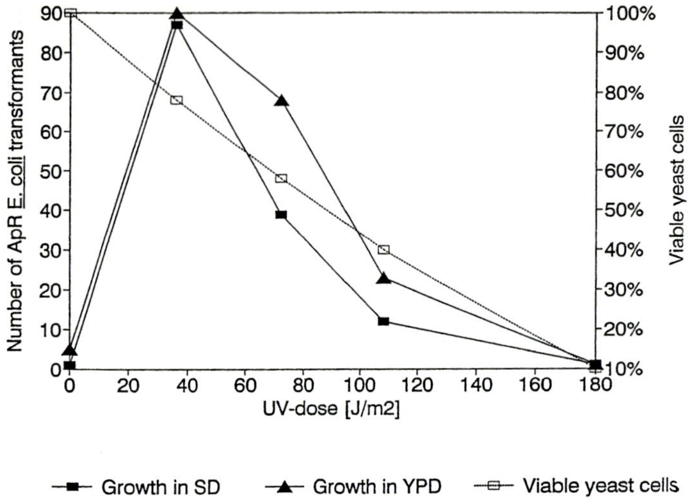

Fig. 1. Effect of ultraviolet (*UV*) irradiation of *H. polymorpha* LR9-4 cells on the frequency of Ap^R transformants of *Escherichia coli* SF8*recA* with the DNA from the irradiated cells. Viable yeast cells were determined by colony formation on *YPD* (broth) medium after 3 days of incubation at 37 °C. Cells were grown in *SD* (minimal) medium or in YPD medium

Conclusions

The results of this study indicate that *H. polymorpha* cells spontanously release segments of chromosomal DNA at low frequency, probably by illegitimate recombinational circle formation. It was calculated that about one circular DNA molecule with the Ap^R marker per 100 cells of *H. polymorpha* LR9-4 respectively per 1000 cells of *H. polymorpha* LR9-1 was present in a fresh culture of the two *H. polymorpha* strains. The aberrant plasmids were derivatives of pFPMT produced by deletions of various length, by rearrangements and probably by insertion of *H. polymorpha* DNA. The data suggest that aberrant plasmids resulted from single and multiple illegitimate recombination events. This conclusion was based on the detailed physical mapping of 14 aberrant plasmids which had sizes between 4 and 17.5 kb. None of these plasmids had a size and genetic structure identical with pFPMT. The apparent insertions of *H. polymorpha* chromosomal DNA had different restriction patterns in the various aberrant plasmids suggesting that the DNA was derived from different locations of a chromosome or even from other chromosomes. Presumably, the release of aberrant plasmids from a chromosome and their reintegration into other loci happened several times comparable to a 'pingpong mechanism'. This model would explain the highly polymorphic population of aberrant plasmids.

Possibly an increasing age of a *H. polymorpha* culture and certainly modest UV irradiation increased the production of aberrant plasmids. The plasmid-like DNA structures identified in this report can enter gene transfer routes and transfer recombinant nucleotide sequences to other microorganisms by natural genetic transformation (see Lorenz and Wackernagel 1994) if they are released from the yeast cells. Such a release can occur during death and lysis of the cells. The released aberrant plasmids would then underlie the same parameters which effect persistence and transfer efficiencies of free natural plasmids. To reduce gene transfer to other microorganisms, it is advisible that recombinant eukaryotic organisms used for applications in biotechnology should not contain prokaryotic origins of replication and possibly also should not contain additionally selective markers such as antibiotic resistance determinants.

Acknowledgements. This work was supported by the Bayer AG (Wuppertal) in the Verbundprojekt Sicherheitsforschung Gentechnik. We are grateful to S. Feldmann, G. Gellissen, R. Roggenkamp and C.P. Hollenberg for providing microorganisms, plasmids and primers.

References

Birnboim HC, Doly J (1979) A rapid alkaline extraction procedure for screening recombinant plasmid DNA. Nucleic Acids Res 7: 1513–1523

Devenish RJ, Newlon CS (1982) Plasmid isolation from yeast. Gene 18: 277–279

Dohmen RJ, Strasser AWM, Höner CB, Hollenberg CP (1991) An efficient transformation procedure enabling long-term storage of competent cells of various yeast genera. Yeast 7: 691–692

Gellissen G, Janowicz ZA, Merckelbach A, Piontek M, Keup U, Weydemann U, Hollenberg CP (1991) Heterologous gene expression in *Hansenula polymorpha*: efficient secretion of glucoamylase. Bio/Technology 9: 291–295

Gellissen G, Janowicz ZA, Weydemann U, Melber K, Strasser AWM, Hollenberg CP (1992) High-level expression of foreign genes in *Hansenula polymorpha*. Biotech Adv 10: 179–189

Hanahan D (1983) Studies on transformation of *Escherichia coli* with plasmids. J Mol Biol 86: 557–580

Lorenz MG, Wackernagel W (1993) Bacterial gene transfer in the environment. In: Wöhrmann W, Tomiuk J (eds) Assessing risks on gene technology. Birkhäuser, Basel, p 43

Lorenz MG, Wackernagel W (1994) Bacterial gene transfer by natural genetic transformation in the environment. Microbiol Rev 58: 563–602

Marmur J (1961) A procedure for the isolation of deoxyribonucleic acid from microorganisms. J Mol Biol 3: 208–218

Roggenkamp R, Hansen H, Eckart M, Janowicz Z, Hollenberg CP (1986) Transformation of the methylotrophic yeast *Hansenula polymorpha* by autonomous replication and integration vectors. Mol Gen Genet 202: 302–308

Romanowski G, Lorenz MG, Wackernagel W (1992) Persistence of free plasmid DNA in soil monitored by various methods, including a transformation assay. Appl Environ Microbiol 58: 3012–3019

Romanowski G, Lorenz MG, Wackernagel W (1993) Use of polymerase chain reaction and electroporation of *Escherichia coli* to monitor the persistence of extracellular plasmid DNA introduced into natural soils. Appl Environ Microbiol 59: 3438–3446

Sherman F, Fink GR, Hicks JB (1986) Methods in yeast genetics. Cold Spring Harbor Laboratory, Cold Spring Harbor

Taylor AF (1988) RecBCD enzyme of *E. coli*. In: Kucherlapati R, Smith GR (eds) Genetic recombination. American Society for Microbiology, Washington

Woodcock DM, Crowther PJ, Doherty J, Jefferson S, DeCruz E, Noyer-Weidner M, Smith SS, Michael MZ, Graham MW (1989) Quantitative evaluation of *Escherichia coli* host strains for tolerance to cytosine methylation in plasmid and phage recombinants. Nucleic Acids Res 17: 3469–3479

19 Instability of Tandem Repetitive DNA in "Natural" and Transgenic Organisms

T. Hankeln[1], A. Winterpacht[2], and E. R. Schmidt[1]

[1]Institut für Molekulargenetik, gentechnologische Sicherheitsforschung und Beratung, Johannes Gutenberg-Universität, 55099 Mainz, Germany
[2]Kinderklinik, Johannes Gutenberg-Universität, 55101 Mainz, Germany

Introduction

Genome research of the last 10 years has forced us to re-evaluate our view of DNA as a relatively stable molecule. Unprecedented levels of DNA instability in germline and soma cells have been observed, associated primarily with tandem repetitive (tr) DNA sequences. We will discuss here briefly the structure and possible functions of trDNA in eukaryotes, the putative mechanisms of mutational change in repeat clusters and the evolutionary dimensions of trDNA instability (for other relevant reviews, see Pardue and Hennig 1990; Vogt 1990; Charlesworth et al. 1994). A special focus will be on the behaviour of trDNA after DNA transfer experiments in transgenic organisms, with reference to our own work in dipteran insects.

Structural Classification of Tandem Repetitive DNA

TrDNA, also referred to as "satellite DNA" due to its isolation by centrifugation techniques, has been classified according to its repeat unit length, cluster length and position in the genome. Classical "macrosatellites" (Fowler et al. 1988) comprise huge blocks of trDNA clusters (hundreds to thousands of kilobasepairs long) which are usually associated with the heterochromatin in the centromere and telomere regions (for reviews, see Beridze 1986; John and Miklos 1979; Brutlag 1980; Singer 1982). Repeat units of macrosatellites may vary from 2 to 2000 basepairs (Dover 1989). "Minisatellite" sequences (Jeffreys et al. 1985) comprise clusters up to a few kilobases in length, with repeat units varying between ten and a few hundred base pairs. Minisatellites are found interspersed all over the genome (Schmidt 1984; Weller et al. 1984; Jeffreys et al. 1985; Nakamura et al. 1987; Hankeln et al. 1994b), with some minisatellite families in

E.R. Schmidt/T. Hankeln (Eds.)
Transgenic Organisms and Biosafety

humans being enriched at the telomeric regions (Royle et al. 1988). "Microsatellites", also called simple tandem repeats (STR), have repeat units of only 2 to about 10 bp. Clusters are usually from a few tens to a few hundreds of bp long and are scattered genome-wide.

A General Function of Tandem Repetitive DNA in Eukaryotes?

Early attempts to ascribe a general genomic function to trDNA, such as regulation of transcription, DNA replication and recombination, have turned out to be mostly invalid (for a detailed discussion, see John 1988; Vogt 1990). Instead, many trDNA sequences may well fit into the concepts of "selfish" DNA (Orgel and Crick 1980; Doolittle and Sapienza 1980) or "ignorant" DNA (Dover 1980). These models describe the presence of trDNA in the genome either as a result of its inherent capability for self-amplification or as a result of other internal genomic forces which amplify trDNA by chance. Notwithstanding these extreme standpoints, tr-clusters (in association with heterochromatin) may have profound position effects on neighbouring DNA regions, e.g. by influencing gene activity or genetic recombination (see below). Additionally, there is growing evidence that some macro-/mini-/microsatellite families are crucially involved in essential cellular functions, some mini- and microsatellites even coding for RNAs or proteins. These functionally interesting trDNA sequences will be important candidates for gene transfer experiments, and their potential instability will be an issue for biosafety considerations. The features of functionally relevant trDNA sequences are discussed in more detail below.

Tandem Repetitive DNA in Centromeres and Telomeres

The centromere is an essential chromosomal domain that acts in conjunction with specialized proteins to mediate proper segregation of chromosomes in meiosis and mitosis (for reviews, see Rattner 1991; Ault and Rieder 1994). While the functional centromeric DNA in the yeast *Saccharomyces cerevisiae* could be delimited to only 125 bp (for review, see Hegemann and Fleig 1993), centromeres in the fission yeast *Schizosaccharomyces pombe* span 40–100 kb of DNA and consist of several families of centromere-specific repeats (Steiner et al. 1993). However, the minimal functional centromeric DNA of *S. pombe* may consist of only a non-repeat 'central core' sequence plus a small part of a K-type repeat (Baum et al. 1994; Steiner and Clarke 1994).

In higher eukaryotes, centromeres are usually associated with large blocks of trDNA. While DNA in the centromeric region may be dominated by one specific satellite family (e.g. in some species of the dipteran genus Chironomus: Rovira et al. 1993; Hankeln et al. 1994a), several different (but sometimes related) classes of tandem repeats have been found in the centromeres of humans (Willard 1990; Grady et al. 1992), mice (Wong and Rattner 1988; Wong et al. 1990; Narayanswami et al. 1992), *Drosophila melanogaster* (Lohe and Brutlag 1987a; Carmena et al. 1993) and also other Chironomus species (our lab; unpublished results). Furthermore, transposable elements and other, yet uncharacterized DNA, have been found within the large centromeric regions of some of these organisms (Lohe and Brutlag 1987b; Hankeln 1990; Wevrick et al. 1992; Le et al. 1995). In spite of the centromere being a heterogeneous mixture of different DNA components, there is growing evidence that satellite DNA is involved in centromere function in humans (Wevrick et al. 1990; Haaf et al. 1992; Tyler-Smith et al. 1993; Brown et al. 1994; Ikeno et al. 1994; Larin et al. 1994), mice (Lica et al. 1986; Radic et al. 1987; Wong and Rattner 1988; but see also Wong et al. 1990) and Drosophila (Murphy and Karpen 1995). It seems possible that special stereophysical characteristics of trDNA such as DNA curvature (Schmidt 1984; Radic et al. 1987; Martinez-Balbas et al. 1990; Hankeln et al. 1994a), foldback structures (Ferrer et al. 1995) or non-Watson/Crick base pairing (Chou et al. 1994) are necessary for its potential function in the centromere (Vig 1994).

Instability of macrosatellite sequences within the centromeres is already visible at the light microscopic level. Heteromorphisms between homologous chromosomes concerning the amount of satellite DNA were for example observed in human metaphase chromosomes (Yurov et al. 1987) and in insect polytene chromosomes (Keyl 1965; Hankeln et al. 1994a), suggesting the activity of amplification and/or deletion mechanisms in trDNA. Large α-satellite clusters in human centromeres were shown to be frequently polymorphic in length (Jabs et al. 1989; Wevrick and Willard 1989; Wevrick et al. 1990). Length variability in centromere-associated trDNA clusters has also been observed in *S. pombe* (Steiner et al. 1993) and in both germline and soma cells of African Green Monkey (Maresca et al. 1984). Owing to its highly repetitious nature, the mechanisms of rearrangement within centromeric trDNA are difficult to investigate, and few approaches have addressed this question experimentally (Wahl et al. 1984; Butner and Lo 1986). Chatterjee and Lo (1989) have analyzed the fate of a thymidine kinase gene insertion embedded in mouse centromeric satellite DNA. They found evidence for deletions in trDNA, possibly mediated by intrachromosomal recombination, and for amplification events, the mechanisms of which remain unclear. It is interesting that in this mouse system such centromeric rearrangements usually do not seem to affect centromere function and chromosome stability unless they are accompanied by changes in the pattern of heterochromatinization (Chatterjee and Lo 1989). In both Drosophila (Murphy and Karpen 1995) and Schizosaccharomyces centromeres (Baum et al. 1994), however, partial deletions of centromeric DNA correlate with malsegregation of chromosomes and thus impair optimal centromere function.

Telomeres are special structures at the chromosomal ends which ensure the proper replication and integrity of linear DNA molecules. The telomeres of eukaryotes as diverse as yeast, protozoa, plants and mammals contain G-rich simple tandem repeats (review: Blackburn 1994). These simple trDNA clusters vary widely in length between species, but often display a narrow length distribution within species (Blackburn 1991). To compensate for gradual loss of DNA during replication, elongation of the simple repeats is accomplished by a specialized ribonucleoprotein complex, the "telomerase", the activity of which is highly regulated (McEachern and Blackburn 1995; and references therein).

In yeast, telomere formation may be accompanied by recombination between different telomeres (Wang and Zakian 1990). Both expansion and contraction events have been observed in telomeric trDNA in unicellular eukaryotes, making telomere length highly variable (Larson et al. 1987). In cultured human fibroblasts, the number of telomere repeats decreases with aging, suggesting a biological role of telomere instability in senescence of somatic tissue (Harley et al. 1990). In contrast, immortal cells show no net loss in telomere length, due to maintenance of telomerase activity (Kim et al. 1994).

Subtelomeric tandem arrays which display length polymorphisms have also been characterized in a number of organisms (Horowitz et al. 1984; Corcoran et al. 1988; Brown et al. 1990). The role of this subtelomeric trDNA is still unclear, as are the mechanisms responsible for its fluctuations.

Different, non-telomerase mechanisms of telomere elongation have been described in the linear mitochondrial DNA of Tetrahymena and in dipteran insects. Chromosome termini in Tetrahymena mtDNA consist of complex, minisatellite-like tandem repeats which are thought to be maintained by unequal crossing-over (Morin and Cech 1986, 1988). A similar mechanism of telomere stabilization may be found in the midge Chironomus. This insect also has complex tandem repeats at the chromosomal ends which might have evolved from simple tandem repeats (Nielsen and Edström 1993). There is fluctuation both in copy number and distribution of repeat sequence variants between chromosomes in Chironomus, arguing for an inherent instability of the telomeric repeat arrays (Cohn and Edström 1992a,b). A third mechanism of telomere stabilization is found in *D. melanogaster* where chromosomal termini are elongated by the site-specific integration of non-LTR transposable elements (review: Mason and Biessmann 1995). Biessmann and co-workers have also described a subtelomeric trDNA cluster in *D. melanogaster* which seems to be polymorphic in length. The potential function of this trDNA is currently being investigated (H. Biessmann, personal communication).

The large amounts of repetitive DNA at centromeres (and often also at telomeres) make molecular analyses rather difficult. Thus, there is a general lack of experimental data on the mechanisms which are responsible for the observed trDNA instability. Many reports invoke general concepts such as unequal crossing over between sister chromatids or homologous chromosomes in mitosis or meiosis, or "saltatory" amplification events to explain trDNA rearrangements (e.g. Marcais et al. 1991). A more specific insight into the turnover of trDNA may therefore be expected from the analysis of the smaller and more easy-to-analyse mini- and microsatellites.

Instability in Minisatellite DNA

Minisatellite clusters are found interspersed in the genomes of many species including humans (Inglehearn and Cooke 1990; and references therein), mice (Jeffreys et al. 1987), insects (Schmidt 1981, 1984; Hankeln 1990; Hankeln et al. 1994b; Hershfield and Swift 1990) and plants (Dallas 1988). Individual minisatellite loci may have been formed in loco, e.g. by expansion from other types of flanking repetitive DNA (Armour et al. 1989; Kelly 1994), or may have spread in the genome by transposition (Schmidt 1984; Hankeln et al. 1994b).

While many minisatellite loci may represent non-functional "junk DNA" (Ohno 1972), recent experiments have provided evidence for a functional importance of some trDNA arrays. Krontiris et al. (1993) were able to demonstrate that rare mutant alleles of a 28 bp minisatellite 3′ of the human H-rAS gene are clearly associated with cancer. On the other hand, the insulin linked polymorphic region (ILPR), a 14 bp trDNA cluster in the 5′-flanking region of the human insulin (INS) gene, has been linked to genetic susceptibility to insulin-dependent diabetes mellitus (IDDM2) (Bennett et al. 1995; Kennedy et al. 1995). It has been shown that ILPR can be bound and activated by transcription factors and that allelic variation of this minisatellite may correlate with an increased level of insulin RNA causing the clinical phenotype. In *D. melanogaster*, a deletion of the Responder satellite locus which contains 120 bp repeat clusters is associated with fitness reduction (Wu et al. 1989). Moreover, in several organisms minisatellite sequences are found being part of genes, coding either for nuclear RNA of unknown function (Hogan et al. 1995) or essential proteins (e.g. Paulsson et al. 1992; Galli and Wieslander 1993; Lindahl et al. 1990; Swallow et al. 1987; Lyons et al. 1988; Simon et al. 1991).

Characteristically, minisatellite arrays show allelic variation in the number of tandem repeat units (VNTR) (Jeffreys et al. 1985; Nakamura et al. 1987). Hypervariability in cluster length caused by gain or loss of repeat units makes minisatellites a valuable tool in linkage analysis and DNA fingerprinting (Armour and Jeffreys 1992). Length variability of some minisatellite loci is not limited to the germline but is also observed in somatic tissues (Kelly et al. 1989; Armour et al. 1989). Not only non-coding minisatellites, but also coding minisatellites show DNA instability (e.g. Paulsson et al. 1992; Simon et al. 1991; Swallow et al. 1987; Lindahl et al. 1990). While some genes containing tandem repeats display enormous length variability (which is then also reflected on the protein level), other coding minisatellites tolerate only limited length differences (Paulsson et al. 1992; Galli and Wieslander 1993). This may depend either on the degree of selection acting on protein structure, or on a sequence-inherent tendency of only some repeat types towards instability.

The hypervariable human minisatellite families originally described by Jeffreys' group display limited sequence similarity to the Chi recombination signal of *E. coli* (Jeffreys et al. 1985). It was therefore originally hypothesized that length variation in minisatellite arrays is predominantly generated by unequal crossing over, and that minisatellites are recombinational hotspots in the human genome (Jeffreys et al. 1985; Royle et al. 1988; Chandley and Mitchell 1988; Wahls et al. 1990). However, linkage of human minisatellites to recombinational hotspots has turned out to be circumstantial, and several hypervariable minisatellites not containing the Chi signal have been described (Jarman and Wells 1989; Dover 1989; Nakamura et al. 1987, but see: Jeffreys et al. 1994). More recently, the development of a technique named minisatellite variant repeat mapping by PCR (MVR-PCR; Jeffreys et al. 1991) has facilitated a detailed structural analysis of newly generated minisatellite alleles and their underlying mechanisms of instability. Studies of highly polymorphic human minisatellites show that instability of trDNA clusters is due to unexpectedly complex events and does not merely comprise changes of array lengths (Jeffreys et al. 1994; Buard and Vergnaud 1994). The results can be summarized as follows:

1. The internal structure of mutant minisatellite alleles differs between sperm and blood, indicating that some processes are germline specific and possibly meiotic.
2. paternal (and possibly also maternal) mutations show a consistent bias towards gain of repeats. This would lead to an expansion of tandem arrays on an evolutionary timescale.
3. Large deletions within clusters are rare (Jeffreys et al. 1990), but may counteract frequent but small gains.
4. Mutations at three different minisatellite loci are highly polar, with rearrangements clustered at one end of the tr array. This implies that instability is regulated by unknown elements outside the trDNA-cluster, and it explains why mutation rates are apparently independent of allele length.
5. In 50% of repeat-gaining mutants from sperm, there is evidence of inter-allelic transfer (between homologous chromosomes) of repeat units. Transfer may be accomplished by a gene conversion-like mechanism rather than by inter-allelic unequal crossing over. There is no exchange of flanking markers during rearrangements (Jeffreys et al. 1994; Wolff et al. 1988, 1989).
6. Many other structural variants can only be explained by complex, multi-step mutation processes, such as allele duplications plus inter-allelic conversion etc. In some cases, anomalous repeats may result from mismatch repair on alternate strands of heteroduplex DNA, formed during conversion.

It remains to be seen whether these data can also explain hypervariability of other minisatellite sequences in other genomes, or even instability of macrosatellites. Still, unequal crossing over between sister chromatids (Smith 1976) and, at least in case of short repeating units, replication slippage (Levinson and Gutman 1987; Coggins and O'Prey 1989) remain viable mechanisms to explain trDNA instability.

Instability of Simple Tandem Repeats in the Human Genome

Simple trDNA in non-coding and also some coding regions is present in most, if not all eukaryotic genomes (Tautz and Renz 1984; Tautz et al. 1986; Epplen 1988; Tautz and Schlötterer 1994; Stallings 1995). The instability of simple repeats within the human genome has lately developed into a field of great interest, since this instability is coupled to severe diseases.

The human genome contains many trDNA sequences, which vary in complexity and are distributed widely throughout the genome. Especially the simple tandem repeats (STRs), which involve di-, tri-, tetra- or pentanucleotides, are hypervariable in number and have become invaluable tools for high resolution mapping of disease genes and for the generation of linkage maps. Recently it has been shown that trinucleotide STRs undergo dynamic mutations, which are associated with several inherited disorders (for reviews, see Richards and Sutherland 1992; Kuhl and Caskey 1993; La Spada et al. 1994; Willems 1994; Künzler et al. 1995; Monckton and Caskey 1995; Sutherland and Richards 1995). These mutations arise through the instability and expansion of certain trinucleotide repeats and are distinguished from conventional mutations by a number of properties:

1. The rate of mutation is related to the copy number and therefore the product of a dynamic mutation has a different risk of undergoing further changes.
2. The number of perfect repeating units of an STR determines the probability of dynamic mutation.
3. The transition from a normal copy number to a clinically significant copy number is not a single event but can involve a number of steps (and generations). In the last few years 12 trinucleotide expansion cases (involving CGG, CTG or CAG) have been associated with specific diseases and/or fragile sites on the chromosomes (Table 1).

The first dynamic mutation was detected in patients with fragile-X syndrome, the most common form of familial mental retardation (Yu et al. 1991; Oberlé et al. 1991). The disease is caused by a CGG expansion in the 5′-untranslated region of the FMR1 gene, an RNA-binding protein widely expressed in embryonic tissue (Hinds et al. 1993; Siomi et al. 1993). While normal X-chromosomes have six to about 50 copies of the repeat, in patients with the full manifestation of the syndrome the repeats have been expanded to 200–1000 copies (Yu et al. 1992; Reiss et al. 1993). Subjects with small size increases (up to 200 copies) are clinically normal and are said to carry a "premutation". The transition from pre to full mutation only occurs on female transmission during oogenesis, while in spermatogenesis the length of the repeats is reduced or remains in the premutation range (Reyniers et al. 1993; Sutherland and Richards 1995). This observation may be explained by a selection in favour of sperm with a premutation versus sperm with a full mutation (Reynier et al. 1993). Moreover, full mutations undergo somatic instability in early development which causes

Table 1. Summary of triplet repeat diseases in humans

Repeat sequence	Condition	Gene	Repeat localisation	Repeat number	
				Normal	Disease
CGG	FRAXA	FMR-1	5′-Untranslated	6–52	200–1000
CGG	FRAXE	?		6–25	200–1000
CGG	FRA11B (Jacobsen syndrome)	CBL2	?	~11	400–800
CAG	SBMA (Kennedy disease)	Androgen receptor	ORF	12-33	<100
CAG	Huntington	Huntington	ORF	9–30	<150
CAG	SCA 1	Ataxin	ORF	9–39	<100
CAG	DRPLA/HRS	Atropin	ORF	9–23	<100
CAG	Machado-Joseph	MJD 1	ORF	16–36	<100
CTG	Myotonic dystrophy	DM kinase (DM-1)	3′–Untranslated	5–40	200–4000

ORF, open reading frame.

heterogeneity in repeat length in different tissues (Wöhrle et al. 1993). Although hypermethylation can be observed as a result of the CGG expansion, it has recently been demonstrated that FMR1 genes with full mutations are transcribed but not translated into protein, indicating a translational suppression by trinucleotide repeat expansion (Feng et al. 1995). Under specific cell culture conditions the CGG expansion (and the resulting methylation) cause fragile sites on the chromosomes. The nature of these fragile sites on the molecular level and the relationship to trinucleotide instability is still unknown. Four other CGG expansions have been reported that are correlated with fragile sites (FRAXE, FRAXF, FRA11B, FRA16A) but only FRAXE and FRA11B are also associated with a clinical phenotype (Hamel et al. 1994; Jones et al. 1995).

Myotonic dystrophy, the most common form of adult muscular dystrophy, is caused by a CTG expansion in the 3’-untranslated region of the gene for myotonin protein kinase (Harley et al. 1992; Buxton et al. 1992; Aslanidis et al. 1992; Brook et al. 1992). The repeat expansion may be extremely large with more than 3000 repeats (with a normal size of five to 35 repeats), although this does not result in the expression of a fragile site (Willems 1994). The massive expansion of triplet repeats is associated with “anticipation”, meaning that the severity of the disease is increased and the age of onset is lowered in successive generations. In females the repeat can increase in copy number from generation to generation without any restraint until it reaches a point which results in (clinical) lethality. In contrast, in males the repeats reach a maximum number of 1000 copies, and mainly a decrease in size upon transmission is observed (Lavedan et al. 1993). It has recently been shown that long stretches of CTG repeats form strong nucleosome positioning signals, suggesting that these repeats may repress transcription of the myotonin protein kinase through the creation of stable nucleosomes (Wang et al. 1994).

CAG expansions have been identified in a group of related neurodegenerative disorders with a late onset expression of the disease. The group comprises Huntington disease (HD), spinobulbar muscular atrophy (SBMA), spinocerebellar ataxia type 1 (SCA1), dentatorubral-pallidoluysian atrophy (DRPLA), Haw River syndrome (HRS) and Machado Joseph disease (MJD) (The Huntington Disease Collaborative Research Group 1993; La Spada et al. 1991; Orr et al. 1993; Koide et al. 1994; Burke et al. 1994). In all cases the CAG repeats are located within the open reading frame of genes and code for polyglutamine stretches. The increase in copy number is low and even a number as little as 5% above the normal level can result in disease (Sutherland and Richards 1995). It has been suggested that reiterated glutamines are involved in protein cross-linking, leading to an aggregation of soluble neuronal proteins (Green et al. 1993; Perutz et al. 1994). On the other hand, number variations of CAG repeats may have a profound effect on the activity of certain transcription factors as suggested for the androgen receptor gene in patients with SBMA (La Spada et al. 1991; Gerber et al. 1994).

It has been shown that a large number of known genes and cloned but uncharacterized cDNAs in humans contain triplet repeats (Riggins et al. 1992; Li et al. 1993; Künzler et al. 1995). Whether these repeats are involved in any form of disease, has yet to be investigated. Cis- and transacting factors may be responsible for STR instability. Meanwhile, strong evidence of a significant founder chromosome effect has been found for myotonic dystrophy (Harley et al. 1991), Huntington disease (MacDonald et al. 1992), fragile X syndrome (Richards et al. 1992; Kunst and Warren 1994) and dentatorubral pallidoluysian atrophy (Koide et al. 1994; Nagafuchi et al. 1994) suggesting the existence of alleles apt to increase the number of tr repeats. The reason for this predisposition may either be a critical repeat number, the homogeneity of the repeats or cis-acting elements in the flanking sequence (Kunst and Warren 1994; Monckton and Caskey 1995). In fragile X syndrome the CGG repeat in normal individuals has single base substitutions leading to an interspersion of CGG and AGG repeats. In contrast, fragile X patients most frequently show uninterrupted clusters of more than 24 perfect CGG repeats (Kunst and Warren 1994; Eichler et al. 1994). Studies of the intraspecific variability of the FMR1 gene in diverse mammalian populations confirms that interruptions of the CGG cluster exert a stabilizing effect (Eichler et al. 1995). Interrupting CATs in the CAG repeat cluster in the spinocerebellar ataxia type 1 (SCA1) gene are present in 98% of the normal alleles, but are absent in unstable mutant alleles (Chung et al. 1993). Although the other known unstable trinucleotide repeat arrays do not contain any interruptions, at least 95% of all normal alleles have less than 25 repeats uninterrupted (Kunst and Warren 1994; Eichler et al. 1995). This indicates that the probability of dynamic mutations of STRs is a function of the copy number of perfect repeat units, a result that has also been predicted for the instability of dinucleotide repeats (Weber 1990).

Once a specific repeat is prone to dynamic mutation there may be different mechanisms that lead to triplet expansion. A very attractive mechanism is the model of Okazaki fragment slippage during lagging strand replication (Richards and Sutherland 1994; Eichler et al. 1994). The average length of an Okazaki fragment is 150–200 bp. For alleles with approximately 70 pure trinucleotide repeats there is an increased likelihood that Okazaki fragment synthesis begins and ends within a continuous tract of repeats. This DNA strand is not anchored by single copy sequences and is able to slip and slide during DNA polymerization, which may lead to a massive introduction of repeats. If we assume a single origin of replication near a repeat, this model would explain the polar variation of either tri- and dinucleotide repeats that has been reported for the FMR1 gene but also for some minisatellite repeat loci (Eichler et al. 1994; Jeffreys et al. 1994; Richards and Sutherland 1994). It would also explain the massive expansion of repeats in DM and FRAXA when transmitted from a parent with greater than 80 repeats (a length of more than approximately one Okazaki fragment) (Richards and Sutherland 1994; Eichler et al. 1994). Nevertheless, gene conversion events and unequal meiotic sister chromatid exchange can not be excluded as mechanisms responsible for STR instability (Jeffreys et al. 1994; Kunst and Warren 1994).

Recent reports have addressed the issue of trans-acting factors (proteins) which might be involved in STR expansion. A genome wide simple repeat sequence instability has been observed in tumors from hereditary nonpolyposis colorectal cancer (HNPCC) kindreds (Ionov et al. 1993; Aaltonen et al. 1993; Thibodeau et al. 1993). It has been shown that mutations in the mismatch repair genes hMSH2, hMLH1, hPMS1 and hPMS2 are associated with most of these cancers (Leach et al. 1993; Bronner et al. 1994; Papadopoulos et al. 1994; Nicolaides et al. 1994; Liu et al. 1995). Mutations in these genes cause a failure to repair replication slippage induced mismatches (Strand et al. 1993). Whether these mutations cause instability of specific target repeats or whether they have a more general effect on different oncogenes and/or tumor suppressor genes is still unclear, but to date there is no indication for involvement of these gene mutations in triplet repeat expansions. There is growing evidence that several other genes are involved in the mismatch repair pathway and the stability of simple repetitive DNA (Johnson et al. 1995) indicating that the components of the mismatch repair system represent key transacting factors in the dynamic mutation process.

Effects of Tandem Repetitive DNA on Chromatin Structure and the Expression of Nearby Genes (Position Effect)

It is a well-known phenomenon that the activity of one and the same gene can be different depending on its position in the genome. Euchromatic genes placed next or into a region of heterochromatin usually become inactivated or show a variegated expression. This phenomenon is called "position effect" or "position effect variegation" (PEV). A variegated gene expression is obtained, when the silencing of the gene is only realized in a subset of the cells in an organism (Muller 1930; Baker 1968; Tartof et al. 1989; Cattanach 1974; Henikoff 1990).

It is widely accepted that the position-dependent silencing of active genes is due to the structure of the heterochromatin, which is obviously more condensed and less genetically active than euchromatin (Heitz 1934; John and Miklos 1988; John 1988; Zhang and Spradling 1995). Tandem repetitive DNA sequences are major DNA components of the "constitutive heterochromatin", which is usually found in pericentromeric regions. However, the role of trDNA in the formation of heterochromatin is poorly understood.

It has been shown that transgenes, when they are organized in a tandemly repeated array, show a variegated expression similar to a typical heterochromatin-induced position effect variegation (Dorer and Henikoff 1994). Also the duplication of genes by unequal crossing over resulting in a tandem arrangement can lead to gene silencing (Matzke et al. 1994; Assaad et al. 1993, Mittelsten-Scheid et al. 1991). The same effect is frequently observed in mammalian transgenic organisms in which the transfected DNA often is organized in large tandem arrays generated either during the integration process (Palmiter and Brinster 1986) or thereafter (Barlow et al. 1995). It has been argued that the tandem arrangement of *any* sequence can lead to the formation of local intercalary heterochromatin by folding of the multiple sequence copies and binding of heterochromatin-specific proteins to such folded DNA (Dorer and Henikoff 1994). This would mean that no specific sequences would be required for heterochromatin formation and gene silencing. In fact, the sequences found in heterochromatin are extraordinarily variable and evolutionarily not conserved (John and Miklos 1988). In accordance with Dorer and Henikoff's model, intercalary heterochromatin has been described which displays all typical features (late replication, underreplication, formation of weak points in polytene chromosomes and C-band staining) – and which contains several AT-rich trDNA families unrelated in sequence (Hägele 1977; Schmidt 1984; Hankeln 1990; Bolshakow et al. 1985).

On the other hand, it is clear that the binding of heterochromatin-specific proteins to tandem repetitive DNA is involved in gene inactivation. A large number of mutations have been isolated, which modify position effect variegation (for a review, see Reuter and Spierer 1992). Some of the responsible genes code for proteins directly involved in heterochromatin formation (James and Elgin 1986; Eissenberg et al. 1990) or for proteins which are general components of the chromatin. These mutations are also modifiers of the silencing effect observed in tandemly repeated transgenes (Dorer and Henikoff 1994), and hence could be structurally involved in formation of local inactive chromatin domains. A model describing how heterochromatin-specific DNA binding proteins and general chromatin proteins (like histones) could interact to build up heterochromatin domains has recently been published by Hecht et al. (1995).

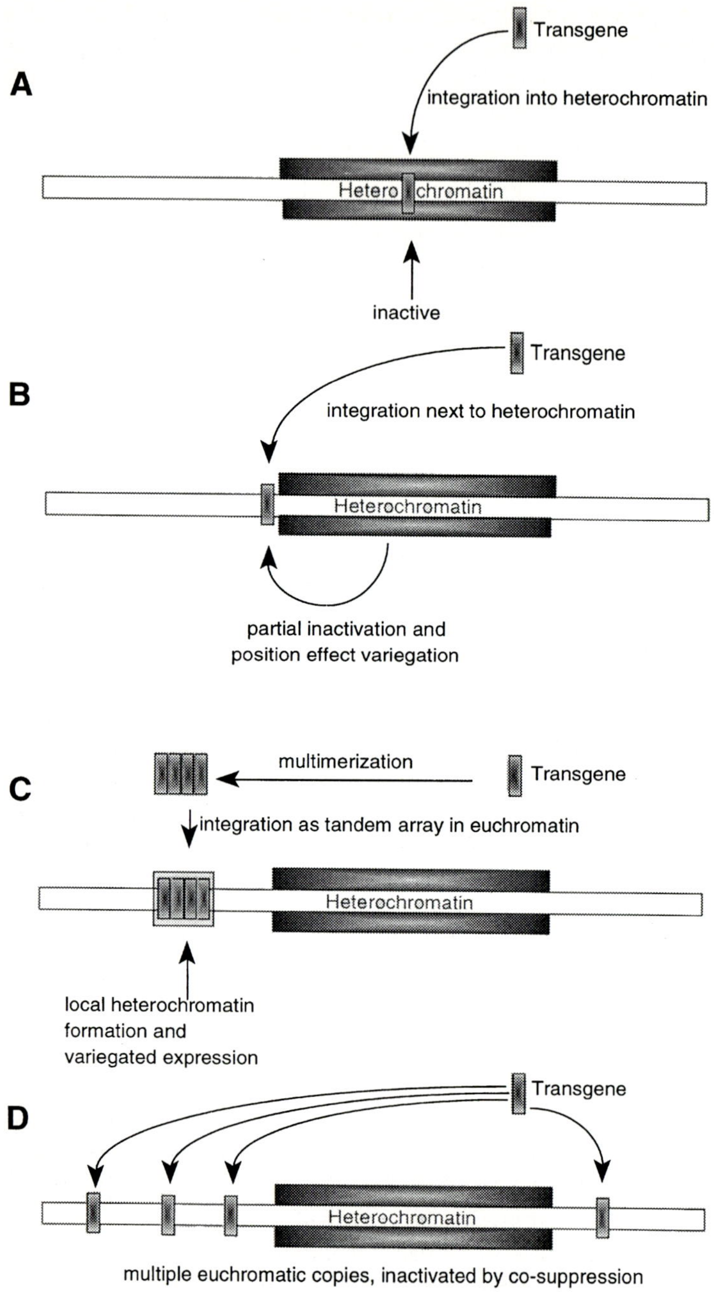
A
Transgene
integration into heterochromatin
Hetero chromatin
inactive
B
Transgene
integration next to heterochromatin
Heterochromatin
partial inactivation and
position effect variegation
C
multimerization
Transgene
integration as tandem array in euchromatin
Heterochromatin
local heterochromatin
formation and
variegated expression
D
Transgene
Heterochromatin
multiple euchromatic copies, inactivated by co-suppression

Additionally, we would like to point out that transgenes present in the host genome in multiple copies but *not* tandemly arranged can also undergo copy number-dependent silencing (see Chap. 20). A similar phenomenon is "co-suppression" which means that a transgene in a genome has a negative effect onto the expression of a homologous or sequence-related gene elsewhere in the genome (Hart et al. 1992; Seymour et al. 1993; Fray and Grierson 1993; Dorlhac de Borne et al. 1994). Co-suppression cannot be explained by a local change of the chromatin structure because the down regulated genes must not be in spatial proximity. We therefore have to assume that different mechanisms exist which lead to a silencing of genes and transgenes, depending on position or copy number. The hypothetical scenarios of gene inactivation are summarized in Fig. 1.

Evolutionary Dimensions of Tandem Repetitive DNA Instability

It has long been noted that even closely related species of animals and plants may differ enormously in their amount, sequence and chromosomal localization of trDNA (e.g. Hennig et al. 1970; Brutlag 1980; Flavell 1986; Hwu et al. 1986; Wong et al. 1990; Charlesworth et al. 1994). The inherent instability of trDNA generated by unequal crossing over, replication slippage or other DNA amplification/deletion mechanisms establishes high levels of repeat sequence similarity within species ("concerted evolution") which are paralleled by higher differences in repeat composition between species (Dover 1982, 1986; Waye and Willard 1989). The grossly different amounts of trDNA present in genomes of related species may contribute significantly to the so-called C-value paradox (Gall 1981; Cavalier-Smith 1985). The degree of instability of trDNA may also differ between species. A comparison of corresponding minisatellite loci in humans and other primates demonstrates that these tr clusters – while being hypervariable in man – are small and stable in most apes and monkeys (Gray and Jeffreys 1991; Ely et al. 1992). The human SCA1 and androgen repeptor (AR) genes contain highly un-

Fig. 1A-D. Different types of inactivation of transgenes. **A** The transgene is integrated into heterochromatic chromosomal regions and is inactive due to the surrounding heterochromatin. **B** The transgene is integrated next to heterochromatin. The inactivating influence of the neighboring heterochromatin is incomplete. This leads to a "variegated" expression, i.e., in certain fractions of the cells in a transgenic organism the transgene is expressed, while in other cells the transgene is silent. **C** A transgene is multimerized during the integration process and integrated into euchromatin as a tandem array. The tandem organization leads to the formation of local heterochromatin resulting in (partial) inactivation of the transgenes. The consequence can be transgene silencing or variegated expression. **D** The transgene is integrated in multiple euchromatic sites. The presence of multiple copies of a transgene can lead to an inactivation of all gene copies by a process called "co-suppression". The mechanism responsible for co-suppression is not yet understood

stable triplet arrays whereas the corresponding mouse genes display either 2 stable (*SCA1*) or no (*AR1*) CAG repeats at the corresponding position (Bingham et al. 1995; Burright et al. 1995). Disorders associated with the expansion of trinucleotide simple repeats have yet to be observed in species other than humans. Recently, Rubinsztein et al. (1995) could show that 33 out of 42 microsatellite arrays are longer in the human genome when compared to orthologous loci in several other primates. Accordingly, the CGG array in the 5' untranslated region of the *FMR1* gene is present in many diverse mammalian orders (and has thus been conserved in position for approximately 150 million years), but only primates show the tendency to expand the cluster length beyond 20 repeats. These data (together with our results in the insect model system; see below) suggest that trDNA instability is a generalized genome-wide phenomenon, which is modulated in a species-specific way.

Stability of Tandem Repetitive DNA in Transgenic Organisms

The transfer of a single defined trDNA array into a foreign genome may simplify its structural and functional analysis, since in the new genomic context the behaviour of the transgene is usually not obscured by the presence of closely related repetitive sequences (see above). Up till now, relatively few investigations have addressed the question of stability of a trDNA transgene, but the data already available promise a wealth of information.

Heartlein et al. (1988) transfected chinese hamster cells with a vector containing monomeric repeats of the human α-macro-satellite. Transformant lines displayed changes in chromosome number and chromosome structure indicating that alphoid DNA might serve as a hot spot for rearrangements. Haaf et al. (1992) transferred a 15.9-kb cluster of human α-satellite repeats into African Green Monkey cells. The calciumphosphate-transfected DNA fragment was obviously ligated into large repeated blocks before integration into the foreign genome and cell lines with single and multiple integration sites were produced. In one multiple integration line, the number, size and chromosomal location of the transgene varied from cell to cell indicating high levels of instability. Larin et al. (1994) introduced a whole YAC clone containing 120 kb of human Y-chromosome α-satellite DNA into hamster and human cells by spheroblast fusion and lipofection respectively. Entire YACs were integrated as intact molecules into the host chromosomes and formed centromere-like structures. In hamster cells, some lines showed limited rearrangements of α-satellite DNA, while others were stable. Lipofection produced multimerized YAC transgenes in human cells, but only single insertions were generated with no gross rearrangements observed. These three studies did not address in detail the potential mechanisms of transgene instability. They indicate, however, that the method of transfection may influence the form of integration and the stability of a transgene. Neil et al. (1990) also reported gross rearrangements of YAC clones containing different human trDNA families upon re-transformation into yeast.

The behaviour of human minisatellite arrays has recently been tested in yeast and mouse genomes. Cederberg et al. (1993) observed length alterations in a human minisatellite MS1 array integrated into *S. cerevisiae* chromosome III. Both gain and loss of repeat units were found in haploid yeast cells, showing that interaction (eg. by unequal crossing-over) between homologous chromosomes is not a prerequisite for formation of new MS1 length alleles. Allen et al. (1994) transferred a human mini-satellite containing 71 repeat units (MS32) into mouse, by micro-injection of embryos with a linear DNA construct. Both single-copy and multi-copy integrates (containing multimerized MS32 tandem arrays plus flanking DNA) were obtained. MS32 transgenes preferentially integrated into mouse major satellite DNA. The reasons for this interesting target-site specificity are not entirely clear. All five single copy integrates had undergone complex rearrangements prior to or during insertion, including loss of repeating units together with flanking DNA, shuffling of repeat variants within the array and disruption of the repeat block into two halves. In four lines, however, the single copy transgene was then stably inherited without showing further instability in the mouse genome. This is in sharp contrast to the hypervariability of MS32 in humans. Mouse lines with multimerized MS32 array integrations, however, showed structural instability of the transgene in pedigrees (Collick et al. 1994). Structural changes included length alterations in individual MS32 arrays, changes in flanking DNA and gross deletions within the MS32 multimere transgenes. It is questionable whether these changes in the multicopy-transgenes equal the 'normal' hypervariability of a single copy MS32 array in the human genome. Most importantly, the reports of Allen et al. (1994) and Collick et al. (1994) demonstrate that human minisatellite transgenes in the mouse genome display a reduced instability, indicating that species may differ in their 'handling' of trDNA. A difference between the human and the mouse genome is also suggested by the behaviour of human simple repeats in transgenic mice. Both normal and expanded CAG trinucleotide arrays of the human androgen receptor gene (Bingham et al. 1995) and the human SCA1 gene (Burright et al. 1995) display a stable transmission as transgenes in the mouse genome.

The Chironomus Model System

Non-biting midges of the dipteran genus Chironomus are preferred objects for studies of genome evolution due to their comparatively small genome and their excellent polytene chromosomes. The two closely related subspecies *C. thummi thummi* and *C. thummi piger* represent an early stage during speciation, but nevertheless differ significantly in genome size (Keyl and Strenzke 1956; Keyl 1965). Much of the 30% additional DNA present in the genome of *C. t. thummi*

(Keyl 1965) is represented by several sequence families of trDNA (Schmidt 1981, 1984; Hankeln 1990). We have characterized 6 unrelated trDNA families from *C. t. thummi* which have repeat units between 48 bp and 298 bp and AT contents between 69% and 80%. Four of the six families display a higher copy number in the genome of *C. t. thummi* compared with the smaller genome of *C. t. piger*. In *C. t. thummi*, all six repeat classes are present at multiple interspersed sites on the chromosomes, a distribution reminiscent of minisatellites (Schmidt 1984; Hankeln 1990; Hankeln et al. 1994b). Individual tandem arrays show the typical length variability (Schmidt et al. 1982; Hankeln et al. 1994b). In *C. t. piger*, however, four of six repeat families are limited in their genomic distribution to the centromeric regions of the chromosomes. One repeat type is not found in *C. t. piger* at all. These data, together with phylogenetical analyses of individual tr arrays (Hankeln et al. 1994b), suggest that several trDNA sequence families have simultaneously been amplified and transposed during divergent evolution of the *C. t. thummi* genome. Similar rearrangements of trDNA either have not happened or were not fixed in the genome of *C. t. piger*. The differences in copy number and distribution of tandem repeats have generated multiple heterozygosities between the genomes of the two subspecies. We suggest that the parallel behaviour of several trDNA families in *C. t. thummi* on the one hand and *C. t. piger* on the other hand reflects a species-specific way to deal with trDNA in the genome. *C. t. thummi* obviously tolerates a lot of trDNA almost everywhere in the genome, whereas *C. t. piger* displays smaller amounts of trDNA within the centromeric regions only. In spite of this, larger clusters of trDNA are also unstable in length in *C. t. thummi*, as seen in the NTS of rDNA (Schmidt et al. 1982) and at other genomic loci (Hankeln et al. 1994b).

To investigate the behaviour of Chironomus tandem repeats as transgenic DNA in a foreign genome, we decided to transfer distinct clusters of trDNA from *C. t. thummi* into the distantly related dipteran fly *D. melanogaster* by P element-mediated germline transformation (Winterpacht 1991). Both small arrays (four repeats) and large arrays (five to 30 repeats) of a 120-bp repeating unit were transferred. The results were rather clear cut. The small arrays did not display any instability in the transgenic genome over several generations of fly breeding. One transgenic Drosophila line with a large cluster (Lig 39) showed a deletion of all but 60 bp of trDNA within the first three generations. The precise elimination of the tandem repeats argues for a sequence-dependent recombination event. Another transgenic line carrying a total of five different integration sites with large clusters (Lig 64) also revealed the rapid elimination of Chironomus repeats from the Drosophila genome. Analysis of single flies over 11 generations invariably showed a gradual shortening of the repeat arrays. Not a single case of cluster growth has been observed. If unequal sister chromatid exchange had been involved (which theoretically generates both shorter *and* longer cluster lengths), one would have to invoke selection for shortened tr clusters to explain the observed bias. Alternatively, intrachromosomal recombination between repeating units generating circular excision products might be responsible for array shortening. Intrachromosomal recombination could in fact be favoured in tandem arrays of *C. t. thummi* repeat families: five out of six sequence families display a sequence-dependent curvature of their DNA helix, which in tandem orientation may generate a highly recombinative solenoid superstructure (Schmidt 1984; Hankeln 1990).

The experiments with transgenic *D. melanogaster* demonstrate that *Drosophila* does not stably accommodate larger tr arrays (more than 4 elements) from Chironomus on its chromosome arms. Like *C. t. piger*, *D. melanogaster* may be a 'low-tolerance' species. This is corraborated by the chromosomal localisation of *D. melanogaster*'s own satellite DNA sequences. The distribution of the simple sequence satellite DNA families of *D. melanogaster* is limited to the centromeric regions. Only rarely, very short clusters are found at interspersed euchromatic positions (Lohe and Roberts 1988). The complex X-chromosomal 359 bp satellite from *D. melanogaster* is also mainly centromeric. Related euchromatic arrays do not exceed the length of four repeating units (DiBartolomeis et al. 1992). A *D. melanogaster* subtelomeric minisatellite cluster also has related arrays at euchromatic positions which do not exceed a very few repeat units (H. Biessmann and co-workers, personal communication). In summary, there might be species-specific barriers towards euchromatically located tr arrays in insects. The genetic basis of such barriers is unknown.

Conclusions

TrDNA is a highly unstable component of the eukaryotic genome. The complex rearrangements of trDNA as well as their underlying genetic mechanisms have still to be elucidated completely. Gene transfer of trDNA arrays into foreign genomes is a valuable tool for investigating these open questions. The behaviour of trDNA within the genome may, however, have been governed by species-specific rules. TrDNA transgenes may therefore display unexpected levels of instability or stability (!) in their new genomic context. Thus, molecular analysis of trDNA in 'natural' and transgenic organisms is an issue for biosafety research.

References

Aaltonen LA, Peltomäki P, Leach FS, Sistonen P, Pylkkänen L, Mecklin J-P, Järvinen H, Powell SM, Jen J, Hamilton SR, Petersen GM, Kinzler KW, Vogelstein B, de la Chapelle A (1993) Clues to the pathogenesis of familial colorectal cancer. Science 260, 812–816

Allen MJ, Jeffreys J, Surani, MA, Barton S, Norris ML, Collick A (1994) Tandemly repeated transgenes of the human minisatellite MS32 (D1S8), with novel mouse gamma satellite integration. Nucl Acids Res 22, 2976–2981

Armour JAL, Jeffreys AJ (1992) Biology and applications of human minisatellite loci. Curr Opin Genet Dev 2, 850–856

Armour JAL, Patel I, Thein SL, Fey MF, Jeffreys AJ (1989) Analysis of somatic mutations at human minisatellite loci in tumors and cell lines. Genomics 4, 328–334

Armour JAL, Wong Z, Wilson V, Royle NJ, Jeffreys AJ (1989) Sequences flanking the repeat arrays of human minisatellites: association with tandem and dispersed repeat elements. Nucl Acids Res 17, 4925–4935

Aslanidis C, Jansen G, Amemiya C, Shutler G, Mahadevan M, Tsilfidis C, Chen C, Alleman J, Wormskamp NGM, Vooijs M, Buxton J, Johnson K, Smeets HJ, Lennon GG, Carrano AV, Korneluk RG, Wieringa B, de Jong PJ (1992) Cloning of the essential myotonic dystrophy region and mapping of the putative defect. Nature 355, 548–551

Assaad FF, Tucker KL, Signer ER (1993) Epigenetic repeat-induced gene silencing (RIGS) in *Arabidopsis*. Plant Mol Biol 22, 1067–1085

Ault JG, Rieder CL (1994) Centrosome and kinetochore movement during mitosis. Curr Opin Cell Biol 6, 41–49

Baker SM, Bronner CE, Zhang L, Plug AW, Robatzek M, Warren G, Elliott EA, Yu J, Ashley T, Arnheim N, Flavell RA, Liskay RM (1995) Male mice defective in the DNA mismatch repair gene *PMS2* exhibit abnormal chromosome synapsis in meiosis. Cell 82, 309–319

Baker WK (1968) Position-effect variegation. Adv Genet 14, 133–169

Barlow C, Meister B, Lendahl U, Vennström B (1995) Altered expression after expansion of a v-erbA transgene in transgenic mice. Transgenic Research 4, 378–387

Baum M, Ngan VK, Clarke L (1994) The centromeric K-type repeat and the central core are together sufficient to establish a functional *Schizosaccharomyces pombe* centromere. Mol Biol Cell 5, 747–761

Bennett ST, Lucassen AM, Gough SCL, Powell EE, Undlien DE, Pritchard LE, Merriman ME, Kawaguchi Y, Dronsfield MJ, Pociot F, Nerup J, Bouzekri N, Cambon-Thomsen A, Roenningen KS, Barnett AH, Bain SC, Todd JA (1995) Susceptibility to human type 1 diabetes at *IDDM2* is determined by tandem repeat variation at the insulin gene minisatellite locus. Nature Genet 9, 284–292

Beridze T (1986) Satellite DNA, Springer-Verlag, Berlin

Bingham PM, Scott MO, Wang S, McPhaul MJ, Wilson EM, Garbern, JY, Merry DE, Fischbeck, KH (1995) Stability of an expanded trinucleotide repeat in the androgen receptor gene in transgenic mice. Nature Genet 9 191–196

Blackburn EH (1991) Structure and function of telomeres. Nature 350, 569–573

Blackburn EH (1994) Telomeres: no end in sight. Cell 77, 621–623

Bolshakov VN, Zharkikh AA, Zhimnler IF (1985) Intercalary heterochromatin in Drosophila. Chromosoma 92, 200–208

Bronner CE, Baker SM, Morrison PT, Warren G, Smith LG, Lescoe MK, Kane M, Earabino C, Lipford J, Lindblom A, Tannergaerd P, Bollag RJ, Godwin AR, Ward DC, Nordenskjoeld M, Fishel R, Kolodner R, Liskay RM (1994) Mutation in the DNA mismatch repair gene homologue hMLH1 is associated with hereditary non-polyposis colon cancer. Nature 368, 258–261

Brook JD, McCurrach ME, Harley HG, Buckler AJ, Church D, Aburatani H, Hunter K, Stanton VP, Thirion J-P, Hudson T, Sohn R, Zemelman B, Snell RG, Rundle SA, Crow S, Davies J, Shelbourne P, Buxton J, Jones C, Juvonen V, Johnson K, Harper PS, Shaw DJ, Housman DE (1992) Molecular basis of myotonic dystrophy: expansion of a trinucleotide (CTG) repeat at the 3′ end of a transcript encoding a protein kinase family member. Cell 68, 799–808

Brown KE, Barnett MA, Burgtorf C, Shaw P, Buckle VJ, Brown WRA (1994) Dissecting the centromere of the human Y chromosome with cloned telomeric DNA. Hum Mol Genet 3, 1227–1237

Brown WRA, MacKinnon PJ, Villasanté A, Spurr N, Buckle VJ, Dobson MJ (1990) Structure and polymorphism of human telomere-associated DNA. Cell 63, 119–132

Brutlag D L (1980) Molecular arrangement and evolution of heterochromatic DNA. Ann Rev Genet 14, 121–144

Buard J, Vergnaud G (1994) Complex recombination events at the hypermutable minisatellite CEB1 (D2S90). EMBO J 13, 3203–3210

Burke JR, Wingfield MS, Lewis KE, Roses AD, Lee JE, Hulette C, Pericak-Vance MA, Vance JM (1994) The haw river syndrome: dentatorubropallidoluysian atrophy (DRPLA) in an African-American family. Nature Genet 7, 521–524

Burright EN, Clark HB, Servadio A, Matilla T, Feddersen RM, Yunis WS, Duvick LA, Zoghbi HY, Orr HT (1995) SCA1 transgenic mice: a model for neurodegeneration caused by an expanded CAG trinucleotide repeat. Cell 82, 937–948

Butner KA, Lo CW (1986) High frequency DNA rearrangements associated with mouse centromeric satellite DNA. J Mol Biol 187, 547–556

Buxton J, Shelbourne P, Davies J, Jones C, van Tongeren T, Aslanidis C, de Jong P, Jansen G, Anvret M, Riley B, Williamson R, Johnson K (1992) Detection of an unstable fragment of DNA specific to individuals with myotonic dystrophy. Nature 355, 547–548

Cabot EL, Doshi P, Wu M-L, Wu C-I (1993) Population genetics of tandem repeats in centromeric heterochromatin: unequal crossing over and chromosomal divergence at the *responder* locus of *Drosophila melanogaster*. Genetics 135, 477–487

Carango P, Noble JE, Marks HG, Funanage VL (1993) Absence of myotonic dystrophy protein kinase (*DMPK*) mRNA as a result of a triplet repeat expansion in myotonic dystrophy. Genomics 18, 340–348

Carmena M, Abad JP, Villasante A, Gonzalez C (1993) The Drosophila melanogaster dodecasatellite sequence is closely linked to the centromere and can form connections between sister chromatids during mitoses. J Cell Sci 105, 41–50

Cattanach BM (1974) Position-effect variegation in the mouse. Genet Res 23, 291–306

Cavalier-Smith T (1985) Eucaryote gene numbers, non-coding DNA and genome size. In: T Cavalier-Smith (Ed) The evolution of genome size (pp 69–103). John Wiley and Sons Ltd, Chichester

Cech TR (1994) Chromosome end games. Science 266, 387–388

Cederberg H, Agurell E, Hedenskog M, Rannug U (1993) Amplification and loss of repeat units of the human minisatellite MS1 integrated in chromosome III of a haploid yeast strain. Mol Gen Genet 238, 38–42

Chandley AC, Mitchell AR (1988) Hypervariable minisatellite regions are sites for crossingover at meiosis in man. Cytogenet Cell Genet 48, 152–155

Charlesworth B, Sniegowski P, Stephan W (1994) The evolutionary dynamics of repetitive DNA in eukaryotes. Nature 371, 215–220

Chatterjee B, Lo CW (1989) Chromosomal recombination and breakage associated with instability in mouse centromeric satellite DNA. J Mol Biol 210, 303–312

Cherdshewasart W, Gharti-Chhetri GB, Negrutiu I (1996) Range of expression and transmission instability of kanamycin – resistance reporter gene in Direct Gene Transfer experiments. This volume

Chou S-H, Zhu L, Reid BR (1994) The unusual structure of the human centromere $(GGA)_2$ motif. Unpaired Guanosine residues stacked between sheared G A Pairs. J Mol Biol 244, 259–268

Chung M, Ranum LPW, Duvick LA, Servadio A, Zoghbi HY, Orr HT (1993) Evidence for a mechanism predisposing to intergenerational CAG repeat instability in spinocerebellar ataxia type I. Nature Genet 5, 254–258

Coggins LW, O'Prey M (1989) DNA tertiary structures formed *in vitro* by misaligned hybridization of multiple tandem repeat sequences. Nucl Acids Res 17, 7417–7427

Cohn M, Edström J-E (1992a) Chromosome ends in *Chironomus palladivittatus* contain different subfamilies of telomere-associated repeats. Chromosoma 101, 634–640

Cohn M, Edström J-E (1992b) Telomere-associated repeats in Chironomus form discrete subfamilies generated by gene conversion. J Mol Evol 35, 114–122

Collick A, Norris ML, Allen MJ, Bois P, Barton SC, Surani MA, Jeffreys AJ (1994) Variable germline and embryonic instability of the human minisatellite MS32 (D1S8) in transgenic mice. EMBO J 13, 5745–5753

Corcoran LM, Thompson JK, Walliker D, Kemp DJ (1988) Homologous recombination within subtelomeric repeat sequences generates chromosome size polymorphism in P. falciparum. Cell 53, 807–813

Dallas JF (1988) Detection of DNA fingerprints of cultivated rice by hybridization with a human minisatellite DNA probe. PNAS 85, 6831–6835

De Wind N, Dekker M, Berns A, Radman M, te Riele H (1995) Inactivation of the mouse *Msh2* gene results in mismatch repair deficiency, methylation tolerance, hyperrecombination, and predisposition to cancer. Cell 82, 321–330

DiBartolomeis SM, Tartof KD, Jackson FR (1992) A superfamily of *Drosophila* satellite related (SR) DNA repeats restricted to the X chromosome euchromatin. Nucl Acids Res 20, 1113–1116

Doolittle WF, Sapienza C (1980) Selfish genes, the phenotype paradigm and genome evolution. Nature 284, 601–603

Dorer DR, Henikoff S (1994) Expansions of transgene repeats cause heterochromatin formation and gene silencing in Drosophila. Cell 77, 993–1002

Dorlhac de Borne F, Vincentz M, Chupeau Y, Vaucheret H (1994) Co-suppression of nitrate reductase host genes and transgenes in transgenic tobacco plants. Mol Gen Genet 243, 613–621

Dover G A (1980) Ignorant DNA? Nature 285, 618–620

Dover GA (1982) Molecular drive: a cohesive mode of species evolution. Nature 299, 111–117

Dover GA (1986) Molecular drive in multigene families: how biological novelties arise, spread and are assimilated. Trend Genet 2, 159–165

Dover GA (1989) Victims or perpetrators of DNA turnover? Nature 342, 347–348

Eichler EE, Holden JJA, Popovich BW, Reiss AL, Snow K, Thibodeau SN, Richards CS, Ward PA, Nelson DL (1994) Length of uninterrupted CGG repeats determines instability in the *FMR1* gene. Nature Genet 8, 88–94

Eichler EE, Kunst CB, Lugenbeel KA, Ryder OA, Davison D, Warren ST, Nelson DL (1995) Evolution of the cryptic *FMR1* CGG repeat. Nat Genet 11, 301–308

Eissenberg JC, James TC, Foster-Hartnett DM, Hartnett T, Ngan V, Elgin SCR (1990) Mutation in a heterochromatin-specific chromosomal protein is associated with suppression of position effect variegation in *Drosophila melanogaster.* Proc Natl Acad Sci 87, 9923–9927

Ely J, Deka R, Chakraborty R, Ferrel RE (1992) Comparison of Five Tandem Repeat Loci between Humans and Chimpanzees. Genomics 14, 692–698

Epplen JT (1988) On simple repeated GAT/CA sequences in animal genomes: a critical reappraisal. J of Heredity 79, 409–417

Feng Y, Zhang F, Lokey LK, Chastain JL, Lakkis L, Eberhart D, Warren ST (1995) Translational suppression by trinucleotide repeat expansion at *FMR1*. Science 268, 731–734

Ferrer N, Azorin F, Villasante A, Gutiérrez C, Abad JP (1995) Centromeric Dodeca-satellite DNA sequences form fold-back structures. J Mol Biol 245, 8–21

Flavell RB (1986) Repetitive DNA and chromosome evolution in plants. Phil Trans R Soc Lond B 312, 227–242

Fowler C, Drinkwater R, Skinner J, Burgoyne L (1988) Human satellite-III DNA: an example of a "macrosatellite" polymorphism. Hum Genet 79, 265–272

Fray RG, Grierson D (1993) Identification and genetic analysis of normal and mutant phytogene synthase genes of tomato by sequencing, complementation and co-suppression. Plant Mol Biol 22, 589–602

Gall JG (1981) Chromosome structure and the c-value paradox. J Cell Biol 91, 3s–14s

Galli J, Wieslander L (1993) A repetitive secretory protein gene of a novel type in *Chir-*

onomus tentans is specifically expressed in the salivary glands and exhibits extensive length polymorphism. J Biol Chem 268, 11888–11893

Gerber H-P, Seipel K, Georgiev O, Höfferer M, Hug M, Rusconi S, Schaffner W (1994) Transcriptional activation modulated by homopolymeric glutamine and proline stretches. Science 263, 808–811

Grady D, Ratliff R, Robinson D, McCanlies E, Meyne J, Moyzis R (1992) Highly conserved repetitive DNA sequences are present at human centromeres. Proc Natl Acad Sci USA 89, 1695–1699

Gray IC, Jeffreys AJ (1991) Evolutionary transience of hypervariable minisatellites in man and the primates. Proc R Soc Lond B 243, 241–253

Green H (1993) Human genetic diseases due to codon reiteration: relationship to an evolutionary mechanism. Cell 74, 955–956

Haaf T, Warburton PE, Willard HF (1992) Integration of human α-satellite DN into Simian chromosomes: centromere protein binding and disruption of normal chromosome segregation. Cell 70, 681–696

Hägele K (1977) Differential staining of polytene chromosome bands in Chironomus by Giemsa banding methods. Chromosoma 59, 207–216

Hamel BC, Smits APT, de Graaff E, Smeets DFCM, Schoute F, Eussen BHJ, Knight SJL, Davies KE, Assman-Hulsmans CFCH, Oostra BA (1994) Segregation of FRAXE in a large family: clinical, psychometric, cytogenetic, and molecular data. Am J Hum Genet 55, 923–931

Hankeln T (1990) Molekulare Analyse phylogenetisch bedeutsamer repetitiver DNA bei Chironomiden. Dissertation zur Erlangung des Grades eines Doktors der Naturwissenschaften an der Ruhr-Universität Bochum

Hankeln T, Fillippova MA, Kiknadze II, Aimanova KG, Schmidt ER (1994a) Centromeric heterochromatin and satellite DNA in the *Chironomus plumosus* species group. Genome 37, 925–934

Hankeln T, Rohwedder A, Weich B, Schmidt ER (1994b) Transposition of minisatellite-like DNA in *Chironomus* midges. Genome 37, 542–549

Harley CB, Futcher AB, Greider CW (1990) Telomeres shorten during aging of human fibroblasts. Nature 345, 458–460

Harley HG, Brook JD, Floyd J, Rundle SA, Crow S, Walsh KV, Thibault M-C, Harper PS, Shaw DJ (1991) Detection of linkage disequilibrium between the myotonic dystrophy locus and a new polymorphic DNA marker. Am J Hum Genet 49, 68–75

Harley HG, Brook JD, Rundle SA, Crow S, Reardon W, Buckler AJ, Harper PS, Housman DE, Shaw DJ (1992) Expansion of an unstable DNA region and phenotypic variation in myotonic dystrophy. Nature 355, 545–546

Hart CM, Fischer B, Neuhaus JM, Meins F Jr (1992) Regulated inactivation of homologous gene expression in transgenic *Nicotiana sylvestris* plants containing a defense-related tobacco chitinase gene. Mol Gen Genet 235, 179–188

Heartlein MW, Knoll JHM, Latt SA (1988) Chromosome instability associated with human alphoid DNA transfected into the Chinese hamster genome. Mol Cell Biol 8, 3611–3618

Hecht A, Laroche T, Strahl-Bolsinger S, Gasser SM, Grunstein M (1995) Histone H3 and H4 N-Termini interact with SIR3 and SIR4 proteins: a molecular model for the formation of heterochromatin in yeast. Cell 80, 583–592

Hegemann JH, Fleig UN (1993) The centromere of budding yeast. BioEssays 15, 451–460

Heitz E (1934) Über α- und β-Heterochromatin sowie Konstanz und Bau der Chromomeren bei *Drosophila*. Biol Zentralbl 54, 588–609

Henikoff S (1990) Position-effect variegation after 60 years. Trends Genet 6, 422–426

Hennig W, Hennig I, Stein H (1970) Repeated sequences in the DNA of *Drosophila* and their localization in giant chromosomes. Chromosoma (Berl) 32, 31–63

Hershfield B, Swift H (1990) Characterization of a tandemly repeated DNA from the fleshfly *Sarcophaga bullata*. Chromosoma (Berl) 99, 125–130

Hinds HL, Ashley CT, Sutcliffe JS, Nelson DL, Warren ST, Housman DE, Schalling M (1993) Tissue specific expression of *FMR-1* provides evidence for a functional role in fragile X syndrome. Nature Genet 3, 36–43

Hogan NC, Slot F, Traverse KL, Garbe JC, Bendena WG, Pardue M-L (1995) Stability of tandem repeats in the *Drosophila melanogaster Hsr-omega* nuclear RNA. Genetics 139, 1611–1621

Horowitz H, Thoburn P, Haber JE (1984) Rearrangements of highly polymorphic regions near telomeres of *Saccharomyces cerevisiae.* Mol Cell Biol 4, 2509–2517

Hwu HR, Roberts JW, Davidson EH, Britten RJ (1986) Insertion and/or deletion of many repeated DNA sequences in human and higher ape evolution. Proc Natl Acad Sci USA 83, 3875–3879

Ikeno M, Masumoto H, Okazaki T (1994) Distribution of CENP-B boxes reflected in CREST centromere antigenic sites on long-range α-satellite DNA arrays of human chromosome 21. Hum Mol Genet 3, 1245–1257

Inglehearn CF, Cooke HJ (1990) A VNTR immediately adjacent to the human pseudoautosomal telomere. Nucl Acids Res 18, 471–476

Ionov Y, Peinado MA, Malkhosyan S, Shibata D, Perucho M (1993) Ubiquitous somatic mutations in simple repeated sequences reveal a new mechanism for colonic carcinogenesis. Nature 363, 558–561

Jabs EW, Goble CA, Cutting GR (1989) Macromolecular organization of human centromeric regions reveals high-frequency, polymorphic macro DNA repeats. Proc Natl Acad Sci 86, 202–206

James TC, Elgin SCR (1986) Identification of a nonhistone chromosomal protein associated with heterochromatin in *Drosophila melanogaster* and its gene. Mol Cell Biol 6, 3862–3872

Jarman AP, Wells AR (1989) Hypervariable minisatellites: recombinators or innocent bystanders? TIG 5, 367–371

Jeffreys AJ, MacLeod A, Tamaki K, Neil DL, Monckton DG (1991) Minisatellite repeat coding as a digital approach to DNA typing. Nature 354, 204–209

Jeffreys AJ, Wilson V, Kelly R, Taylor BA, Bulfield G (1987) Mouse DNA 'fingerprints': analysis of chromosome localization and germ-line stability of hypervariable loci in recombinant inbred strains. Nucl Acids Res 15, 2823–2836

Jeffreys AJ, Wilson V, Thein SL (1985) Hypervariable minisatellite regions in human DNA. Nature 314, 67–73

Jeffreys AJ, Neumann R, Wilson V (1990) Repeat unit sequence variation in minisatellites: a novel source of DNA polymorphism for studying variation and mutation by single molecule analysis. Cell 60, 473–485

Jeffreys AJ, Tamaki K, MacLeod A, Monckton DG, Neil DL, Armour JAL (1994) Complex gene conversion events in germline mutation at human minisatellites. Nature Genet 6, 136–145

John B (1988) The biology of heterochromatin. In: RS Verma (Ed) Heterochromatin (pp 1–147) Cambridge University Press, New York

John B, Miklos G L G (1979) Functional aspects of satellite DNA and heterochromatin. Int Rev Cytol 58, 1–114

John B, Miklos GLG (1988) The eukaryote genome in development and evolution. Allen and Unwin, Boston

Johnson RE, Kovvali GK, Prakash L, Prakash S (1995) Requirement of the yeast RTH1 5' to 3' exonuclease for the stability of simple repetitive DNA. Science 269, 238–240

Jones C, Penny L, Mattina T, Yu S, Baker E, Voullaire L, Langdon W Y, Sutherland G R, Richards R I, Tunnacliffe A (1995) Association of a chromosome deletion syndrome with a fragile site within the proto-oncogene CBL2. Nature 376, 145–149

Karpen GH, Spradling AC (1992) Analysis of subtelomeric heterochromatin in the *Drosophila* minichromosome Dp1187 by single P element insertional mutagenesis. Genetics 132, 737–753

Kawaguchi Y, Okamoto T, Taniwaki M, Aizawa M, Inoue M, Katayama S, Kawakami H, Nakamura S, Nishimura M, Akiguchi I, Kimura J, Narumiya S, Kakizuka A (1994) CAG expansions in a novel gene for Machado-joseph disease at chromosome 14q32.1. Nature Genet 8, 221–227

Kelly RG (1994) Similar origins of two mouse minisatellites within transposon-like LTRs. Genomics 24, 509–515

Kelly R, Bulfield G, Collick A, Gibbs M, Jeffreys AJ (1989) Characterization of a highly unstable mouse minisatellite locus: evidence for somatic mutation during early development. Genomics 5, 844–856

Kennedy GC, German MS, Rutter WJ (1995) The minisatellite in the diabetes susceptibility locus *IDDM2* regulates insulin transcription. Nature Genet 9, 293–298

Keyl H-G (1965) Duplikationen von Untereinheiten der chromosomalen DNS während der Evolution von *Chironomus thummi*. Chromosoma (Berl) 17, 139–180

Keyl H-G, Strenzke K (1956) Taxonomie und Cytologie von zwei Subspezies der Art *Chironomus thummi*. Z Naturforschg 11b, 727–735

Kim NW, Piatyszek MA, Prowse KR, Harley CB, West MD, Ho PLC, Coviello GM, Wright WE, Weinrich SL, Shay JW (1994) Specific association of human telomerase activity with immortal cells and cancer. Science 266, 2011–2014

Koide R, Ikeuchi T, Onodera O, Tanaka H, Igarashi S, Endo K, Takahashi H, Kondo R, Ishikawa A, Hayashi T, Saito M, Tomoda A, Miike T, Naito H, Ikuta F, Tsuji S (1994) Unstable expansion of CAG repeat in hereditary dentatorubral-pallidoluysian atrophy (DRPLA). Nature Genet 6, 9–13

Krontiris TG, Devlin B, Karp DD, Robert NJ, Risch N (1993) An association between the risk of cancer and mutations in the hras1 minisatellite locus. J of Medicine 329, 517–523

Kuhl DPA, Caskey T (1993) Trinucleotide repeats and genome variation. Curr Opin Genet Development 3, 404–407

Kunst CB, Warren ST (1994) Cryptic and polar variation of the fragile X repeat could result in predisposing normal alleles. Cell 77, 853–861

Künzler P, Matsuo K, Schaffner W (1995) Pathological, physiological, and evolutionary aspects of short unstable DNA repeats in the human genome. Biol. Chem Hoppe-Seyler 376, 201–211

La Spada AR, Paulson HL, Fischbeck KH (1994) Trinucleotide repeat expansion in neurological disease. Annals of Neurology 36, 814–822

La Spada AR, Wilson EM, Lubahn DB, Hardling AE, Fischbeck KH (1991) Androgen receptor gene mutations in X-linked spinal and bulbar muscular atrophy. Nature 352, 77–79

Larin Z, Fricker MD, Tyler-Smith C (1994) De novo formation of several features of a centromere following introduction of a Y alphoid YAC into mammalian cells. Hum Mol Genet 3, 689–695

Larson DD, Spangler EA, Blackburn EH (1987) Dynamics of telomere length variation in *Tetrahymena thermophila*. Cell 50, 477–483

Lavedan C, Hofmann-Radvanyi H, Shelbourne P, Rabes J-P, Duros C, Savoy D, Dehaupas I, Luce S, Johnson K, Junien C (1993) Myotonic dystrophy: size- and sex-dependent dynamics of CTG meiotic instability, and somatic mosaicism. Am J Hum Genet 52, 875–883

Le M-H, Duricka D, Karpen GH (1995) Islands of complex DNA are widespread in Drosophila centric heterochromatin. Genetics 141, 283–303

Leach FS, Nicolaides NC, Papadopoulos N, Liu B, Jen J, Parsons R, Peltomäki P, Sistonen P, Aaltonen LA, Nyström-Lahti M, Guan X-Y, Zhang J, Meltzer PS, Yu J-W, Kao F-T,

Chen DJ, Cerosaletti KM, Fournier REK, Todd S, Lewis T, Leach RJ, Naylor SL, Weissenbach J, Mecklin J-P, Järvinen H, Petersen GM, Hamilton SR, Green J, Jass J, Watson P, Lynch HT, Trent JM, de la Chapelle A, Kinzler KW, Vogelstein B (1993) Mutations of a *mutS* homolog in hereditary nonpolyopsis colorectal cancer. Cell 75, 1215–1225

Levinson G, Gutman GA (1987) Slipped-strand mispairing: a major mechanism for DNA sequence evolution. Mol Biol Evol 4, 203–221

Li S-H, McInnis MG, Margolis RL, Antonarakis SE, Ross CA (1993) Novel triplet repeat containing genes in human brain: cloning, expression, and length polymorphisms. Genomics 16, 572–579

Lica LM, Narayanswami S, Hamkalo BA (1986) Mouse satellite DNA, centromere structure, and sister chromatid pairing. J Cell Biol 103, 1145–1151

Lindahl G, Gersdorf E, Menzel HJ , Seed M, Humphries S, Utermann G (1990) Variation in the size of human apolipoprotein A is due to a hypervariable region in the gene. Hum Genet 84, 563–567

Liu B, Nicolaides NC, Markowitz S, Willson JKV, Parsons RE, Jen J, Papadopolous N, Peltomäki P, de la Chapelle A, Hamilton SR, Kinzler KW, Vogelstein B (1995) Mismatch repair gene defects in sporadic colorectal cancers with microsatellite instability. Nature Genet 9, 48–55

Lohe AR, Brutlag DL (1987a) Identical satellite DNA sequences in sibling species of *Drosophila.* J Mol Biol 194, 161–170

Lohe AR, Brutlag DL (1987b) Adjacent satellite DNA segments in *Drosophila*: structure of junctions. J Mol Biol 194, 171–179

Lohe AR, Roberts P (1988) Evolution of satellite DNA sequences in *Drosophila.* In: RS Verma (Ed) Heterochromatin (pp 148–186) Cambridge University Press, New York

Lyons KM, Stein JH, Smithies O (1988) Length polymorphisms in human proline-rich protein genes generated by intragenic unequal crossing over. Genetics 120, 267–278

MacDonald ME, Novelletto A, Lin C, Tagle D, Barnes G, Bates G, Taylor S, Allitto B, Altherr M, Myers R, Lehrach H, Collins FS, Wasmuth JJ, Frontali M, Gusella JF (1992) The huntington's disease candidate region exhibits many different haplotypes. Nature Genet 1, 99–103

Marcais B, Charlieu JP, Allain B, Brun E, Bellis M, Roizès G (1991) On the mode of evolution of alpha satellite DNA in human populations. J Mol Evol 33, 42–48

Maresca A, Singer MF, Lee TNH (1984) Continuous reorganization leads to extensive polymorphism in a monkey centromeric satellite. J Mol Biol 179, 629–649

Martinez-Balbas A, Rodriguez-Campos A, Garcia-Ramirez M, Sainz J, Carrera P, Aymami J, Azorin F (1989) Satellite DNAs contain sequences that induce curvature. Biochemistry 29, 2342–2348

Mason JM, Biessmann H (1995) The unusual telomeres of *Drosophila.* TIG 11, 58–62

Matzke M A, Matzke AJM, Mittelsten-Scheid O (1994) Inactivation of repeated genes – DNA-DNA interaction. In: Homologous recombination and gene silencing in plants. Paszkowski J (ed) pp 271–307. Kluwer Academic Publishers, Netherlands

McEachern MJ, Blackburn EH (1995) Runaway telomere alongation caused by telomerase RNA gene mutations. Nature 376, 403–408

McLean JW, Tomlinson JE, Kuang W-J, Eaton DL, Chen EY, Fless GM, Scanu AM, Lawn RM (1987) cDNA sequence of human apolipoprotein(a) is homologous to plasminogen. Nature 300, 132–140

Mittelsten-Scheid O, Paszkowski J, Potrykus J (1991) Reversible inactivation of a transgene in *Arabidopsis thaliana.* Mol Gen Genet 228, 104–112

Monckton DG, Caskey CT (1995) Unstable triplet repeat diseases. Circulation 91, 513–519

Morin GB, Cech TR (1986) The telomeres of the linear mitochondrial DNA of *Tetrahymena thermophila* consist of 53 bp tandem repeats. Cell 46, 873–883

Morin GB, Cech TR (1988) Mitochondrial telomeres: surprising diversity of repeated telomeric DNA sequences among six species of Tetrahymena. Cell 52, 367–374

Muller HJ (1930) Types of viable variations induced by X-rays in Drosophila. J Genet 22, 299–334

Murphy TD, Karpen GH (1995) Localization of centromere function in a *Drosophila* minichromosome. Cell 82, 599–609

Nagafuchi S, Yanagisawa H, Sato K, Shirayama T, Ohsaki E, Bundo M, Takeda T, Tadokoro K, Kondo I, Murayama N, Tanaka Y, Kikushima H, Umino K, Kurosawa H, Furukawa T, Nihei K, Inoue T, Sano A, Komure O, Takahashi M, Yoshizawa T, Kanazawa I, Yamada M (1994) Dentatorubral and pallidoluysian atrophy expansion of an unstable CAG trinucleotide on chromosome 12p. Nature Genet 6, 14–18

Nakamura Y, Leppert M, O'Connell P, Wolff R, Holm T, Culver M, Martin C, Fujimoto E, Hoff M, Kumlin E, White R (1987) Variable number of tandem repeat (VNTR) markers for human gene mapping. Science 235, 1616–1628

Narayanswami S, Doggett NA, Clark LM, Hildebrand CE, Weier H-U, Hamkalo BA (1992) Cytological and molecular characterization of centromeres in *Mus domesticus* and *Mus spretus*. Mammalian Genome 2, 186–194

Neil DL, Villasante A, Fisher RB, Vetrie D, Cox B, Tyler-Smith C (1990) Structural instability of human tandemly repeated DNA sequences cloned in yeast artificial chromosome vectors. Nucl Acids Res 18, 1421–1428

Nicolaides NC, Papdopoulos N, Liu B, Wei Y-F, Carter KC, Ruben SM, Rosen CA, Haseltine WA, Fleischmann RD, Fraser CM, Adams MD, Venter JC, Dunlop MG, Hamilton SR, Petersen GM, de la Chapelle A, Vogelstein B, Kinzler KW (1994) Mutations of two PMS homologues in hereditary nonpolyposis colon cancer. Nature 371, 75–80

Nielsen L, Edström J-E (1993) Complex telomere-associated repeat units in members of the genus *Chironomus* evolve from sequences similar to simple telomeric repeats. Mol Cell Biol 13, 1583–1589

Oberlé I, Rousseau F, Heitz D, Kretz C, Devys D, Hanauer A, Boué J, Bertheas MF, Mandel JL (1991) Instability of a 550-base pair DNA segment and abnormal methylation in fragile X syndrome. Science 252, 1097–1102

Ohno S (1972) So much 'junk' in our genome. In: Evolution of genetic systems (Smith HH ed) Brookhaven Symp. No. 26, Gordon and Breach, New York 1972, pp. 366–370

Orgel LE, Crick FHC (1980) Selfish DNA: the ultimate parasite. Nature 284, 604–607

Orr HT, Chung M, Banfi S, Kwiatkowski Jr. TJ, Servadio A, Beaudet AL, McCall AE, Duvick LA, Ranum LPW, Zoghbi HY (1993) Expansion of an unstable trinucleotide CAG repeat in spinocerebellar ataxia type 1. Nature Genet 4, 221–226

Palmiter RD, Brinster RL (1986) Germ line transformation of mice. Ann Rev Genet 20, 465–499

Papadopoulos N, Nicolaides NC, Wei Y-F, Ruben SM, Carter KC, Rosen CA, Haseltine WA, Fleischmann RD, Fraser CM, Adams MA, Venter JC, Hamilton SR, Petersen GM, Watson P, Lynch HT, Peltomäki P, Mecklin J-P, de la Chapelle A, Kinzler KW, Vogelstein B (1994) Mutation of a *mutL* homolog in hereditary colon cancer. Science 263, 1625–1629

Pardue ML, Hennig W (1990) Heterochromatin: junk or collectors item? Chromosoma 100, 3–7

Paulsson G, Höög C, Bernholm K, Wieslander L (1992) Balbiani ring 1 gene in *Chironomus tentans*. Sequence organization and dynamics of a coding minisatellite. J Mol Biol 225, 349–361

Perutz MF, Johnson T, Suzuki M, Finch JT (1994) Glutamine repeats as polar zippers: their possible role in inherited neurodegenerative diseases. Proc Natl Acad Sci USA 91, 5355–5358

Pimpinelli S, Berloco M, Fanti L, Dimitri P, Bonaccorsi S, Marchetti E, Caizzi R, Caggese C, Gatti M (1995) Transposable elements are stable structural components of *Drosophila melanogaster* heterochromatin. Proc Natl Acad Sci 92, 3804–3808

Radic MZ, Lundgren K, Hamkalo BA (1987) Curvature of mouse satellite DNA and condensation of heterochromatin. Cell 50, 1101–1108

Rattner JB (1991) The structure of the mammalian centromere. BioEssays 13, 51–56

Reiss AL, Freund L, Abrams MT, Boehm C, Kazazian H (1993) Neurobehavioral effects of the fragile X premutation in adult women: a controlled study. Am J Hum Genet 52, 884–894

Reuter G, Spierer P (1992) Position effect variegation and chromatin proteins. Bioessays 14, 605–612

Reyniers E, Vits L, De Boulle K, Van Roy B, Van Velzen D, de Graaff E, Verkerk AJMH, Jorens H ZJ, Darby JK, Oostra B, Willems PJ (1993) The full mutation in the FMR-1 gene of male fragile X patients is absent in their sperm. Nature Genet 4, 143–146

Richards RI, Holman K, Friend K, Kremer E, Hillen D, Staples A, Brown WT, Goonewardena P, Tarleton J, Schwartz C, Sutherland GR (1992) Evidence of founder chromosomes in fragile X syndrome. Nature Genet 1, 257–260

Richards RI, Sutherland GR (1992) Dynamic mutations: a new class of mutations causing human disease. Cell 70, 709–712

Richards RI, Sutherland GR (1994) Simple repeat DNA is not replicated simply. Nature Genet 6, 114–116

Riggins GJ, Lokey LK, Chastain JL, Leiner HA, Sherman SL, Wilkinson KD, Warren ST (1992) Human genes containing polymorphic trinucleotide repeats. Nature Genet 2, 186–191

Rovira C, Beermann W, Edström J-E (1993) A repetitive DNA sequence associated with the centromeres of *Chironomus pallidivittatus.* Nucl Acids Res 21, 1775–1781

Royle NJ, Clarkson RE, Wong Z, Jeffreys AJ (1988) Clustering of hypervariable minisatellites in proterminal regions of human autosomes. Genomics 3, 352–360

Rubinsztein DC, Amos W, Leggo J, Goodburn S, Jain S, Li S-H, Margolis RL, Ross CA, Ferguson-Smith MA (1995) Microsatellite evolution – evidence for directionality and variation in rate between species. Nature Genet 10, 337–343

Schmidt ER (1981) The development of a 120 base-pair repetitive DNA sequence in *Chironomus thummi* is correlated to the duplication of defined chromosomal segments. FEBS Lett. 129, 21–24

Schmidt ER (1984) Clustered and interspersed repetitive DNA family of Chironomus. J Mol Biol 178, 1–15

Schmidt ER, Godwin EA, Keyl H-G, Israelewski N (1982) Cloning and analysis of ribosomal DNA of *Chironomus thummi thummi* and *Chironomus thummi* piger. Chromosoma (Berl) 87, 389–407

Seymour GB, Fray RG, Hill P, Tucker GA (1993) Down-regulation of two non-homologous endogenous tomato genes with a single chimaeric sense gene construct. Plant Mol Biol 23, 1–9

Simon M, Philieps M, Green H (1991) Polymorphism due to variable number of repeats in the human involucrin gene. Genomics 9, 576–580

Singer MF (1982) Highly repeated sequences in mammalian genomes. Int Rev Cytol 76, 67–112

Siomi H, Siomi MC, Nussbaum RL, Dreyfuss G (1993) The protein product of the fragile X gene, *FMR1*, has characteristics of an RNA-binding protein. Cell 74, 291–298

Smith GP (1976) Evolution of repeated DNA sequences by unequal crossover. Science 191, 528–535

Stallings RL (1995) Conservation and evolution of $(CT)_n/(GA)_n$ microsatellite sequences at orthologous positions in diverse mammalian genomes. Genomics 25, 107–113

Steiner NC, Clarke L (1994) A novel epigenetic effect can alter centromere function in fission yeast. Cell 79, 865–874

Steiner NC, Hahnenberger M, Clarke L (1993) Centromeres of the fission yeast *Schizosaccharomyces pombe* are highly variable genetic loci. Mol Cell Biol 13, 4578–4587

Strand M, Prolla TA, Liskay RM, Petes TD (1993) Destabilization of tracts of simple repetitive DNA in yeast by mutations affecting DNA mismatch repair. Nature 365, 274–276

Sutherland GR, Richards RK (1995) Simple tandem DNA repeats and human genetic disease. Proc Natl Acad Sci 92, 3636–3641

Swallow DM, Gendler S, Griffiths B, Corney G, Taylor-Papadimitriou J, Bramwell M E (1987) The human tumour-associated epithelial mucins are coded by an expressed hypervariable gene locus PUM. Nature 328, 82–84

Tartof KD, Bishop C, Jones M, Hobbs CA, Locke J (1989) Towards an understanding of position effect variegation. Dev Genet 10, 162–176

Tautz D, Renz M (1984) Simple sequences are ubiquitous repetitive components of eukaryotic genomes. Nucl Acids Res 12, 4127–4138

Tautz D, Schlötterer C (1994) Simple sequences. Curr Opin Genet Development 4, 832–837

Tautz D, Trick M, Dover GA (1986) Cryptic simplicity in DNA is a major source of genetic variation. Nature 322, 652–656

The Huntington's Disease Collaborative Research Group (1993) A novel gene containing a trinucleotide repeat that is expanded and unstable on Huntington's disease chromosomes. Cell 72, 971–983

Thibodeau SN, Bren G, Schaid D (1993) Microsatellite instability in cancer of the proximal colon. Science 260, 816–819

Tyler-Smith C, Oakey RJ, Larin Z, Fisher RB, Crocker M, Affara NA, Ferguson-Smith MA, Muenke M, Zuffardi O, Jobling M A (1993) Localization of DNA sequences required for human centromere function through an analysis of rearranged Y chromosomes. Nature Genet 5, 368–375

Vig BK (1994) Do specific nucleotide bases constitute the centromere? Mutation Res 309, 1–10

Vogt P (1990) Potential genetic functions of tandem repeated DNA sequence blocks in the human genome are based on a highly conserved "chromatin folding code". Hum Genet 84, 301–336

Wahl GM, de Saint Vincent R, DeRose ML (1984) Effect of chromosomal position on amplification of transfected genes in animal cells. Nature 307, 516–520

Wahls WP, Wallace LJ, Moore PD (1990) Hypervariable minisatellite DNA is a hotspot for homologous recombination in human cells. Cell 60, 95–103

Wang S-S, Zakian VA (1990) Telomere-telomere recombination provides an express pathway for telomere acquisition. Nature 345, 456–460

Wang Y-H, Amirhaeri S, Kang S, Wells RD, Griffith J D (1994) Preferential nucleosome assembly at DNA triplet repeats from the myotonic dystrophy gene. Science 265, 669–671

Waye JS, Willard H (1989) Concerted evolution of alpha satellite DNA: Evidence for species specificity and a general lack of sequence conservation among alphoid sequences of higher primates. Chromosoma 98, 273–279

Weber JL (1990) Informativeness of human $(dC\text{-}dA)_n \cdot (dG\text{-}dT)_n$ polymorphisms. Genomics 7, 524–530

Weller P, Jeffreys AJ, Wilson V, Blanchetot A (1984) Organization of the human myoglobin gene. EMBO J 3, 439–446

Wevrick R, Earnshaw WC, Howard-Peebles PN, Willard HF (1990) Partial deletion of alpha satellite DNA associated with reduced amounts of the centromere protein CENP-B in a mitotically stable human chromosome rearrangement. Mol Cell Biol 10, 6374–6380

Wevrick R, Willard HF (1989) Long-range organization of tandem arrays of α satellite DNA at the centromeres of human chromosomes: high-frequency array-length polymorphism and meiotic stability. Proc Natl Acad Sci 86, 9394–9398

Wevrick R, Willard VP, Willard HF (1992) Structure of DNA near long tandem arrays of alpha satellite DNA at the centromere of human chromosome 7. Genomics 14, 912–923

Willard HF (1990) Centromeres of mammalian chromosomes. TIG 6, 410–416

Willems PJ (1994) Dynamic mutations hit double figures. Nature Genet 8, 213–215

Winterpacht A (1991) Instabilität tandem-repetitiver Elemente aus *Chironomus thummi* nach Transfektion in Escherichia coli und in *Drosophila melanogaster.* Dissertation zur Erlangung des Grades eines Doktors der Naturwissenschaften an der Ruhr-Universität Bochum

Wöhrle D, Hennig I, Vogel W, Steinbach P (1993) Mitotic stability of fragile c mutations in differentiated cells indicates early post-conceptional trinucleotide repeat expansion. Nature Genet 4, 140–142

Wolff RK, Nakamura Y, White R (1988) Molecular characterization of a spontaneously generated new allele at a VNTR locus: no exchange of flanking DNA sequence. Genomics 3, 347–351

Wolff RK, Plaetke R, Jeffreys AJ, White R (1989) Unequal crossingover between homologous chromosomes is not the major mechanism involved in the generation of new alleles at VNTR loci. Genomics 5, 382–384

Wong AKC, Biddle FG, Rattner JB (1990) The chromosomal distribution of the major and minor satellite is not conserved in the genus Mus. Chromosoma (Berl) 99 190–195

Wong AKC, Rattner JB (1988) Sequence organization and cytological localization of the minor satellite of mouse. Nucl Acids Res 16, 11645–11661

Wu C-I, True JR, Johnson N (1989) Fitness reduction associated with the deletion of a satellite DNA array. Nature 341, 248–251

Yu S, Mulley J, Loesch D, Turner G, Donnelly A, Gedeon A, Hillen D, Kremer E, Lynch M, Pritchard M, Sutherland G R, Richards R I (1992) Fragile-X syndrome: unique genetics of the heritable unstable element. Am J Hum Genet 50, 968–980

Yu S, Pritchard M, Kremer E, Lynch M, Nancarrow J, Baker E, Holman K, Mulley JC, Warren ST, Schlessinger D, Sutherland GR, Richards RI (1991) Fragile X genotype characterized by an unstable region of DNA. Science 252, 1179–1181

Yurov YB, Mitkevich SP, Alexandrov IA (1987) Application of cloned satellite DNA sequences to molecular-cytogenetic analysis of constitutive heterochromatin heteromorphisms in man. Hum Genet 76, 157–164

Zhang P, Spradling C (1995) The Drosophila salivary gland chromocenter contains highly polytenized subdomains of mitotic heterochromatin. Genetics 139, 659–670

Zhong N, Dobkin C, Brown WT (1993) A complex mutable polymorphism located within the fragile X gene. Nature Genet 5, 248–252

III Expression of Transgenes

20 Range of Expression and Transmission Instability of the Kanamycin-Resistance Reporter Gene in Direct Gene Transfer Experiments

W. Cherdshewasart[1], G.B. Gharti-Chhetri[2], and I. Negrutiu[3]

[1]Biology Department, Chulalongkorn University, Bangkok 10330, Thailand
[2]Agronomy Department, IAAS, Rampur, Chitwan, Nepal
[3]Ecole Normale Supérieure de Lyon, Reconnaissance Cellulaire et Amélioration des Plantes. 46, Allée d'Italie, 69364 Lyon, France

Introduction

Two decades of breakthroughs in plant transformation have consolidated a technology that today represents one of the key tools in basic and applied research:

1. Genetic modification and agronomic improvement of plants. Current experiments and field tests involve traits such as pathogen (viral, fungal, insect etc.)-resistance, herbicide-resistance, protein quality enhancement, fructan, trehalose etc. production, enhancement of vegetative growth, sugar production etc., genetic castration or improved shelf-life.
2. Cloning of tagged genes and their functional analysis, including the generation of additional or even novel classes of mutants by antisense, overexpression and "illegitimate" expression of the cloned genes.
3. Risk assessment of horizontal transgene dissemination according to virulence/pathogenic criteria, public perception and policy decision making.

This report will consider elements from items 1 and 3 above, taking into account the following consideration:

- The foreign genes are integrated at random, as single or multiple copies.
- The transgene is generally inherited in a Mendelian fashion as a single or as multiple loci.
- (High) copy number as a critical trigger for transgene suppression/inactivation. The nature and the extent at which involved molecular mechanisms operate are not properly defined at present.
- Field trials are conducted at a rapid pace in various parts of the world in order to perform standard breeding evaluations, to provide some assessment for potential environmental risks and in view of commercial release.

E.R. Schmidt/T. Hankeln (Eds.)
Transgenic Organisms and Biosafety

The present state of knowledge with respect to the fate of foreign genes in the host plant genome, the combined use of a relatively reduced range of selectable marker genes and of an increasing number of genes of interest, the integration of transgenic lines in standard breeding schemes and the implicit economic expectations they create, all point towards the necessity of improving means of predicting and controlling both integration events and expression patterns of foreign genes.

Choice of Experimental Design

Within this context, we have defined an experimental design which attempts to evaluate the spectrum of instability events following direct gene transfer into *Nicotiana plumbaginifolia* ($2n = 20$) under typical breeding conditions:

1. Analysis of a large enough population of transgenic plants that was previously screened in Ro for average to high levels of resistance to the selectable marker gene NPT(neomycin phosphotransferase), leading to kanamycin-resistance (Km^R).
2. Following segregation analysis in R_1, the batch of transgenic plants was further analysed according to the number of segregating loci. Most of the tests were performed with the monogenic group and concerned
   a) The stability of transgene expression in successive meiotic generations
   b) Linkage analysis
   c) The systematic use in Ro of a backcross to the wild type as a means of "stabilizing" and standardizing "in vitro" regenerants before their use in various types of crosses
3. Crosses between independent transgenic lines and evaluation according to the initial number of segregating loci.
4. Comparisons with respect to transgene integration profile, transmission and expression stability in simple versus co-transformation experiments.

Terminology

The following terms need to be defined according to the established experimental design and the described instability events (see also Cherdshewasart et al. 1993).

- *Integrative transformation* is used in place of stable transformation which we consider as inappropriate in general.
- The *integration pattern* describes the migration profile in Southern blot analysis of a transgene in an individual transformant and its successive meiotic generation(s) following restriction with an appropriate restriction

enzyme which reveals the expected functional fragments of the transgene. It is assumed that the pattern remains unchanged within the plant and in successive generations. The analysis gives an estimate of the copy number. A simple integration pattern corresponds to the presence of a major hybridization band at the expected size of a functional copy plus one or two minor bands (the junction fragments). A complex integration pattern (rearranged, scrambled integration) is produced whenever three or more major bands are observed and exhibit unexpected size values.

- The *segregation pattern* is the result of segregation tests performed in R_1 and/or BC_1 (F_1) of a given transgenic line. It establishes the number of expressed segregating loci for the selectable marker gene. It is assumed that the pattern remains unchanged in successive self-generations and contributes the established number of loci in defined cross-combinations.
- *Expression instability* of the transgene results from deviations from the observed segregation pattern in R_1/BC_1 and can be due to inactivation or activation of one or several loci in later meiotic generations. Discordance between the number of expected loci in selfing versus cross-combinations is defined *discordant segregations.*
- *Partial inactivation* is characterized by non-Mendelian ratios, while *complete inactivation* corresponds to loss of resistance from one meiotic generation to the other. Complete inactivation may also occur between Ro and R_1 generations (first meiosis) in which case the initial number of expressed loci remains unknown.
- *Genomic disorders* produce early seed abortion in crosses between transformants with multiple NPT loci.

The Nicotiana Direct Gene Transfer System: Evidence for DNA Uptake and Integration Conditions

Optimization studies for DNA uptake were performed in a transient gene expression system. It was concluded that plant protoplasts took up DNA under similar conditions with bacterial cells (e.g. dication type and concentration, PEG concentration). The uptake was shown to be an instantaneous process, all protoplasts being able to take up and express the foreign DNA under defined experimental parameters. Since leaf protoplasts, which are highly vacuolated and have the nucleus compressed between the plasmalemma and the tonoplast, exhibited higher levels of transient gene expression as compared to cell suspension protoplasts, we hypothesized a possible correlation between the apposition of the nucleus onto the plasmalemma and a more efficient transport of the DNA into the nucleus.

With respect to integrative transformation, we were able to demonstrate that treatments which stimulate DNA repair (UV irradiation) or delay DNA re-

Table 1. Segregation patterns for Km^R in transformant populations produced from haploid and diploid batches of *N. plumbaginifolia* protoplasts

Specification	Haploid (224 transformants)		Diploid (166 transformants)	
	Heterozygous (%)	Homozygous (%)	Heterozygous (%)	Homozygous (%)
Total	62	38	98	2
Monogenics	43	95	43	–
Digenics	29	0	20	–
Multigenics				
Instability events	11.5	0	12	–

The initial segregations correspond to the R_1/BC_1 generation. The "haploid" batch was subdivided in two groups, according to the initial segregation ratios. The homozygous group identified in R_1 was further analysed by segregation tests in BC_1 (F_1) and F_2 generations. The "instability events" class includes complete or partial loss of Km^R, as well as "discordant Segregations" (see Table 3).

plication (3-aminobenzamide) significantly increase the transformation efficiency (Gharti-Chhetri et al. 1990). It is interesting to note that using haploid protoplasts for transformation we could better characterize the timing of DNA integration (Table 1): integration before or during the first DNA replication cycle resulted in homozygous transformants which almost exclusively integrated and inherited the DNA as a single locus. This class of transformants appeared particularly stable in both the transmission and the expression of the transgene. The other class of transformants was heterozygous for the trait and exhibited a high proportion of multiple insertion loci and instability events.

Simple Transformation: Fate of the Selectable Marker Gene According to the Reproduction Mode

The integration and the segregation pattern with respect to the expression (in)stability was monitored in self versus cross-pollination tests.

The *integration pattern* is evaluated in Table 2. It can be seen that the first passage through meiosis can generate a certain level of DNA rearrangement. This was significantly higher in transformed lines in which an altered Km^R expression was observed.

The *segregation pattern* was evaluated in parallel in R_1 and BC_1 generations (Table 1). It can be seen that roughly half of the transformants transmitted the trait as a single locus. All instability classes together represented 12% of the analysed population and was considered as the background level of expression instability.

Table 2. Integration profile of NPT gene in *N. plumbaginifolia* transformants following "simple" transformation versus co-transformation

Meiotic generation	Cases analysed	Copy number range NPT	Samples with complex integration pattern (%)
Simple transformation			
Ro	98	1–10	15.3
R_1	36	1–10	19.5
$R_{1(M)}$	32	1–10	26
Co-transformation			
Ro	46	1–30 (50)	52

In the case of simple transformation, the comparison was extended to the R_1 generation. Copy number range and the extent of DNA rearrangement were assessed. More than three major bands were considered as complex rearrangements in restrictions theoretically generating functional bands. $R_{1(M)}$ corresponds to a particular class of transformants in which loss of Km^R was observed between the Ro and R_1 generations.

A comparison between Tables 1 and 2 indicates that there is some correlation between the level of rearrangement and the extent of expression instability. However, many lines with rearranged NPT sequences expressed appropriate levels of Km^R before and after the first meiotic generation; the opposite does not seem to be true, as the frequency of lines with rearrangements increases when belonging to the class of transformants exhibiting Km^R alterations in R_1 generation. Such comparisons suggest that part of the expression instability observed in the first meiotic generation is generated by rearrangement events.

Backcrossing Reveals the Existence of Silenced Loci

Up to 9% of the evaluated transformants in R_1/BC_1 generations revealed "discordant" segregations (Table 3). These are individual transformants in which there is a discrepancy between the number of loci scored either by selfing (R_1) or by backcrossing (BC_1). The selfed R_1 group exhibits a majority of digenic transformants, with an important fraction of inactivated lines (17%). The corresponding BC_1 group reveals instead a high proportion of multigenics (62%) and a low level of inactivated lines (4%).

Table 3. Distribution of "discordant" segregation classes according to the number of segregating loci in R_1 versus BC_1 generations

Reproduction mode	Number of transformants	Monogenics (%)	Digenics (%)	Multigenics (%)	Loss/ inactivation (%)
R_1	51	18	41	24	17
F_1	51	16	18	62	4

The comparison strongly indicates that in genuine Ro transformants and their selfed progeny R_1 as many as 51% of the lines [(62–24)+(17–4)] contain non-expressed NPT loci. The BC seems to activate them, suggesting that the major process involved here is co-suppression. Less plausible, but not excluded, is the occurrence of gene conversion events following outcrossing. Since discordant segregations affect a defined fraction of the transgenic plant population, we speculated that this corresponded to a particular physical configuration of the transgene loci/inserts within the host genome.

Linkage Analysis: While Random Integration Is the Rule, Linked Loci Generate More Instability Events

More than 200 independent transgenic lines which were classified as monogenic in BC_1 progeny tests were crossed among them in order to evaluate the randomness of integration events (Cherdshewasart et al. 1993). Only 5% of the crosses identified linkage events. Within the two main linkage groups, instability events ocurred in half of the cross-combinations analysed, affecting 20% of the individual plants tested in the F_2 generations. Further tests in F_3 generation demonstrated that the instability was continously generated within the "normal" fraction of a transgenic family affected by partial or complete inactivation in previous meiotic generations.

Crosses Between Independent Transgenic Lines: Genomic Disorders Occur when Transgene Loci Are in Excess and Depend on the Relative "Configuration" of the Transgene Within the Chromosomes

Km^R clones are produced in protoplast cultures at high rates (10^{-3} to 10^{-4}). Only a small fraction is characterized in the end, i.e. the one usually exhibiting high levels of Km^R and good regeneration response. Such transgenic plants usually contain one to ten copies of the gene integrated in one to several loci in the genome and are self-fertile. The transgene is said to be in a "balanced configuration".

When such independent transformed lines are crossed among them, the level and type of instability depends on the number of loci present in each of the genomes, although there is no linear correlation as such.

With two to three loci involved, the alterations in segregation patterns indicate both activation of silent sites and inactivation of expressed loci (Table 4).

With four loci, a threshold level seems to be attained: each cross-combination tested exhibits at least one inactivated locus, with as much as 60% of the analysed cases having two loci inactivated simultaneously. With six loci, four to five of them appear to be systematically inactivated (Table 4).

Table 4. Extent of expression instabilities in crosses between independent transformants as a function of the number of NPT loci

Specification	Monogenics × monogenics		Monogenics × digenics		Monogenics × trigenics		Digenics × digenics		Trigenics × trigenics	
	(%)	Loci involved	(%)	Loci involved	(%)	Loci involved	(%)	Loci involved	(%)	Loci involved
Expected segregation class	75	–	78	–	0	–	0	–	0	–
Activation or gene conversion events	10	1	4	At least one	0	–	0	–	0	–
Inactivation events	15	One or two	18	One to three (one locus, 7%; two or more, 11%)	100	One to four (one locus, 18%; two, 64%; three or more, 18%)	100	Two (57%) or three (43%)	100	Four (50%) or five (50%)

The percentage of cases producing the expected class of segregants is compared to those showing alterations in segregation ratios. Both activation of silent loci and inactivation of active loci are evaluated.

Table 5. Genomic disorders as determined by the number of combined transgene loci in diallel crosses between digenic or trigenic transformed lines

Specification	Digenics × digenics(*n*)	Trigenics x trigenics(*n*)
Independent transformants evaluated	7	7
With seed setting in all combinations tested	1	0
With seed setting in some of the combinations tested	3	1
Total crosses performed	42	42
Early abortions	30 (71%)	37 (88%)

From these results we concluded that whenever the combined number of loci in crosses between transformed lines was four or higher, only one to two active loci were maintained in the progeny plants.

Diallel crosses between digenic or trigenic lines (Table 5) clearly demonstrated the existence of a positive and negative transgene configuration: genomic disorders were produced or not as a function of the direction of the cross and the number of transgene loci (also see Cherdshewasart et al. 1993). The more loci involved, the higher the proportion of cross-pollinations that aborted. "Positive configuration" of NPT sequences at high locus number appeared as a rare event (approximately 10%, see Table 5).

It can perhaps be argued that several of the results presented in this chapter may have resulted from polyploidy/aneuploidy problems. As far as the linkage tests and discordant segregation analyses are concerned, ploidy/aneuploidy can be excluded. For example, the fact that BC_1 series (wild type pollen was always used as male partner) revealed additional Km^R loci speaks against ploidy as the main reason. If it were, only resistant progeny would be obtained. The same reasoning applies for the genomic disorder results (Table 5): ploidy differencies can not explain the high abortion rates and the "positive configuration" cases observed in certain line combinations.

Co-transformation Experiments: Presence of Two Distinct Types of Sequences and Absence of Selection Pressure for the Non-selectable Gene Result in Excessive Transgene Rearrangement and Increased Expression Instability

Co-transformation experiments have been performed by using the selectable gene (Km^R) and the non-selectable gene (the gene of interest) either in the same construct (linked form) or on separate plasmids (unlinked form).

The NPT gene integration profile and transmission pattern were compared in simple and co-transformation experiments. The results on the integration profile are summarized in Table 2 and demonstrate a spectacular increase in average copy number and a strong tendency towards complex rearrangements in the co-transformation system. The differences were less pronounced when comparing the transmission patterns: a relatively similar distribution of various segregation categories was observed with, however, a fourfold increase in the rates of expression instability (from 5% to 20%).

Comparisons were also made between the integration profile of the NPT gene and the non-selectable gene in corresponding transgenic plants. A random association of simple and complex integration profiles was observed for the marker gene and the gene of interest. However, the non-selectable gene was integrated as complex rearranged fragments two times more frequently than the selectable gene.

From these and other observations not described here, several conclusions were drawn:

1. The observed rearrangements have most likely occurred due to recombination before integration into the genome.
2. Enhanced rearrangements seem to be favoured in more complex mixtures of transgene constructs.
3. The lack of selection pressure on the non-selectable gene results in an increased tendency to transgene rearrangement and high copy number.
4. No significant differences were noticed between linked and unlinked constructs for the above mentioned trends.

All the above results concern observations made on Ro and R_1 transgenic plants. We expect that the instability levels reported in the simple transformation system (transgene configuration, copy number, reproduction mode in successive meiotic generations etc.) would be further increased in the co-transformation experiments.

Conclusions

Alterations in transgene expression and transmission pattern have been reported in several plant species and in various transformation systems. A common characteristic of transformants that exhibit modifications in transgene expression is a complex integration pattern with multiples of complete and rearranged gene copies. Duplicated sequences in tandem repeats were more frequently inactivated than unlinked repeats (Mittelsten Scheid et al. 1991).

We present molecular evidence that supports such data. In addition, we show that alterations in initial segregation rates result from the reduction or the increase in the number of segregating loci as a function of the number of insertion sites and the mode of sexual transmission. Co-suppression appeared as the main process involved. There was a clear-cut release of suppression in crosses to wild type. On the other hand, the number of expressed loci was systematically reduced to one or two loci in crosses between transformants which accumulated four or more active NPT loci.

In addition and as we reported elsewhere (Cherdshewasart et al. 1993), gene inactivation also appeared at unpredictable rates in successive meiotic generations. Once established, the process was transmitted to the next generations in a cumulative manner.

Thus, the phenotypic loss of Km^R occurred at various stages of transformant ontogeny and phylogeny. Physical loss of transgene sequences, rearrangement and amplification, gene conversion, methylation or co-suppression all account – to variable degrees – for the expression instability events described (also see Cherdshewasart et al. 1993 and the references therein). They systematically recall situations described in animal cells (Brenig and Brem 1993), which can be summarized as follows: known features characteristic of active or inactive chromatin (such as methylation pattern, DNase I hypersensitive sites or nucleosome positioning of transgenes) are not sufficient to predict the transcriptional activity of transgenes. One envisaged strategy is the inclusion of nuclear matrix attachment site sequences in the constructs to enhance the expression stability of the transgenes.

So far we have discussed the fate of a selectable transgene in the host genome for which – in addition to the initial selection procedure – the selection pressure can be applied at least after each meiotic generation on germinating seedlings. We have seen that in the case of a non-selectable transgene, which in general is supposed to represent the gene of interest, the lack of selection constraints and the presence of the selectable marker gene sequences are conditions that increase the instability level.

Although our results are preliminary, they confirm the findings on simple transformation and are in agreement with results available in animal cells (Clark et al. 1992).

Thus in mice co-transformed genes were integrated at any one locus as a mixture of interspersed and non-interspersed arrays of the two transgene sequences (one to 50 or more copies per transgene). In many of the samples multiple fragments were generated indicating the complex nature of the rearranged sequences.

The above discussion is meant to demonstrate the present limits and limitations of our art in predicting and controlling integration events and expression patterns of transgenes. It looks like the host genome has the means to "recognize" in space and time transgene sequences and is capable of altering their expression. When there is a high copy number of a transgene, but mainly when the number of insertion loci of a transgene is higher than three or four, i.e. when the transgene has several characteristics of a gene family without beeing identical to it, the genome of a true diploid species such as *N. plumbaginifolia* seems to be constrained to adopt some drastic measures. It is either able to control the situation and renders simultaneously silent active loci with the exception of one or two, or the loci are in such excess and/or configuration that the whole genome suffers severe disorders with, as a consequence, genome imbalance and death.

On the whole, our observations and their interpretation should raise concern mainly with respect to the applied and commercial aspects of direct gene transfer. This is particularly true in species resistant to transformation/regeneration where the number of transformed clones is limited and extensive screening for transformed clones with defined properties (one copy/one locus, high level of expression, stability under field conditions, etc.) is problematic.

In addition, our work with co-transformed genes strongly suggests restricting the transformation procedure to the gene of interest and developing appropriate means of early detection (e.g. PCR screening) and evaluation of its expression. Eliminating the need for marker genes will reduce certain breeding constraints and solve some matters of dispute with respect to environmental issues.

References

Brenig B, Brem G (1993) Comparative analysis of transgenic expression in mice. Chemicaoggi 1–2:33–40

Cherdshewasart W, Gharti-Chhetri GB, Saul MW, Jacobs M, Negrutiu I (1993) Expression instability and genetic disorders in transgenic *N. plumbaginifolia*. Transgenic Res 2: 307–320

Clark AJ, Cowper A, Wallace R, Wright G, Simons JP (1992) Rescuing transgene expression by co-integration. Biotechnology 10: 1450–1454

Gharti-Chhetri GB, Cherdshewasart W, Paszkowski J, Jacobs M, Negrutiu I (1990) Hybrid genes in the analysis of transformation conditions. 3. Temporal/spatial fate of NPTII gene integration, its inheritance and factors affecting these processes in *N. plumbaginifolia*. Plant Mol Biol 14: 687–696

Mittelsten Scheid O, Paszkowski J, Potrykus I (1991) Reversible inactivation of a transgene in *Arabidopsis thaliana*. Mol Gen Genet 228: 104–112

21 Variability of Organ-Specific Expression of Reporter Genes in Transgenic Plants

J. Landsmann, C. van der Hoeven, and A. Dietz-Pfeilstetter

Biologische Bundesanstalt für Land- und Forstwirtschaft, Institut für Biochemie und Pflanzenvirologie, Messeweg 11/12, 38104 Braunschweig, Germany

Introduction

Expression of introduced genes in transgenic plants varies between individual regenerants. Horsch et al. (1985) assumed influences of the surrounding DNA or the chromatin structure at the insertion site on the foreign genes and called this phenomenon "position effect". Integration of the introduced genes via *Agrobacterium* into the plant genome occurs randomly. The number of integrated gene copies cannot be influenced at the time of transformation. These uncertainties hamper the predictability of transformation results (Deroles and Gardner 1988a,b; Dunwell and Paul 1990).

We have fused the par promoter from the hemoglobin gene of the Indonesian elm *Parasponia andersonii* to the gus, the pat and the cat genes and cloned these into the binary vector pLX222 (Landsmann et al. 1988), containing the nos-nptII gene. Plants were transformed via *Agrobacterium tumefaciens* LBA4404 and selected on kanamycin. GUS (β-D-glucuronidase), PAT (phosphinothricin acetyltransferase), CAT (chloramphenicol acetyltransferase) and NPTII (neomycin-phosphotransferase II) were biochemically quantified in a large number of transgenic tobacco and petunia plants. Sampling was standardized as much as possible.

The use of the root specific par promoter (Landsmann et al. 1986) made it possible to study not only the level of gene expression but also the leakiness of organ specificity.

Variability of Expression Between Individual Transformants

GUS expression of 140 in vitro tobacco plants from a single transformation with the gus gene was assayed at two to four different time points over a period of 18

E.R. Schmidt/T. Hankeln (Eds.)
Transgenic Organisms and Biosafety

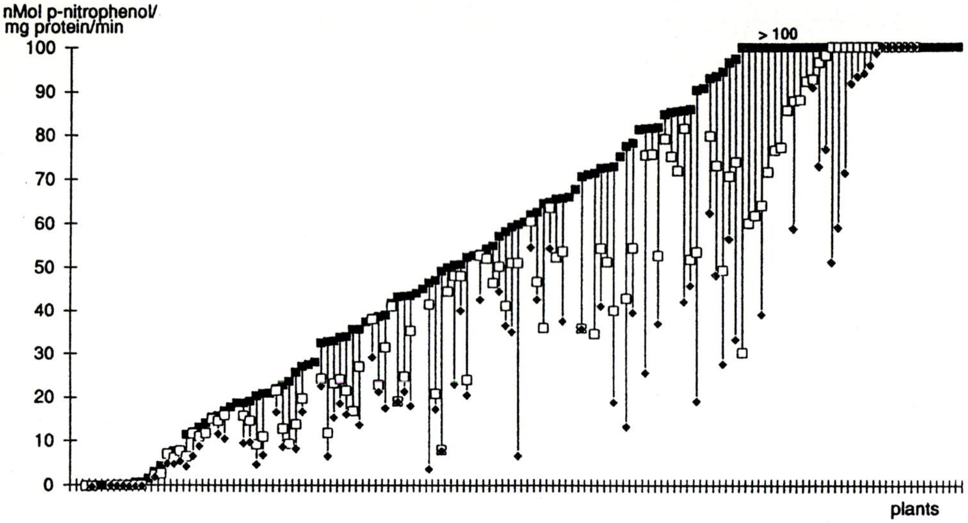

Fig. 1. GUS expression of roots from individual PARGUSH transformants at three different time points (different symbols, stacked); ordered according to highest (*black squares*) root GUS results (in case of nearly identical values, symbols may be masked by other symbols)

months (van der Hoeven et al. 1994a). A total of 128 plants showed GUS expression in the roots varying from 2 to over 100 nmol *p*-nitrophenol/mg protein/min. Levels above 100 nmol *p*-nitrophenol/mg protein/min appeared in 35 cases at least at one assay date. There was a continuous spectrum from weakly to highly expressing plants (Fig. 1). Ten plants did not express the gus gene at all. 46 of the plants expressing the gus gene in the roots also showed GUS expression in the leaves in the range of 2 to 98 nMol p-nitrophenol/mg protein/min. Generally, the level of gene expression in leaves was much lower than in roots.

Leaves and roots of 30 in vitro tobacco plants transformed with the pat gene (Eckes et al. 1989) were assayed two to three times (at 4- to 8-week intervals). All transformants showed the expected root expression (ranging from 3 to 304 nmol acetyl-ppt/mg protein/h). A total of 21 of these plants at times also expressed the enzyme in leaves (varying from 2 to 135 nmol acetyl-ppt/mg protein/h).

Seventy-three tobacco and 13 petunia plants transformed with the cat gene expressed the reporter gene in the roots. In only three of the tobacco plants, but in nine of the petunia plants high CAT expression was detected in leaves (Dietz et al. 1994).

Plants with high expression from the par driven reporter genes in leaves always showed high expression in roots, but high root expression was no indication of coordinated high leaf expression at the time of sampling.

Co-transferred together with the gus or pat or cat genes into the plants was the nptII gene driven by the nos promoter. Ninety in vitro plants were analysed for NPTII activity two or three times. Nearly all plants showed leaf as well as root

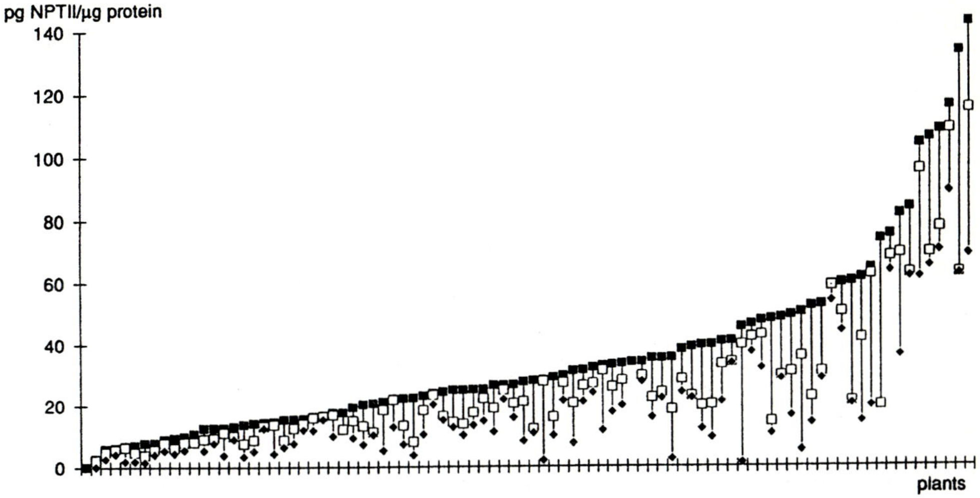

Fig. 2. NPTII expression of roots from individual PARGUSH transformants at three different time points (different symbols, stacked); ordered according to highest (*black squares*) root NPTII results (in case of nearly identical values, symbols may be masked by other symbols)

expression at all time points, ranging from 2 to 142 pg NPTII/μg protein (Fig. 2). There were more plants with high root expression than with high leaf expression. In general the plants expressed the nptII gene stronger in the roots but in 18% of the assays the relationship was reversed.

Variability of Expression Between Different Leaves Within a Plant

Samples were standardized by taking the central section of comparable leaves (sixth from the top) for the inter-individual comparisons. Assays of the third to 11th leaf (counted from the apex) generally showed lower expression at the top of the plant, increasing towards the middle and decreasing or increasing further to the bottom of the plant (Fig. 3). Some plants expressed the reporter gene equally in all leaves tested.

A total of 28 secondary leaf regenerants from two tobacco plants expressing GUS in the leaves were tested in roots and leaves. Variations of GUS expression between these regenerants were comparable to the variations observed between different time points within the respective primary transformants.

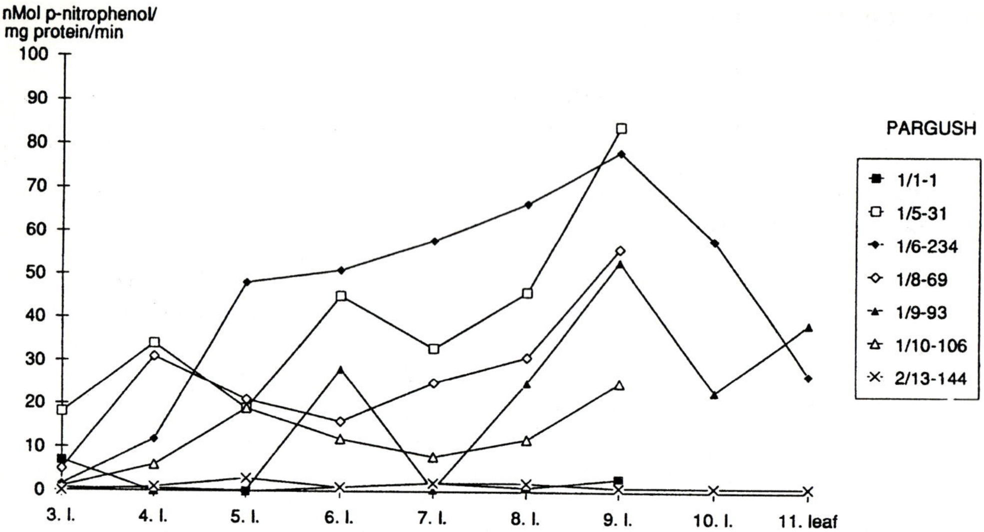

Fig. 3. GUS expression in different leaves of seven individual PARGUSH tobacco transformants; leaves counted from the apex

Variability of Expression Over Time

Repeated assays of the same in vitro plant at different time points after transformation showed fluctuating reporter gene expression from date to date (see Figs. 1,2). This concerned roots as well as leaves. No continuous increase or decrease of gene expression could be seen in the course of plant cultivation.

134 tobacco plants (comprising 23 original transformants spanning the entire range of low to high PAT expression plus 111 secondary regenerants) were transferred from tissue culture to soil in the greenhouse. They were tested for resistance to the non-selective herbicide BASTA. It was assumed that different PAT expression levels (as determined from the in vitro plants) would be manifested in different resistance levels to spray applications of the herbicide. However, there was no difference between transformants neither in appearance or growth nor in resistance to the herbicide applications.

PAT assays were performed with eight of the herbicide treated greenhouse plants before and after the herbicide spraying date. In these assays several leaves from all nodes of the plants and root samples were tested. All roots assayed showed PAT activity comparable to the in vitro data. However, in none of the leaves was PAT activity above detection level, not even in those plants which had shown very high PAT activity in the leaves during their tissue culture phase.

Gene Copy Numbers

Thirty-six tobacco plants transformed with the gus gene were submitted to Southern blot analyses with the 1-kb par promoter fragment as hybridization probe. In 32 of these plants positive hybridization signals indicating stable integration of the foreign DNA in the plant genome could be detected. Eight plants had obtained a single copy, seven plants two copies, eight plants three copies, seven plants four copies, and two plants more than four copies.

Three plants without detectable GUS activity had multiple gene copies. Two of the four plants without clearly detectable Southern hybridization signals with the par probe showed GUS and NPTII expression, possibly indicating loss of restriction sites resulting in insufficient blotting of large DNA fragments.

A positive correlation of a high gene copy number with high NPTII expression in leaves as well as in roots could be verified (Fig. 4). Out of 17 plants with three or more integrated gene copies 12 showed high GUS activity. Four of the eight plants with a single gene copy also showed high GUS activity.

Discussion

Transgenic plants obtained from *Agrobacterium*-mediated transformations showed gross variations in several characters. Individuals from single transfor-

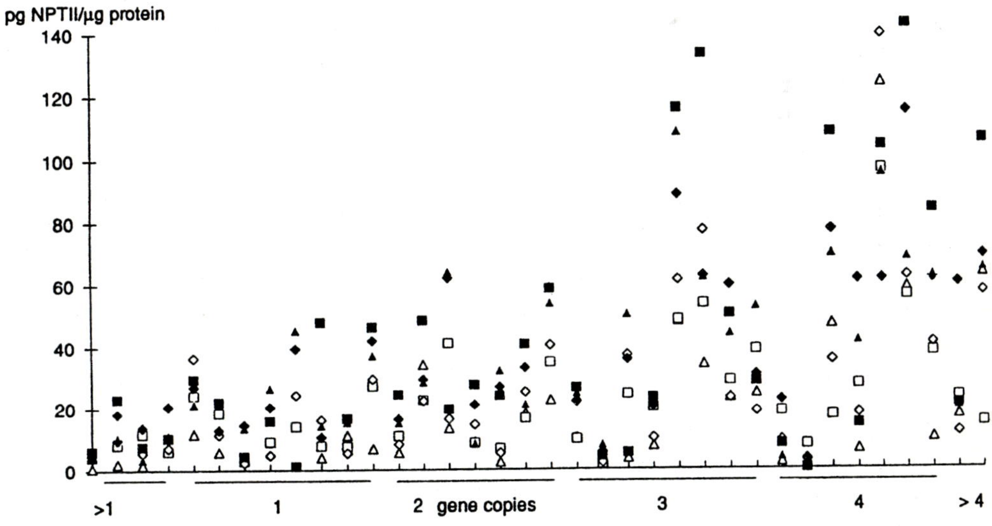

Fig. 4. NPTII expression of roots (*black symbols*) and leaves (*white symbols*) from individual PARGUSH transformants at different time points (differently shaped symbols, stacked); ordered according to highest transgene copy number

mation experiments differed in copy number of transgenes, corresponding expression levels and tightness of organ/tissue specificity. This was shown for different gene constructs with the par promoter and, to a lesser extent, for the nos-nptII gene. However, some general conclusions can be drawn from the variations observed.

Organ specificity was physiologically and/or developmentally modulated in time. The root specificity was not regulated by a switch on or off mechanism. There were more or less continuous gradients between on and off states. Physiologically different plant tissue may result in local or temporal differences in gene activity, e.g. as a consequence of secondary DNA modifications (methylation) (Matzke et al. 1989; Linn et al. 1990; Meyer et al. 1992; Cherdshewasart et al. 1993).

The general influence on gene activity of physiological conditions was also indicated by only weak activity of the endogenous nitrate reductase in young tobacco leaves at the tip of the plants, not unlike the introduced GUS or PAT activity (Fig. 5). Physiological stability of the transcripts and/or the translation products may be major factors determining the effectiveness of the introduced trait and the variability observed. This might explain the herbicide resistance of tobacco plants – transformed with the par-pat gene – in the absence of detectable PAT activity (van der Hoeven et al. 1994b) and the lack of cat mRNA in tobacco leaves with high CAT activity (Landsmann et al. 1988).

Organ and tissue specificity is mainly a function of promoter sequences. Leakiness (i.e. leaf expression) of par regulation was mostly temporal and spacial and can thus be assumed not to be a result of promoter mutations. The leakiness

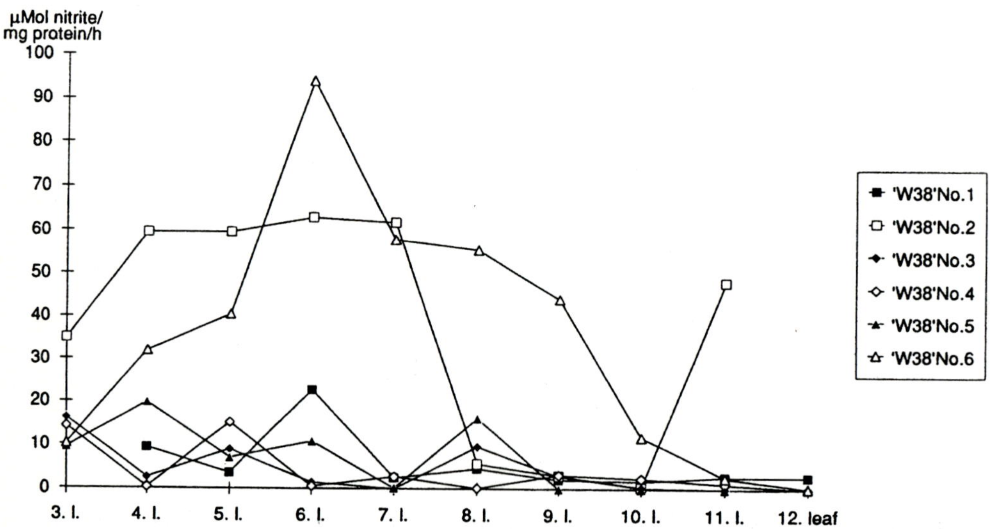

Fig. 5. Expression of endogenous nitrate reductase in different leaves of six untransformed tobacco plants "W38"; leaves counted from the apex

may be a result of an overall increase of expression as indicated by coincidences of high root expression and leaf expression. Activity of the co-transferred genes was not generally synchronized. However, an individual (position dependent) effect on the overall level of expression of the introduced genes was detectable.

Gene copy numbers could not be correlated to a particular expression rate. Multiple gene copies increased chances of high expression, but were not sufficient for a continuously high expression level. The level of gene expression did not significantly decrease or increase over months of in vitro propagation of the transgenic plants, so that gene instability was not considered a major factor with the transgenes. Variations of expression within a plant (in different organs/tissues, at different assay time points) were in the same range as between somatic regenerants from the same transgenic plant. Regenerants from a single transformation experiment may be regarded as largely comparable to somaclonal variants (Landsmann and Uhrig 1985).

Acknowledgements. We would like to thank Katrin Schulze and Birgit Effinghausen-Büssing for excellent technical assistance. This work was supported by BMFT grant BCT-0319–486A.

References

Cherdshewasart W, Gharti-Chhetri GB, Saul MW, Jacobs M, Negrutiu I (1993) Expression instability and genetic disorders in transgenic *Nicotiana plumbaginifolia* L plants. Transgenic Res 2: 307–320

Deroles SC, Gardner RC (1988a) Expression and inheritance of kanamycin resistance in a large number of transgenic petunias generated by *Agrobacterium*-mediated transformation. Plant Mol Biol 11: 355–364

Deroles SC, Gardner RC (1988b) Analysis of the T-DNA structure in a large number of transgenic petunias generated by *Agrobacterium*-mediated transformation. Plant Mol Biol 11: 365–377

Dunwell JM, Paul EM (1990) Impact of genetically modified crops in agriculture. Outlook Agriculture 19(2): 103–109

Dietz A, Kay V, Schlake T, Landsmann J, Bode J (1994) A plant scaffold attached region detected close to a T-DNA integration site is active in mammalian cells. Nucleic Acids Res 22: 2744–2751

Eckes P, Vijtewaal B, Donn G (1989) Synthetic gene confers resistance to the broad spectrum herbicide L-phosphinothricin in plants. J Cell Biochem [Suppl 13D]

Horsch RB, Fry JE, Hoffmann NL, Eichholtz D, Rogers SG, Fraley RT (1985) A simple and general method for transferring genes into plants. Science 227: 1229–1231

Landsmann J, Dennis ES, Higgins TJV, Appleby CA, Kortt AA, Peacock WJ (1986) Common evolutionary origin of legume and non-legume plant haemoglobins. Nature 324: 166–168

Landsmann J, Llewellyn D, Dennis ES, Peacock WJ (1988) Organ regulated expression of the *Parasponia andersonii* haemoglobin gene in transgenic tobacco plants. Mol Gen Genet 214: 68–73

Landsmann J, Uhrig H (1985) Somaclonal variation in *Solanum tuberosum* detected at the molecular level. Theoret Appl Gen 71: 500–505

Linn F, Heidmann I, Saedler H, Meyer P (1990) Epigenetic changes in the expression of the maize A1 gene in *Petunia hybrida*: Role of numbers of integrated gene copies and state of methylation. Mol Gen Genet 222: 329–336

Matzke MA, Primig M, Trnovsky J, Matzke AJM (1989) Reversible methylation and inactivation of marker genes in sequentially transformed tobacco plants. EMBO J 8(3): 643–649

Meyer P, Linn F, Heidmann I, Meyer ZAH, Niedenhof I, Saedler H (1992) Endogenous and environmental factors influence 35S promoter methylation of a maize A1 gene construct in transgenic petunia and its colour phenotype. Mol Gen Genet 231: 345–352

van der Hoeven C, Dietz A, Landsmann J (1994a) Variability of organ specific gene expression in transgenic tobacco plants. Transgenic Res 3: 159–165

van der Hoeven C, Dietz A, Landsmann J (1994b) Expression of phosphinothricin acetyltransferase from the root specific par promoter in transgenic tobacco plants is sufficient for herbicide tolerance. Plant Cell Reports 14: 165–170

22 Heat-Induced Loss of Transgene Activity Detected and Analysed in Several Transgenic *Nicotiana tabacum* Lines

K. Neumann[1,2], W. Dröge-Laser[1,3], S. Köhne[1], and I. Broer[1]

[1]Universität Bielefeld, Fakultät für Biologie, Lehrstuhl für Genetik, Universitätsstrasse, Postfach 100131, 33501 Bielefeld, Germany
[2]*Current address:* Universität Rostock, Fachbereich Biologie, Doberanerstrasse 143, 18051 Rostock, Germany
[3]*Current address:* The Salk Institute for Biological Studies, Plant Biology Laboratory, 10010 North Torrey Pines Road, La Jolla, CA 92037, USA

Abstract

Heat treatment (37 °C) of various transgenic tobacco (*Nicotiana tabacum*) plants obtained from three different transformation experiments resulted in a temporary loss of transgene activity. During the heat treatment, almost no transgene-encoded protein is detectable. In this work, the influence of methylation on the observed reaction has been analysed applying two different approaches: Firstly, in vivo incubation with 5-azacytidine was carried out by various methods to induce a widespread demethylation in the plant genome. Secondly, Southern analysis with methylation-sensitive restriction endonucleases was performed using DNA isolated from heat-treated and non-heat-treated plants, respectively. With this method, the methylation status of single C residues could be examined. However, no evidence could be obtained that DNA methylation is involved in the heat-induced strong reduction of the transgene-encoded phenotype.

Introduction

The efficient long-term use of transgenic plants either in research or in agriculture demands a continued stable expression of the recombinant genes the plants were transformed with. Many of the transgenic plants analysed up to now inherited and expressed the integrated constructs according to the Mendelian laws (e.g. Otten et al. 1981). Still, more and more cases are reported in which transgenes physically present in the plant genome are inactivated (for a review, see Jorgensen 1990, 1991). There are probably different reasons for the various

E.R. Schmidt/T. Hankeln (Eds.)
Transgenic Organisms and Biosafety

inactivation events reported: One mechanism often described seems to involve inactivation at the transcriptional level frequently accompanied by changes in the methylation patterns of transgenes (Matzke et al. 1989; Matzke and Matzke 1993; Hobbs et al. 1993; Meyer et al. 1993). Usually, hypermethylation of DNA is linked to the loss of gene activity (e.g. John and Amasino 1989; Weber et al. 1990; Renckens et al. 1992). Furthermore, the chromosomal region into which the transgene integrates seems to be of some importance for its activity. A transgene integrated into a hypermethylated chromosomal environment became hypermethylated. The same construct remained hypomethylated and transcriptionally active after integration into an unmethylated region (Pröls and Meyer 1992). On the other hand, transgene-specific de novo methylation of this construct could also be found sometimes in spite of integration into a hypomethylated region (Meyer and Heidmann 1994).

Transcriptional inactivation accompanied by DNA methylation is not only found in transgenes, but also in endogenous genes (Watson et al. 1987a; Bianchi and Viotti 1988; Ngernprasitsiri et al. 1989) and transposons (Martin et al. 1989; Dennis and Brettell 1990; Peschke et al. 1991). The multitude of cases in which gene inactivation was found together with methylation indicates the regulatory importance of this mechanism, although it still remains unclear whether methylation is cause or consequence of the observed inactivation event.

Another inactivation mechanism often found is the homology-dependent gene silencing termed as "co-suppression" (Jorgensen 1990). Here, the introduction of multiple copies of a transgene or of a transgene with homology to an endogenous gene into the plant genome can result in silencing of some or all of the homologous genes (reviewed, e.g., by Jorgensen 1991; Matzke and Matzke 1993). The suppression of gene activity frequently happens only in a part of the transformed plants. For the progeny of each transformant, the ratio of silenced to active plants remains constant over subsequent generations (Dorlhac de Borne et al. 1994). The molecular mechanisms of this homology-triggered inactivation seem to differ from case to case: Sometimes a transcriptional inactivation often accompanied by methylation appears to be involved (e.g. Matzke et al. 1989; Matzke and Matzke 1993; Hobbs et al. 1993; Meyer et al 1993). In contrast, other cases seem to be post-transcriptional; no DNA modifications like methylation could be found to date (de Carvalho et al. 1992; Kooter and Mol 1993). In general, the results indicate that distinct chromosomal regions differ in their susceptibility for homology-triggered inactivation; some are inherently unstable. This attribute seems to be amplified when the transgene integrated into such an unstable locus is homozygous and/or when an unlinked homologous gene is present (Neuhuber et al. 1994; Dorlhac de Borne et al. 1994).

Furthermore, environmental conditions like growth in the greenhouse or in closed containers (Hart et al. 1992) and developmental state of the plant material have been shown to influence the stability of transgene expression (Hart et al. 1992; Meyer et al. 1992). Therefore, we decided to study the stability of transgene expression under conditions of environmental stress. Recently, we described a heat-induced inactivation of a transgene (*pat* gene) (Walter et al. 1992) conferring resistance to the herbicide phosphinothricin (Pt) (Bayer et al. 1972; Lea et al. 1984) in transgenic suspension cultures of *Medicago sativa*. The *pat* gene, encoding a phosphinothricin-*N*-acetyl transferase (PAT) which inactivates Pt by acetylation, was isolated from *Streptomyces viridochromogenes* (Strauch et al. 1988) and transferred to various plant species (e.g. Wohlleben et al. 1988; Broer et al. 1989; Dröge et al. 1992). To rule out the possibility that the artificial suspension culture system used for *M. sativa* might have a negative influence on transgene expression, we studied the stability of the *pat* gene in transgenic *Nicotiana tabacum* plants. As described in Neumann et al. (in preparation), a 37 °C, heat-treatment of tobacco plants carrying the *pat* gene resulted in a loss of the transgene-encoded phenotype. Plants treated this way show a strong reduction in PAT activity detected by using a PAT assay (Dröge et al. 1992, Dröge-Laser et al. 1994) or western blot analysis (Vinnemeier et al., 1995), respectively. The reaction could be observed in two different transformants analysed so far. The heat-induced inactivation occurred in plants of the S1 and the S2 generation. After shifting the culture temperature back to normal (24 °C), reactivation of the transgene encoded phenotype occurred.

Moreover, to exclude the possibility that the *pat* coding region itself was causing the observed reaction, other sequences like the luciferase coding region (*luc*) obtained from *Photinus pyralis* (Ow et al. 1986), and the coding region of the kanamycin resistance gene (*npt*II) from Tn*5* (Herrera-Estrella et al. 1983) were tested for their ability to be inactivated in plants by a heat treatment.

Here we describe the influence of methylation on the observed heat-induced reaction of transgene activity in several transgenic *Nicotiana tabacum* lines carrying the *pat* and the *npt*II gene or the *luc* and the *npt*II gene, respectively.

Results and Discussion

Construction and Characterization of Transgenic Tobacco Plants

In order to analyse the stability of transgene expression in *Nicotiana tabacum* plants at 37 °C, various promoter-reporter-gene constructs were tested in transgenic tobacco obtained by *Agrobacterium tumefaciens*-mediated leaf-disc transformation. The promoter of the *A. tumefaciens* nopaline synthase gene (NOS_{pro}) (Bevan 1984) fused to the neomycin-phosphotransferase (*npt*II) coding region was cotransformed with a construct consisting of the 35S cauliflower mosaic virus promoter ($35S_{pro}$) (Baulcombe et al. 1986) combined with the phosphinothricin-*N*-acetyl transferase (*pat*) coding region (Wohlleben et al. 1988). The resulting plants were termed as C plants. The NOS_{pro}-*npt*II construct transformed together with a *Photinus pyralis* luciferase (*luc*) coding region (Ow et al. 1986) fused to a promoter derived from the small subunit-ribulose bisphosphate carboxylase gene (SSU_{pro}) resulted in the B plants, or, when fused to the $35S_{pro,}$ in the A plants, respectively (Quandt et al. 1992).

Various independent transformants of the A, B, or C plants were tested for their copy number and the homo-/hemizygous state of the transgenes in the S1 generation mostly used for the experiments. All the different plants used here show a 37 °C-induced diminution of the transgene-encoded activity. The NPTII or LUC activities are nearly lost in all transgenic plants during a heat treatment with only slight activities remaining sometimes, while the Pt-resistance phenotype has a stronger rest activity: The amount of the herbicide Pt necessary to induce sensitivity at 37 °C is still higher in any transgenic compared to wildtype plants, but at 24 °C transgenic plants survive this treatment without any problems. In all plants analysed so far the observed strong reduction of the transgene-encoded phenotype was reversible: Renewed cultivation at 24 °C after applying a heat treatment resulted in a restoration of transgene activity (Neumann et al. in preparation).

One mechanism often discussed in connection with a loss of gene activity is the hypermethylation of the silenced DNA and sometimes of the surrounding chromosomal region as well. Therefore, we decided to test the influence of methylation on the heat-induced diminution of the transgene-encoded phenotype.

Influence of Methylation on the Reaction to Heat Treatment

Two different approaches were used to analyse the involvement of DNA methylation: In a first approach, the plants were grown in the presence of the cytidine analogon 5-azacytidine (5-aza). 5-aza is incorporated into the DNA instead of cytidine, but, in contrast to the latter, cannot be methylated. Interestingly, 5-aza induces demethylation of nearby C residues (Watson et al. 1987b). Therefore, only relatively few C residues have to be exchanged with 5-aza to create a widespread demethylation.

If a heat-induced hypermethylation of the transgene and/or its integration site causes the observed reaction, incubation with 5-aza should reduce the effect and a continued stable expression in spite of a 37 °C treatment should result. The experiments were conducted using various A or B transformants, since their LUC activity can be detected easily and fast. Three experimental set-ups were applied for in vivo incorporation of 5-aza:

1. 100 µl of a 25 m*M* 5-aza solution were applied on approximately 8 cm^2 leaf segment of an aseptically grown plant. Since 5-aza is not light stable, the plants were kept in the dark for one day to ensure its intact incorporation. After one, two or three days of further growth in the light to allow demethylation to spread throughout the genome, the plants were heat-treated and afterwards tested for their LUC activity.
2. The plants were grown on MS medium supplemented with 1–20 µ*M* 5-aza. In order to protect 5-aza from light, carbon (15 g/l) was added to the medium. After 14 days, the plants were heat-treated and tested for their LUC activity.

3. The third experiment was carried out using seeds germinating on 5-aza according to Burn et al. (1993), who have been successful in activating a gene previously silenced by methylation: The freshly sterilized seeds were put on filters soaked with a 50 μM 5-aza solution and kept in the dark. Each day, filter and 5-aza solution were changed to provide the seeds with sufficient active 5-aza. After germination, the seeds were put on MS medium, heat-treated and tested for their LUC and NPTII activity.

The results obtained with these three experimental set-ups were always the same: The heat-induced loss of the transgene-encoded phenotype occurred just as observed in plants cultivated at 37 °C without 5-aza. Hence either there is no heat-induced, inactivating hypermethylation of the transgenes or the region in which they are inserted, or this method is not sensitive enough to detect the methylation of single important methylated C residues.

Therefore, we decided to try another approach to analyse the possible influence of methylation patterns. Various methylation-sensitive restriction endonucleases were used to cut DNA isolated from different heat-treated and control transgenic A, B and C plant lines. By this method, the state of methylation of five *Msp*I/*Hpa*II restriction sites (CCGG) and a *Pst*I site (CTGCAG) located in the *npt*II coding region was analysed. Furthermore, a *Msp*I/*Hpa*II site located in the $35S_{pro}$, which was found to be methylated in the silenced A1 gene in transgenic *Petunia hybrida* (Meyer et al. 1992), was examined.

In no case a heat-correlated change of the methylation patterns could be detected. Analysis of a homo- and a hemizygous C transformant revealed differences in their methylation pattern, but they were identical in their reaction to a 37 °C treatment. Nevertheless, although this indicates that the sensitivity of the method used here was sufficient to detect differences in DNA methylation, no connection could be found between state of methylation and gene silencing. Neither a general hypermethylation (as analysed using 5-aza) nor the methylation of single C residues as shown in other cases seem to be involved in the loss of transgene activity observed here. Furthermore, the quick reactivation of the transgene-encoded activity occurring after the end of a heat treatment would have been rather unusual for genes inactivated by methylation, therefore we conclude that methylation is not likely to be involved in the silencing mechanism analysed here.

Since there is no hypermethylation detectable, the heat-induced diminution of the transgene-encoded phenotype might be due to another mechanism. Co-suppression seems unlikely, since some of the plants used here carried only a single copy of the transgene. It is now important to investigate the actual inactivation mechanism which should lead to further interesting conclusions on the plant's gene regulation concerning fast reactions to environmental changes. Although it has not been tested up to now, it cannot be excluded that a similar inactivation might occur when transferring genetically engineered plants to the field. The identification of plant lines stable at 37 °C might be an important step in the design of plants optimized for the use in agriculture.

Acknowledgment. We wish to thank Prof. Dr. A. Pühler for the opportunity to work in Bielefeld and for his helpful comments and interest in the subject.

References

Baulcombe DC, Saunders GR, Bevan MW, Mayo AM, Harrison BD (1986) Expression of biological active viral satellite RNA from nuclear genome of transformed plants. Nature 321: 446–229

Bayer E, Gugel KH, Hagele K, Hagenmaier H, Jessipow S, König WA, Zäher A (1972) Stoffwechselprodukte von Mikroorganismen. Phosphinothricin und Phosphinothricyl-Alanyl-Alanin. Helv Chim Acta 55: 224–239

Bevan M (1984) Binary *Agrobacterium* vectors for plant transformation. Nucleic Acid Res 12: 8711–8721

Bianchi MW, Viotti A (1988) DNA methylation and tissue specific transcription of the storage protein genes in maize. Plant Mol Biol 11: 203–214

Broer I, Arnold W, Wohlleben W, Pühler A (1989) The phosphinothricine-*N*-acetyltransferase gene as a selection marker for plant genetic engeneering. In: Galling G (ed) Proceedings of the Braunschweig Symposium on Applied Plant Molecular Biology, Technische Universität Braunschweig, Germany, pp 240–246

Burn JE, Bagnall DJ, Metzger JD, Dennis ES, Peacock WJ (1993) DNA methylation, vernalization, and the initiation of flowering. Proc Natl Acad Sci USA 90: 287–291

De Carvalho F, Gheysen G, Kushnir S, Van Montagu M, Inzé D, Castresana C (1992) Suppression of -1,3-glucanase transgene expression in homozygous plants. EMBO J 11: 2595–2602

Dennis ES, Brettell RIS (1990) DNA methylation of maize transposable elements is correlated with activity. Phil Trans R Soc Lond B326: 217–229

Dorlhac de Borne F, Vincentz M, Chupeau Y, Vaucheret H (1994) Co-suppression of nitrate reductase host genes and transgenes in transgenic tobacco plants. Mol Gen Genet 243: 613–621

Dröge W, Broer I, Pühler A (1992) Transgenic plants containing the Phosphinothricin-*N*-acetyltransferase gene metabolize the herbicide L-phosphinothricin (glufosinate) differently from untransformed plants. Planta 187: 142–151

Dröge-Laser W, Siemling U, Pühler A, Broer I (1994) The metabolites of the herbicide L-phosphinothricin (glufosinate). Plant Physiol. 105: 159–166

Hobbs SLA, Warkentin TD, DeLong CMO (1993) Transgene copy number can be negatively associated with transgene expression. Plant Mol Biol 21: 17–26

John MC, Amasino RM (1989) Extensive changes in DNA methylation patterns accompany activation of a silent T-DNA *ipt* gene in *Agrobacterium tumefaciens*-transformed plant cells. Mol Cell Biol 9: 4298–4303

Jorgensen R (1990) Altered gene expression in plants due to *trans* interactions between homologous genes. Trends Biotechnol 8: 340–344

Jorgensen R (1991) Silencing of plant genes by homologous transgenes. Ag Biotech News and Information 4,9: 256N–273N

Kooter JN, Mol JNM (1993) *Trans*-inactivation of gene expression in plants. Curr Opin Biotechnol 4: 166–171

Lea PJ, Joy KW, Ramos JL, Guerrero MG (1984) The action of the 2-amino-4-(methylphosphinyl)-butanoic acid (phosphinothricin) and its 2-oxo-derivate on the metabolism of cyanobacteria and higher plants. Phytochemistry 23: 1–6

Martin C, Prescott A, Lister C, MacKay S (1989) Activity of the transposon *Tam3* in *Anterrhinum* and tobacco: possible role of DNA-methylation. EMBO J 8: 997–1004

Matzke MA, Primig M, Trnovsky J, Matzke AJM (1989) Reversible methylation and inactivation of marker genes in sequentially transformed tobacco plants. EMBO J 8: 643–649

Matzke MA, Matzke JM (1993) Genomic inprinting in plants: parental effects and *trans*-inactivation phenomena. Annu Rev Plant Mol Biol 44: 53–76

Meyer P, Linn F, Heidmann I, Meyer zu Altenschildesche H, Niedenhof I, Saedler H (1992) Endogenous and environmental factors influence 35S promoter methylation of a maize A1 gene construct in transgenic petunia and its colour phenotype. Mol Gen Genet 231: 345–352

Meyer P, Heidmann I, Niedenhof I (1993) Differences in DNA-methylation are associated with a paramutation phenomenon in transgenic petunia. Plant J 4: 89–100

Meyer P, Heidmann I (1994) Epigenic variants of a transgenic petunia line show hypermethylation in transgene DNA: an indication for specific recognition of foreign DNA in transgenic plants. Mol Gen Genet 243: 390–399

Neuhuber F, Park Y-D, Matzke AJM, Matzke MA (1994) Susceptibility of transgene loci to homology-dependent gene silencing. Mol Gen Genet 244: 230–241

Neumann K, Dröge-Laser W, Vinnemeier J, Broer I. Heat treatment induced instability of transgene encoded herbicide resistance in tobacco (*Nicotiana tabacum*) plants (in preparation)

Ngernprasirtsiri J, Kobayashi H, Akazawa T (1989) Transcriptional regulation and DNA methylation of nuclear genes for photosynthesis in nongreen plant cells. Proc Natl Acad Sci USA 86: 7919–7923

Otten L, de Greve H, Hernalsteens JP, Van Montagu M, Schieder O, Straub J, Schell J (1981) Mendelian transmission of genes introduced into plants by the Ti plasmids of *Agrobacterium tumefaciens*. Mol Gen Genet 183: 209–213

Ow DW, Wood KW, De Luca M, De Wet JR, Helinski DR, Howel SH (1986) Transient and stable expression of the firefly luziferase gene in plant cells and plants. Science 234: 856–859

Peschke VM, Phillips RL, Gengenbach BG (1991) Genetic and molecular analysis of tissue-culture-derived *Ac* elements. Theor Appl Genet 82: 121–129

Pröls F, Meyer P (1992) The methylation patterns of chromosomal integration regions influence gene activity of transferred DNA in *Petunia hybrida*. Plant J 2: 465–475

Quandt HJ, Broer I, Pühler A (1992) Tissue-specific activity and light-dependant regulation of a soybean *rbc*S promoter in transgenic tobacco plants monitored with the firefly luciferase gene. Plant Mol Biol 82: 59–70

Renckens S, De Greve H, Van Montagu M, Hernalsteens JP (1992) *Petunia* plants escape from negative selection against a transgene by silencing the foreign DNA via methylation. Mol Gen Genet 233: 53–64

Strauch E, Wohlleben W, Pühler A (1988) Cloning of a phosphinothricin-*N*-acetyltransferase gene from *Streptomyces viridochromogenes* Tü 494 and its expression in *Streptomyces lividans* and *Escherichia coli*. Gene 63: 65–74

Vinnemeier J, Dröge-Laser W, Pistorius EK, Broer I. Purification and partial characterization of the *Streptomyces viridochromogenes* Tü494 phosphinothricin-*N*-acetyltransferase mediating resistance to the herbicide phosphinothricin in transgenic plants. Naturforsch. 50C: 796–805

Walter C, Broer I, Hillemann D, Pühler A (1992) High frequency heat inactivation of the phosphinothricin resistance gene in transgenic single cell suspension cultures of *Medicago sativa*. Mol Gen Genet 235: 189–196

Watson GC, Kaufman LS, Thompson WF (1987a) Developemental regulation of cytosin methylation in nuclear ribosomal RNA genes of *Pisum sativum*. J Mol Biol 193: 15–26

Watson JD, Hopkins NH, Roberts JW, Argetsinger Steitz J, Weiner AM (1987b) Molecular Biology of the Gene. 4th ed. Benjamin/Cummings, Menlo Park, p 695

Weber H, Ziechmann C, Graessmann A (1990) In vitro DNA methylation inhibits gene expression in transgenic tobacco. EMBO J 9: 4409–4415

Wohlleben W, Arnold W, Broer I, Hillemann D, Strauch E, Pühler A (1988) Nucleotide sequence of the phosphinothricin-*N*-acetyl-transferase gene from *Streptomyces viridochromogenes* Tü 494 and its expression in *Nicotiana tabacum*. Gene 70: 25–37

23 Quantification of Heat Treatment-Induced Transgene Silencing Using Single Cells of *Nicotiana tabacum*

S. Köhne[1], K. Neumann[1,2], W. Dröge-Laser[1,3], and I. Broer[1]

[1]Universität Bielefeld, Fakultät für Biologie, Lehrstuhl für Genetik, Universitätsstrasse, Postfach 100131, 33501 Bielefeld, Germany
[2]*Current address:* Universität Rostock, Fachbereich Biologie, Doberanerstr. 143, 18051 Rostock, Germany
[3]*Current address:* The Salk Institute for Biological Studies, Plant Biology Laboratory, 10010 North Torrey Pines Road, La Jolla, CA 92037, USA

Abstract

The growth of transgenic phosphinothricin-resistant single cell cultures under unusually high temperatures (37 °C) led to the loss of herbicide resistance in more than 90% of the cells (Walter et al. 1992). A strong reduction of herbicide resistance could be observed in transgenic tobacco plants after an incubation for 1 to 10 days at 37 °C (Neumann et al. 1996). In this study we tried to quantify the phenomenon analysing the reduction of the transgene encoded phenotype in single cells of different tobacco lines. The transgenic lines used carried transgenes with different promoters, coding regions and terminator sequences. In the single cell system, a reversible loss of the transgene encoded phenotype was detectable during the heat treatment. After the reduction of the growth temperature to 24 °C, the number of cells which lost the transgene encoded phenotype decreased, reaching the initial level after approximately 6 days. On the single cell level, evidence for DNA methylation associated with the phenomenon could not be found by in vivo incubation of heat-treated plants with 5-azacytidine, a compound preventing DNA methylation. Transient expression of a *gus* A gene transferred to protoplasts of transgenic plants was not disturbed by a heat treatment of cells, while the integrated transgenes were still inactivated in the same experiment.

E.R. Schmidt/T. Hankeln (Eds.)
Transgenic Organisms and Biosafety

Introduction

The ability to introduce foreign genes and their subsequent expression in plants has become an important tool for the use of plants biotechnology in science and industry. In the USA as well as in Europe, a number of field tests using transgenic plants are planned or have already been carried out. An example of a gene used in such experiments, is the gene conferring resistance to the herbicide phosphinothricin (Pt), which is produced as phosphinothricyl-alanyl-alanine (Ptt) by *Streptomyces hygroscopicus* and *Streptomyces viridochromogenes*. Pt is a structural analogue of glutamic acid which competitively inhibits bacterial and plant glutamine synthetases. Resistance genes (*bar, pat*) were isolated from the producer strains and transferred to a variety of plant species. Both genes code for a phosphinothricin-*N*-acetyl-transferase (PAT) which inactivates Pt. The *pat* gene was tranferred to *Nicotiana tabacum* SRI (Wohlleben et al. 1988) to give rise to Pt-resistant plants. Herbicide-resistant plants are proposed for release into the environment and use in agriculture. It is consequentially of importance to ascertain the stability of transgene expression in the transgenic plant in relation to changes in the environmental conditions.

As reviewed by Jorgensen (1992), the silencing of transgenes present in more than one copy in the plant was frequently observed. It has been assumed that the occurrence of transgene silencing was the result of mutation or suppression of the transgene. The copy number or the homo/hemizygous state of the foreign gene can influence the transgene expression (Jorgensen 1990, 1992; Hobbs et al. 1993). The inactivation of transgenes by an additional partially homologous copy of the T-DNA (*trans*-inactivation) has been reported (Matzke et al. 1989; Matzke and Matzke 1990, 1991). The introduction of extra copies of an endogenous gene to boost expression, led to a coordinate silencing not only of the introduced transgenes but also of the endogenous gene. This phenomenon, termed co-suppression, was described for the petunia chalcone synthase (*chs*) gene, encoding one step in floral pigment biosynthesis. The gene was introduced into a deep violet flowering line of *Petunia hybrida*. Surprisingly up to 42% of the transgenic plants produced flowers which were completely white or had variegated flowers with white or pale sectors on the normal colored background (Mol et al. 1990; Napoli et al. 1990; van der Krol et al. 1990). Co-suppression effects could be observed for many different transgenes (van der Krol 1990; Smith et al. 1990; Neuhaus et al. 1991; de Carvalho et al. 1992; Hart et al. 1992). Several experiments investigating the mRNA level indicated that the co-suppression is correlated with a post-transcriptional event (Mol et al. 1991; Kooter and Mol 1993; Favell 1994). Generally, co-suppression does not appear to be asscociated with methylation of the suppressed genes (Mol et al. 1991; Hart et al. 1992; Goring et al. 1991) and is reversible both somatically and upon segregation of the transgene (Neuhaus et al. 1991; Hart et al. 1992; Goring et al. 1991), indicating that the repressed state might be unstable. Nevertheless, in other cases DNA metylation has been found to be involved in transgene inactivation events (Gelvin et al. 1983; Matzke and Matzke 1991; Renckens et al. 1992; Ottaviani et al. 1993). The phenomenon does not occur in all transgenic plants containing extra copies of an endogenous gene (Mol et al. 1990; Napoli et al. 1990; van der Krol 1990; Lagrimini et al. 1990). Other factors, such as the level of transcription of the ectopic copies and/or chromosomal location, seem to be important.

Environmental factors may also play a role in the transgene suppression. One of the best studied cases of transgene inactivation induced by environmental changes is the silencing of the single copy chimeric maize A1 gene in a white flowering mutant of *Petunia hybrida* (Meyer et al. 1987, 1992). The transgene suppression was correlated with methylation of the cauliflower mosaic virus 35 S promoter which was fused to the A1 coding region (Linn et al. 1990; Meyer et al. 1992). Another example is the silencing of the tobacco chitinase gene that was dependent on the early rearing conditions of the plant (Hart et al. 1992).

Most of the changes in the transgene expression in the plant mentioned above, occurred before or during meiosis. Since in the field the meiotic stability of a transgene is of less importance than its somatic stability, we analysed the stability of transgene encoded activities during a heat treatment period using herbicide resistant plants. A reducible reduction of the transgene encoded phenotype was observed in homo- and hemizygous descendants of different primary transformants (Neumann et al. 1996). In order to quantify this effect, we tested single cells of transgenic *Nicotiana tabacum* plants concerning their transgene expression after the heat treatment. We established a method to quantify inactivation events counting the number of cells that lost transgene expression in individual leaves of transgenic plants after changes in incubation temperatures and analysed the influence of different transgenes, promoters, copy number and insertion loci on the amount of inactivated cells.

Results and Discussion

Using a newly established single cell culture system, we demonstrated that even a single copy of a herbicide-resistance gene in cells of a transgenic *Medicago sativa* suspension culture was inactivated, at high frequency, when the growth temperature increased. The growth rate of the culture was not influenced by the heat treatment. This effect occurred during somatic cell division and was quantified under controlled conditions (Walter et al. 1992). Later we found that reduction of transgene encoded activity also occurs in transgenic plants when grown at 37 °C for at least 2 days. The activity was regained after some days of culture at 24 °C (Neumann et al. 1996). In order to determine whether transgenes are silenced in all cells of a heat treated plant or whether there are only certain amounts of cells involved, the expression of transgenes has to be analysed in single plant cells

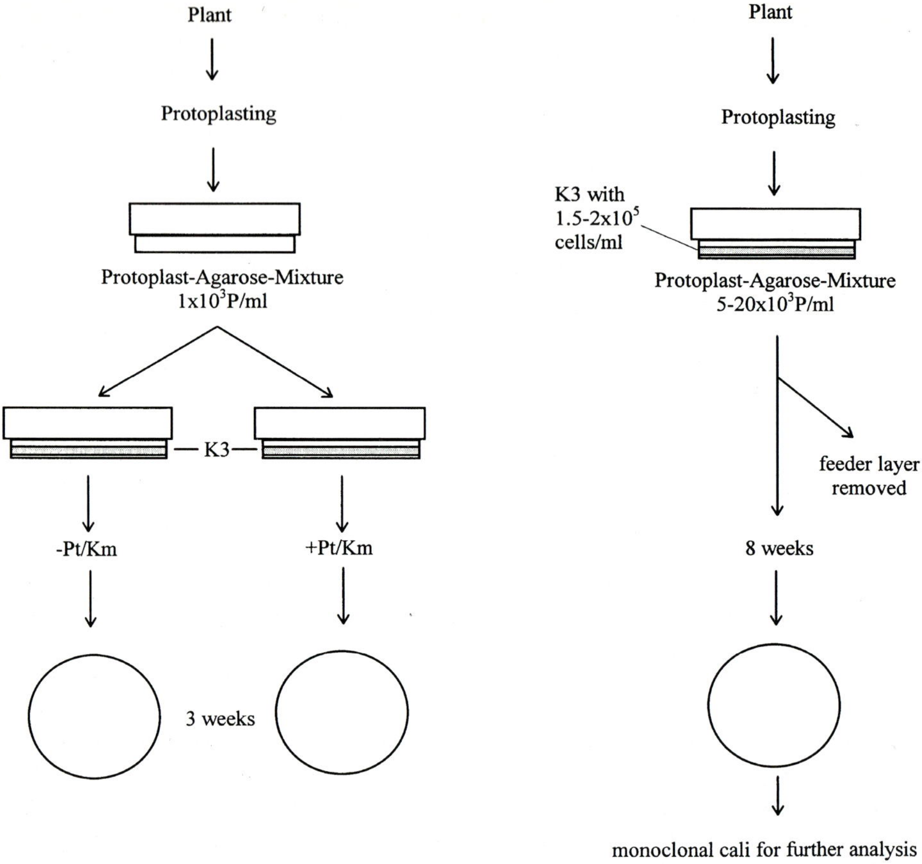

Fig. 1. The two methods used to analyse transgene expression in single cells of *N. tabacum*. *Left*, method 1; *right*, method 2. *Pt*, phosphinothricin. See text for details

which have to be regenerated to monoclonal calli. Different transgenic tobacco plant lines, similar in age, height and physiological state were incubated at normal and heat treatment temperatures for up to 10 days. Leaves obtained from the same positions of the plant were protoplasted at each day of the heat treatment as well as the day before and after. We used two different methods (Fig. 1) to analyse the transgene encoded activity:

- *Method 1* allowed quantitative analysis of the amount of resistant and sensitive cells in a tissue. Protoplasts were embedded in agarose in the presence or absence of selective agents. The number of resistant calli was compared to the total number of regenerating cells in the absence of the herbicide. With this method we received different percentages of sensitive cells in the individual transgenic tobacco plant lines.
- *Method 2* allowed the regeneration of monoclonal calli. Protoplasts were embedded in agarose without selective pressure. To ensure growth at low

densities, the cells were overlayed with a high density protoplast culture and incubated for 8 weeks. Monoclonal calli could then be analysed individually.

Using the methods described above we tried to answer the following questions:

1. *Is the reversible strong reduction of the transgene encoded phenotype also detectable in single cells of heat treated plants?* In order to determine whether the reduction of transgene encoded activity is detectable on single cell level, we incubated different S1 descendants of primary transformants at 37 °C for 1–10 days. Leaves of the plants were digested with cell wall degrading enzymes and the isolated protoplasts were embedded as described in method 1. Due to the experimental set up, first selection could be applied one day after heat-treatment. Depending on the plant line used and the transgene analysed, more than 40% of the transgenic protoplasts lost the herbicide resistance. When the plants returned to normal growth temperatures (24 °C), the number of sensitive cells decreased, reaching the initial amount after up to 6 days. Hence, cells present in the plant leaf show the same reversible diminution of the transgene encoded activity as observed in the plant. Since the number of sensitive cells in the heat treated plant decreases after the reduction of the growth temperatures, the addition of selective pressure one day after the heat treatment may not reflect the actual percentage in the plant during the treatment. This is underlined by the fact, that the decrease in the number of sensitive cells goes along with the increase in the time after the heat treatment (Fig. 2). Nevertheless, although the percentage of sensitive cells might be higher then 40% in the leaf, we can not expect that 100% of the cells show this phenotype. Therefore, the leaf cells do not all display the same level of reduction. It has to be ascertained whether this is based on different changes in the cells genome. In order to analyse the cause of this phenomenon on the molecular level, we had to analyse changes in single plant cells. This can only be achieved when the reduced transgene encoded phenotype is stably maintained in the callus phase where the individual can multiply.
2. *What are the molecular reasons for the heat induced reduction of herbicide resistance?* The instability of gene expression after heat treatment in the transgenic plant may be due to lasting protein or mRNA instabilities or to modifications on DNA level. Mutations in single cells are only detectable when these cells proliferate and produce enough monoclonal tissue to isolate effective amounts of DNA for further analysis. We therefore regenerated monoclonal calli using method 2 without selective pressure, hoping that in contrast to the plant, the changed phenotype is stably maintained in the callus tissue as observed in *Medicago sativa* (Walter et al. 1992). All of the calli tested, displayed a resistant phenotype. No sensitive monoclonal calli could be regenerated. Hence, stable mutations on the DNA level can not be expected. One phenomenon often correlated with reversible gene silencing is the methylation of transgenes. We therefore incubated the transgenic plants on medium containing the cytosin analogon 5-azacytidine. The addition of this compound

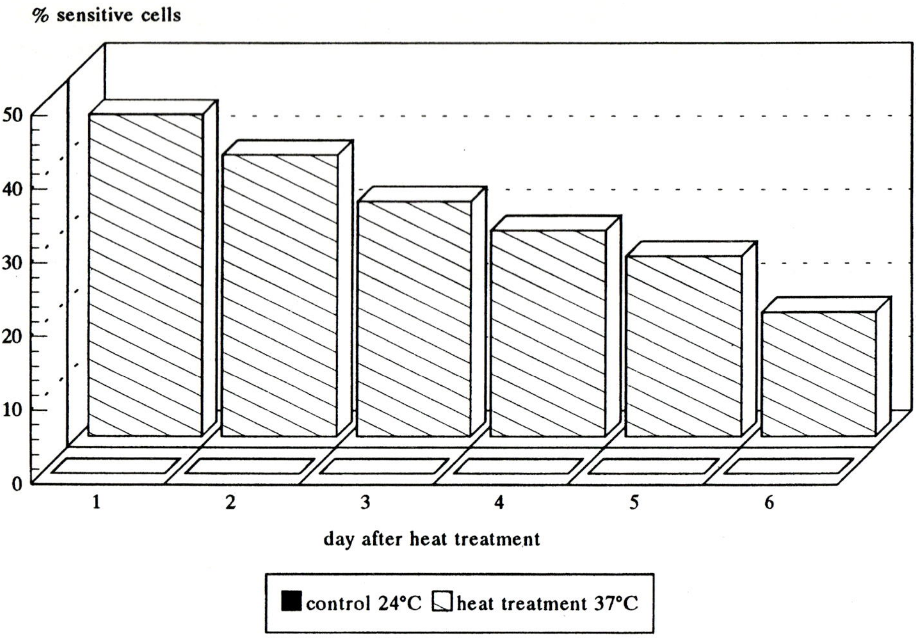

Fig. 2. Analysis of the percentage of sensitive cells after the end of the heat treatment

prevents methylation and should therefore influence the amount of sensitive cells in the plant, if methylation is involved. Neither transgene activity nor the percentage of sensitive cells measured with method 1 were influenced by the presence of 5-azacytidine. Nevertheless, an impact on the regeneration of calli was detectable in the presence and in the absence of antibiotics, proving the disturbance of regulatory processes necessary for callus regeneration by 5-azacytidine. Therefore, although 5-azacytidine seemed to be incorporated, it did not prevent transgene silencing. The analysis of transgene-DNA isolated from heat treated plants with methylation sensitive restriction enzyms additionally proved that the silenced transgenes were not hypermethylated (Neumann et al. 1996). In conclusion, neither a stable mutation nor methylation is involved in the observed phenomenon. Other regulatory influences have to be predicted.

3. *Is the reduction of the transgene encoded activity influenced by a trans-acting factor?* Since the degree of reduction of transgene expression seems to differ in the leaf cells, it may be assumed that transgene expression is regulated by a plant cell-specific regulatory factor. Gene expression can not only be reduced by methylation, but also by *cis-* or *trans*-acting elements present in the plants genome. *Trans*-acting factors interfering with sequences present on the transgene itself should also interact with transgenes not integrated into the plant genome but transiently present in protoplasts. Therefore, transient *gus*A expression was analysed in heat treated protoplasts of a line that efficiently

inactivates transgenes under heat stress conditions. The line that inactivated the integrated transgenes under heat treatment conditions in the protoplast system was not able to inhibit the GUS activity *in trans* (Fig. 3).This result does not support the idea of a *trans*-acting factor that is involved in the regulation of transgene expression on DNA-level. It therefore has to be ascertained whether *cis* -acting elements, present near the insertion locus of the transgene, influence the degree of transgene silencing. In this case, position effects should lead to different levels of sensitive cells in the individual transformants. In all transgenic plants we analysed to date, different levels of sensitive cells could be traced. In addition, a divergent degree of inactivation of different transgenes can be observed in the same plant. Hence, it can not be excluded that *cis*-acting elements influence heat treatment induced transgene silencing.

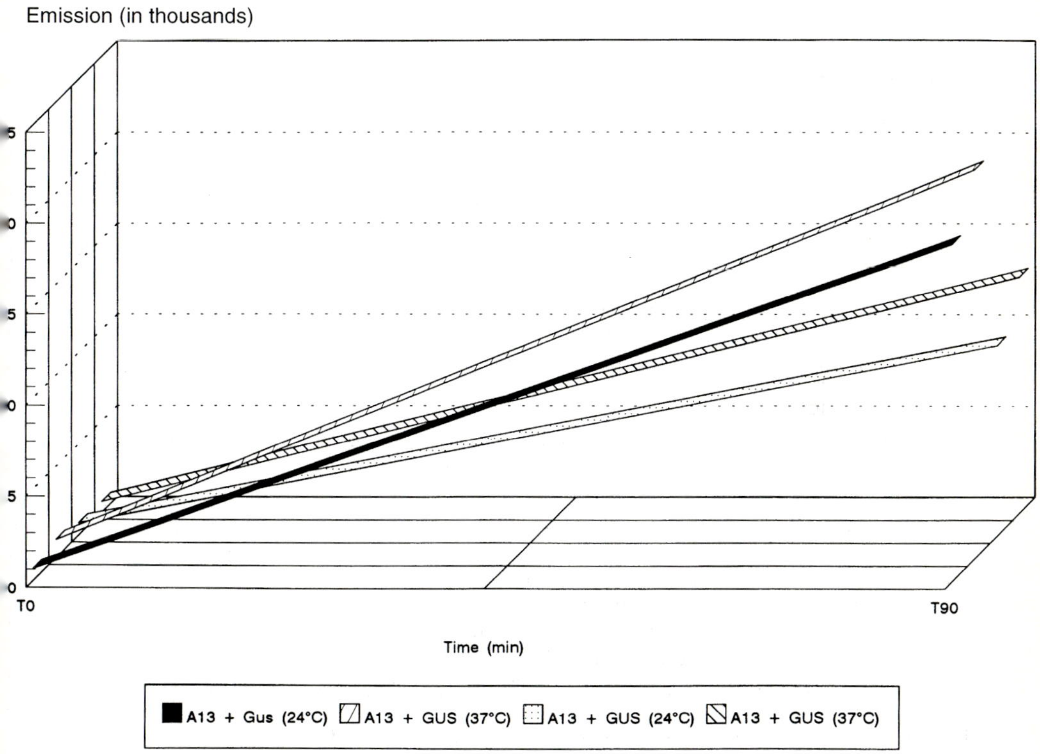

Fig. 3. Measurement of the transient *gus*A expression in protoplasts of a transgenic plant which inactivates transgenes under heat stress conditions

References

De Carvalho F, Gheysen G, Kushnir S, Van Montagu M, Inze D, Castresana C (1992) Suppression of β-1,3-glucanase transgene expression in homozygous plants. EMBO J 11/7: 2595–2602

Favell RB (1994) Inactivation of gene expression in plants as a consequence of specific sequence duplication. Proc Natl Acad Sci USA 91: 3490–3496

Gelvin SB, Karcher SJ, DiRita VJ (1983) Methylation of the T-DNA in *Agrobacterium tumefaciens* and in several crown gall tumors. Nucleic Acids Res 11: 159–174

Goring DR, Thomson L, Rothstein SJ (1991) Transformation of a partial nopaline synthase gene into tobacco suppresses the expression of a resident wild-type gene. Proc Natl Acad Sci USA 88: 1770–1774

Jorgensen R, Synder C, Jones JDG (1987) T-DNA is organized predominantly in inverted repeat structures in plants transformed with *Agrobacterium tumefaciens* C58 derivates. Mol Gen Genet 207: 471–477

Jorgensen R (1990) Altered gene expression in plants due to *trans* interactions between homologous genes. TIBTECH 8: 340–344

Jorgensen R (1992) Silencing of plant genes by homologous transgenes. AgBiotech News Inform 4: 265N–273

Hart CM, Fischer B, Neuhaus J-M, Meins FJ (1992) Regulated inactivation of homologous gene expression in *Nicotiana sylvestris* plants containing a defence-related tobacco chitinase gene. Mol Gen Genet 235: 179–188

Hobbs S, Warkentin TD, DeLong CMO (1993) Transgene copy number can be postively or negatively associated with transgene expression. Plant Mol Biol 21: 17–26

Kooter JM, Mol NM (1993) *Trans*-inactivation of gene expression in plants. Curr Opin Biotechnol 4: 166–171

Lagrimini LM, Bedford S, Rothstein S (1990) Peroxidase-induced wilting in transgenic tobacco plants. Plant Cell 2: 7–18

Linn F, Heidmann I, Saedler H, Meyer P (1990) Epigenetic changes in the expression of the maize A1 gene in *Petunia hybrida*: role of numbers of integrated gene copies and state of methylation. Mol Gen Genet 222: 329–336

Matzke MA, Primig M, Trnovsky J, Matzke AJM (1989) Reversible methylation and inactivation of marker genes in sequentially transformed tobacco plants. EMBO J 8/3: 643–649

Matzke M, Matzke AJM (1990) Gene interactions and epigenic variation in transgenic plants. Dev Genet 11: 214–223

Matzke M, Matzke AJM (1991) Differential inactivation and methylation of a transgene in plants by two suppressor loci containing homologous sequences. Plant Mol Biol 16: 821–830

Meyer P, Heidmann I, Forkmann G, Saedler H (1987) A new petunia flower colour generated by transformation of a mutant with a maize gene. Nature 330: 677–678

Meyer P, Linn F, Heidmann I, Meyer zu Altenschildesche H, Niedenhof I, Saedler H (1992) Endogenous and environmental factors influence 35S promotor methylation of a maize A1 gene construct in transgenic petunia and its colour phenotype. Mol Gen Genet 231: 345–352

Mol JNM, van der Krol AR, van Tunen AJ, van Blokland R, de Lange P, Stuitje AR (1990) Regulation of plant gene expression by antisense RNA. FEBS Lett 268: 427–430

Mol JNM, van Blokland R, Kooter J (1991) More about co-suppression. TIBTECH 9: 182–183

Napoli C, Lemieux C, Jorgensen R (1990) Introduction of a chimeric chalcone synthase gene into petunia results in reversible co-suppression of homologous genes *in trans*. Plant Cell 2: 279–289

Neuhaus J-M, Ahl-Goy P, Hinz U, Flores S, Meins FJ (1991) High level expression of a tobacco chitinase gene in *Nicotiana sylvestris*. Susceptibility of transgenic plants to *Cercospora nicotianae* infection. Plant Mol Biol 16: 141–150

Neumann K, Dröge-Laser W, Löhne S, Broer I (1996) The heat induced loss of transgene activity detected and analysed in several transgenic *Nicotiana tabacum* lines. Proceedings of the Biosafety Workshop, Mainz (1994) Springer, Berlin Heidelberg New York

Ottaviani M-P, Smits T, Hänisch ten Cate CH (1993) Differential methylation and expression of the β-glucuronidase and neomycin phosphotransferase genes in transgenic plants of potato cv. Bintje. Plant Sci 88: 73–81

Renckens S, de Greve H, Van Montagu M, Hernalsteens JP (1992) *Petunia* plants escape from negative selection against a transgene by silencing the foreign DNA via methylation. Mol Gen Genet 233: 53–64

Smith CJS, Watson CF, Bird CR, Ray J, Schuch W, Grierson D (1990) Expression of a truncated tomato polygalacturonase gene inhibits expression of the endogenous gene in transgenic plants. Mol Gen Genet 224: 477–481

Van der Krol AR, Mur LA, Beld M, Mol JNM, Stuitje AR (1990) Flavonoid genes in *Petunia*: addition of a limited number of gene copies may lead to a suppression of gene expression. Plant Cell 2: 291–299

Walter C, Broer I, Hillemann D, Pühler A (1992) High frequency, heat treatment-induced inactivation of the phosphinothricin resistance gene in transgenic single cell suspension cultures of *Medicago sativa*. Mol Gen Genet 235: 189–196

Wohlleben W, Arnold W, Broer I, Hillemann D, Strauch E, Pühler A (1988) Nucleotide sequence of the Phosphinothricin-*N*-acetyltransferase gene from *Streptomyces viridochromogenes* Tü 494 and its expression in *Nicotiana tabacum*. Gene 70: 25–37

24 Genetic Stability and Expression of Foreign Genes in Transgenic Lines of the Legume *Vicia narbonensis*

M. Meixner, S. Brinkmann, O. Schieder, and T. Pickardt

Free University of Berlin, Department of Biology, Institute for Applied Genetics, Albrecht-Thaer-Weg 6, 14195 Berlin, Germany

Abstract

We have transformed the grain legume *Vicia narbonensis* via *Agrobacterium*-mediated gene transfer and regenerated fertile plants carrying the T-DNA of the construct pGV3850Hygro which confers the characters hygromycin resistance and nopalin production. Initial transformants harbouring a single copy integration of the T-DNA were identified and the progeny of three such clones was analysed in detail. The transgene was inherited in the first generation according to Mendelian laws as shown by biochemical and Southern analyses. Through self pollinations homozygous as well as hemizygous sublines were obtained and further characterized. In order to follow the transmission of the two foreign genes for the hygromycin phosphotransferase (hph) and the nopalin synthase (nos) for a long period, biochemical and molecular analyses were performed up to the fifth generation.The genetic stability as well as the stable expression of both characters was demonstrated for homozygous sublines and their progenies. For the hemizygous sublines sometimes unexpected results were obtained, indicating that in some cases genetic instability and/or modification of parts of the integrated foreign DNA might have occurred within the hemizygous offspring.

Introduction

The transformation of plants is a key technique in modern biotechnology and plant molecular genetics. Although these are very recent developments, the production of transgenic plants is a routine procedure for a spectrum of plant species especially including the families of the solanaceae and the crucifers. For an increasing number of agronomically important plant species the introduction of foreign genes is possible today and several lines or cultivars are shortly before

E.R. Schmidt/T. Hankeln (Eds.)
Transgenic Organisms and Biosafety

or ready to use in agriculture. Transgenic (crop) plants will only be of agronomic importance, if their engineered phenotype is faithfully transmitted to the sexual offspring through many generations or is stably passed to vegetative populations for a long time (Finnegan and McElroy 1994). Normally one would expect that transgenes integrated into the plant nuclear genome are stably maintained in the genome and are inherited according to Mendelian laws. This has been reported to a large extent in many publications for plants transformed by means of the *Agrobacterium*-mediated gene transfer (e.g. Otten et al. 1981) as well as by the methods of direct gene transfer or DNA delivery (e.g. Potrykus et al. 1985) for the first few generations. But there is also a large number of publications, describing the unexpected variation of expression levels and/or the instability of foreign genes in transgenic plants and their offspring.

The expression levels of foreign genes in plants are generally unpredictable and vary strongly between single individual transformants. These variations are often described, when analysing large numbers of independent transformants and are usually explained by one or more of the following:

- Different positions of the transgenes within the plant genomes (e.g. Pröls and Meyer 1992)
- Differences in copy numbers and arrangements (e.g. Hobbs et al. 1993)
- Incomplete/unexpected integration patterns of the transgenes caused for instance by integration mechanisms during direct gene transfer methods (e.g. Schmidt-Rogge et al. 1993)
- Early modifications of the introduced DNA (e.g. Mittelsten-Scheid et al. 1991)
- Unexpected modulation of gene expression (e.g. Linn et al. 1990)

However, these particular problems can be overcome by the production of many independent transformants and subsequent screening for the most suitable individuals. More difficult is the situation concerning:

- The variation of alien gene expression sometimes observed within a single generation (e.g. Meyer et al. 1992)
- The loss of characters following selection for linked selectable marker genes (e.g. Ottaviani et al. 1993)
- The unstable transmission of transgenes in the sexual offspring not following the Mendelian laws (e.g. Deroles and Gardner 1988)
- The unexpected behaviour of transgenes in crossing experiments and the variation of gene expression values in the progeny (e.g. Hobbs et al. 1993, Matzke and Matzke 1993)

When investigating this phenomenon in more detail at the molecular level it was found, that sometimes the loss/variation of gene function may be a result of illegitimate recombination events or mutations (especially deletions). But in most cases the loss/variation of the traits was due to suppression of expression of the foreign genes rather than to loss of the transgenes themselves (e.g. Matzke et al. 1989; Mittelsten-Scheid et al. 1991) and has been correlated with hypermethylation of the foreign DNA (epigenetic changes; e.g. Meyer et al. 1992), high number of transgene copies (cis-/trans-inactivation; e.g. van der Krol et al. 1990), interference of different transgenes with a similar construction scheme (trans-inactivation; e.g. Matzke and Matzke 1993) or other modification phenomena. In addition, the expression of foreign genes may be affected by environmental conditions too (Meyer et al. 1992; Walter et al. 1992). This reflects the situation in nature, where for instance transposition frequencies of transposable elements seem to be regulated by their methylation status (Martin et al. 1989) or by environmental conditions (Carpenter et al. 1987) and where endogenous genes undergo changes in their methylation pattern during the transition from an active into an inactive form (Bianchi and Viotti 1988).

Although there has been a lot of work invested in the field of transgene inactivation and gene silencing within the last few years, many unanswered questions remain and the published literature mostly deals with a few solanaceous species or crucifers only (for reviews, see Matzke et al. 1994; Finnegan and McElroy 1994). Finnegan and McElroy (1994) recently estimated that among more than 30 companies involved in the commercialization of transgenic crop plants "almost all ... had observed some level of transgene inactivation. Many of those questioned indicated that most cases of transgene inactivation never reach the literature." Companies give only part of their data to the scientific community because of economic interests or the preservation of patent rights.

These were some of the reasons why we started a detailed progeny analysis with transgenic lines of *Vicia narbonensis*. In contrast to most other grain legumes, *Vicia narbonensis* can be transformed routinely and fertile transgenic plants were regenerated (Pickardt et al. 1991). This legume species has been employed as a model grain legume in our laboratory for several years.

Results and Discussion

We took a randomly selected set of six hygromycin resistant clones that were transformed by *Agrobacterium*-mediated gene transfer using the strain C58C1 with the cointegrative vector pGV3850Hygro (Kreuzaler and Kato, unpublished; see Fig. 1).

These clones were analysed in more detail using the following methods:

- A tissue culture method in order to confirm the hygromycin resistance
- A biochemical assay for nopalin production
- Southern blot analysis in order to prove the integration of the transgene into the plant genome and to determine the copy number as well (for description of methods, see also Pickardt et al. 1991; Meixner et al. 1994)

Southern blot analyses revealed the presence and the integration of the T-DNA into the plant genome in five out of the six clones (Fig. 2, clones 1, 2, 4, 9 and 18). All five clones produced the opin nopalin. Clone no. 5 in Fig. 2, which was resistant to hygromycin but failed to produce nopalin, did not give rise to a

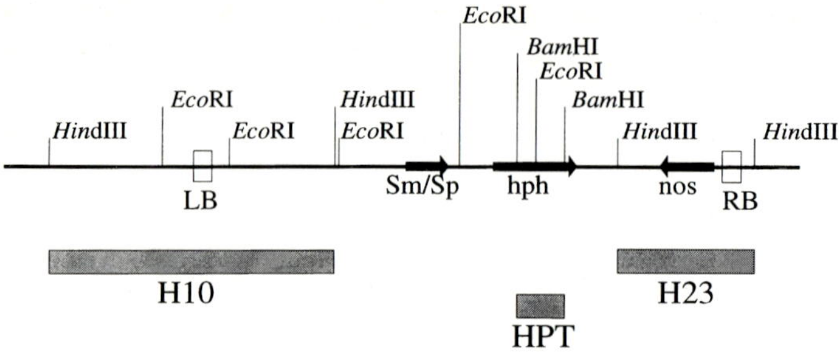

Fig. 1. T-DNA region from the plasmid pGV3850Hygro. *LB*, *RB*, left and right border sequences of the T-DNA; *Sm/Sp*, bacterial streptomycin/spectinomycin resistance; *hph*, plant gene cassette for hygromycin B resistance; *nos*, nopaline synthase gene; *H10*, 6.6-kbp *Hin*dIII fragment, containing the left border and neighbouring sequences; *H23*, 3.2-kbp *Hin*dIII fragment harboring the right border and neighbouring sequences; *HPT*, 1.1-kbp fragment with the amino acid coding sequence of the hph gene cassette

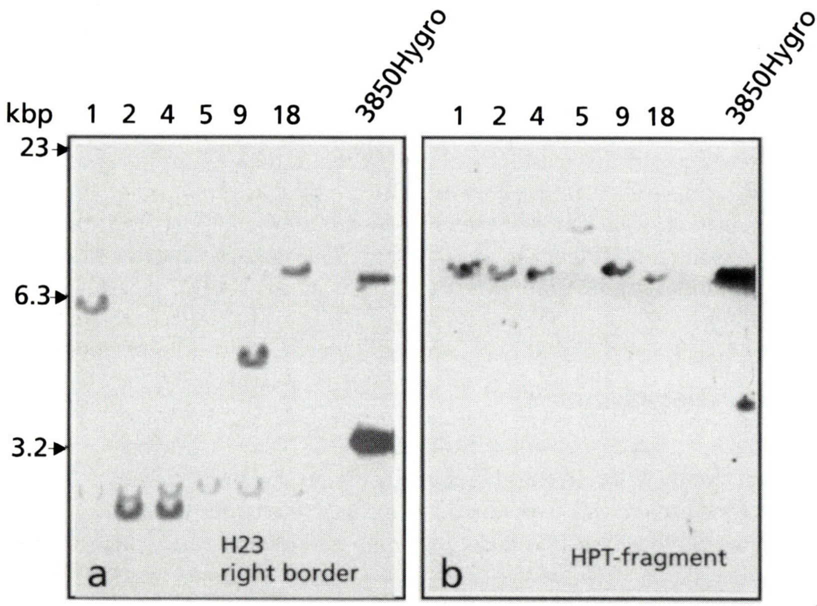

Fig. 2a,b. Southern blot analysis of primary transformants of *Vicia narbonensis*, harboring the T-DNA of the vector pGV3850Hygro. Genomic DNA from the in vitro clones 1, 2, 4, 5, 9, and 18 was digested with *Hin*dIII and probed with **a** the H23 right border fragment and **b** the 1.1-kbp HPT fragment. (3850Hygro, DNA from *A. tumefaciens* C58C1/pGV3850Hygro)

hybridization signal for the right border and neighbouring sequences (Fig. 2a) and showed a week signal for the 7.5-kbp internal *Hin*dIII-fragment of the T-DNA with the hph gene at approximately 10 kbp only (Fig. 2b). Possibly the loss of parts of the transgene was due to incomplete integration or partial deletion of the T-DNA.

The *Hin*dIII-digest of the DNA from all other initial transformants presented in Fig. 2 revealed the integration of a single copy of the transgene into the plant DNA as indicated by the presence of single junction fragments for the right border region (Fig. 2a) as well as for the left border (data not shown). In addition, in all five clones the single copy of the T-DNA seems to contain its complete 7.5-kbp internal *Hin*dIII-fragment (Fig. 2b) including the hph gene cassette (see Fig. 1).

For the progeny analyses of the three transgenic clones 4, 9 and 18, individual in vitro shoots were grown up to mature plants. The flowering and fruiting initial transgenics were named 4.1, 9.1 and 18.1 (R0 in Table 1). From these R0 plants, seven, 28 and 21 seeds respectively were obtained that were capable of germination. As shown in Table 1, the resistance to the antibiotic hygromycin and the nopalin production, both conferred by the integrated T-DNA, were stably transmitted to the three R1 generations. Moreover, in the first generation the inheritance of both characters followed the Mendelian laws in all three lines, as evaluated by methods especially modified for the statistical analysis of small progenies (test with Yates correction, see Strickberger 1985). Because of the small numbers of seeds per single plant that were harvested under greenhouse conditons, the following R2 progenies were also tested. The phenotypical 3:1 distribution was confirmed and the genotype segregated in a 1:2:1 fashion (Figs. 3a,b, 4).

For the initial transgenic lines 4.1 and 9.1, an extended progeny analysis up to the fourth respectively the fifth generation was performed regarding the expression of the nos gene (Fig. 3a,b) and for 18.1 in view to the expression of both the hph as well as the nos gene (Fig. 4). The segregation patterns principally followed the Mendelian laws. In the homozygous offspring (underlined plant numbers in Figs. 3, 4), the characters were stably maintained in all three lines. From approximately 300 homozygous individuals obtained in the offspring of the three clones none failed the expression of the nos gene. When analysing the DNA of randomly selected plants from the progenies of the three clones (Fig. 5), the nos gene (which is part of the H23 right border fragment that was used as a molecular probe in Fig. 5a) was found to be stably transmitted throughout all generations. For all plants investigated specific hybridization signals were ob-

Table 1. Segregation analysis of three transgenic R0 lines after self-pollination, based on hygromycin resistance and nopalin production

R0 line	R1 hyg^r:hyg^s	nop$^+$:nop$^-$	ratio[a]	χ^2 [b]	P[c]
4.1	6:1	6:1	6:1	0.0476	>0.7
9.1	23:5	23:5	4.6:1	0.482	>0.4
18.1	17:4	17:4	4.2:1	0.141	>0.7

[a] 3:1 expected
[b] χ^2 values calculated using the Yates correction term as described by Strickberger (1985)
[c] $P=\chi^2$ probability with 1 d.f.

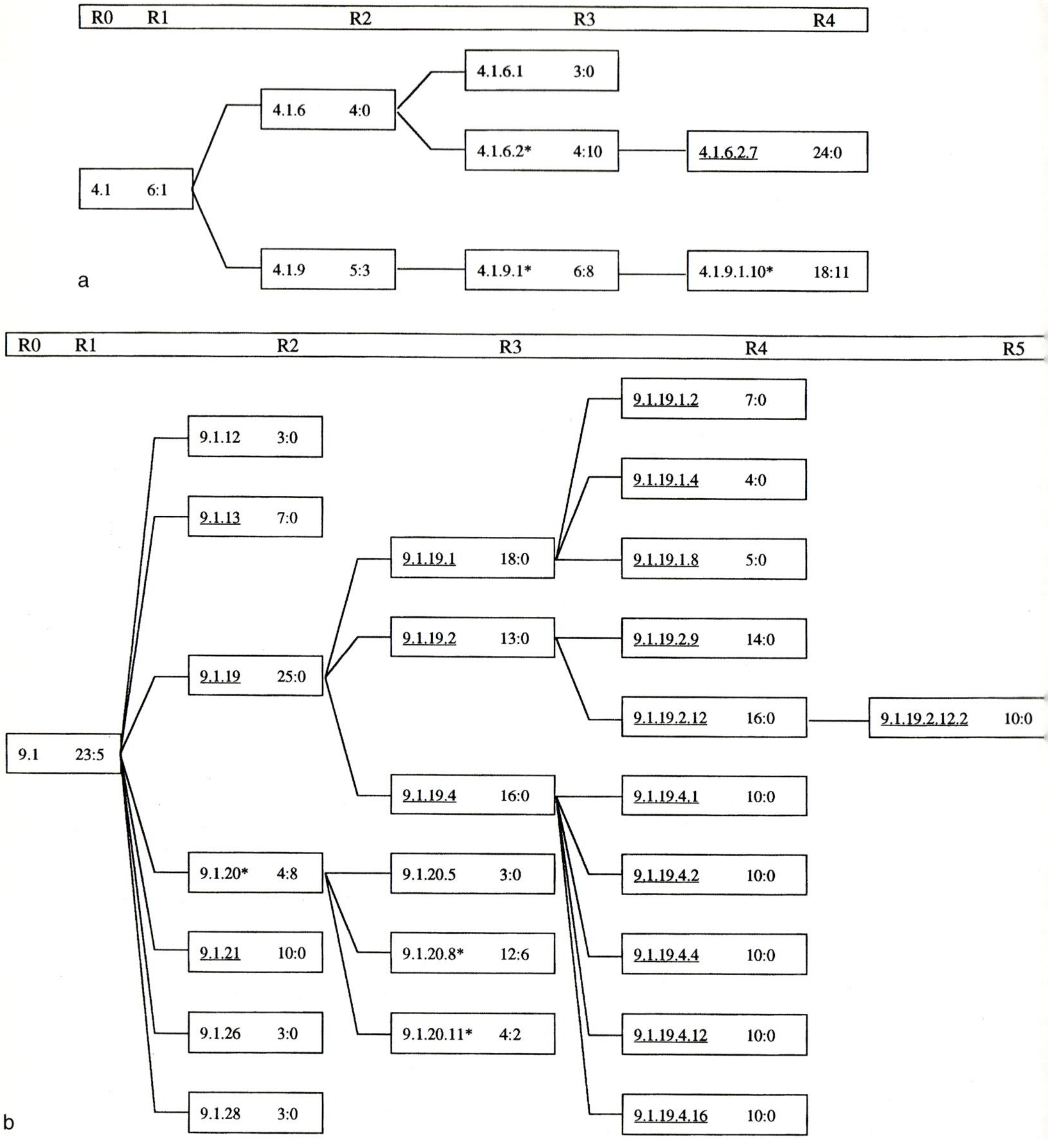

Fig. 3a,b. Segregation analysis in the progeny of the transgenic lines 4.1 (**a**) and 9.1 (**b**) regarding nopalin production. Candidates with a possible deviation from the normal segregation pattern are indicated by an *asterisk*. *Underlined* plant numbers represent homozygous individuals

tained at approximately 2.6 kbp for the R2 to the R4 of clone 4, 5.0 kbp for the R2 to the R4 of clone 9 and 6.5 kbp for the R2 to the R3 of clone 18, respectively. These positions are identical to the fragment size of the appropriate initial transformants (see Fig. 2a). The same DNA blot was reprobed with the 1.1 kbp

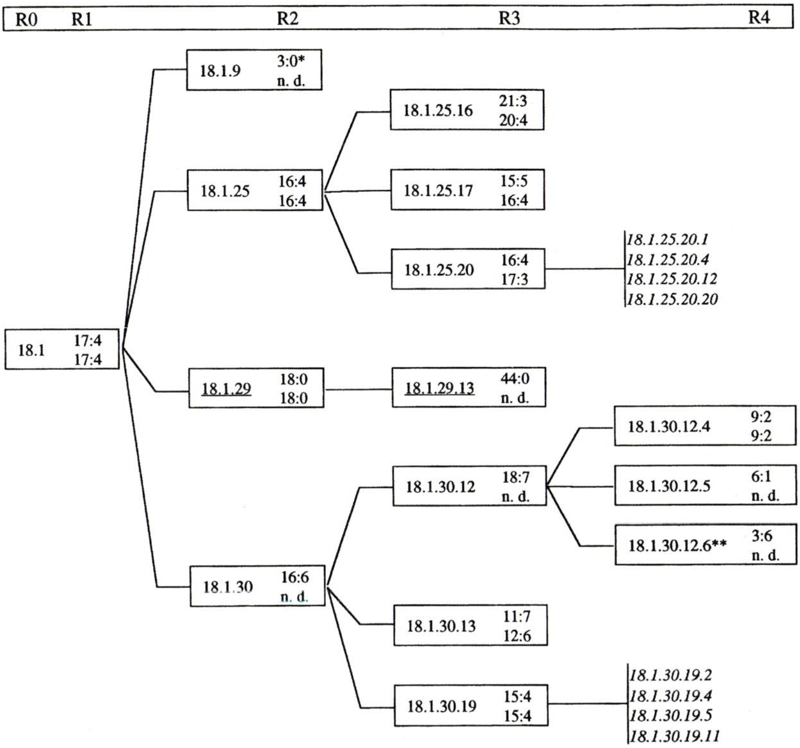

Fig. 4. Cosegregation of the *nos* gene and the *hph* gene in the progeny of the transgenic line 18.1. Numbers at the *top* in each frame represent the segregation ratio for nopalin production, at the *bottom* the ratio for hygromycin resistance. *n.d.*, not determined. *Two asterisks* indicate candidates with a possible deviation from the normal segregation pattern. *Underlined* plant numbers represent homozygous individuals. *Italics* indicate individuals in which DNA was subjected to Southern blot analysis (see Fig. 6)

fragment containing the coding region of the hph gene (Fig. 5b). The presence, the right size and no qualitative changes of the 7.5 kbp long internal *Hin*dIII-fragment (carrying amongst others the hph gene cassette) were shown in the offspring of the three clones. This indicates the stable integration of the T-DNA into the plant genome as well as the conservation of the transgene during at least four generations in the homozygous offspring.

The situation seems to be more complicated when computing the results for the hemizygous plant material (not underlined in Figs. 3,4). Although the low number of seeds that could be obtained per single plant (usually five to 30) was insufficient for a statistically sound evaluation, there are indications for a more frequent loss of gene function in hemizygous plants when compared to homozygous sublines. Candidates that possibly showed a deviation from the expected 3:1 segregation pattern of one of the two traits are marked by asterisks in Fig. 3 and 4. In order to follow up the hypothesis of abnormal loss of gene function in hemizygous plant material, we began to study the cosegregation of the two traits in

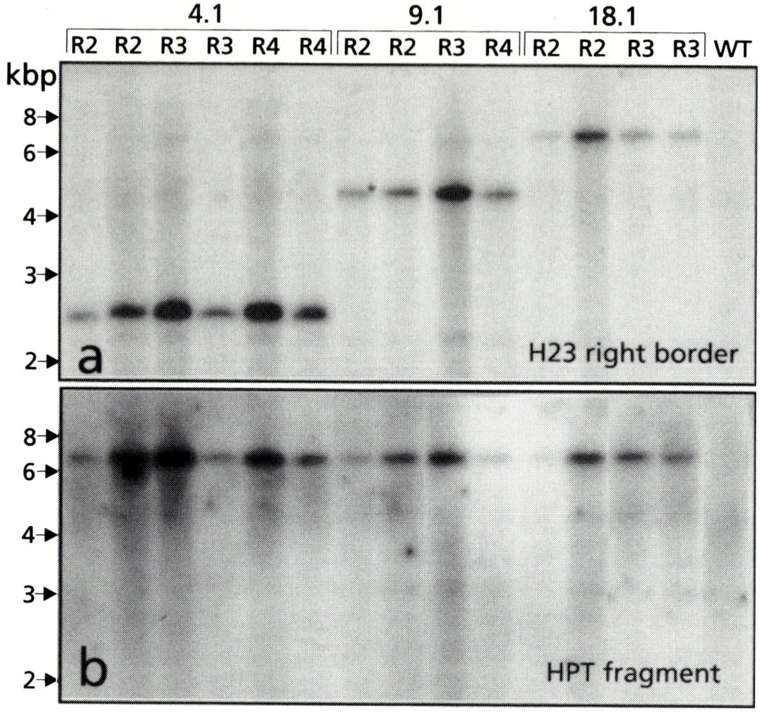

Fig. 5a,b. Southern blot analysis of individual plants from the progeny of the three initial transformants 4.1, 9.1 and 18.1 (generations R2–R4). Genomic DNA was digested with *Hin*dIII, subjected to Southern blotting and probed with the H23 right border fragment (**a**) and the 1.1-kbp HPT fragment (**b**). *WT*, DNA from an untransformed plant

some homozygous and hemizygous pools (Fig. 4, Table 2). As shown in Fig. 4, the cosegregation in the homozygous population seems to be perfect, whereas differences in the cosegregation patterns were observed in the hemizygous plant material. This is illustrated in more detail in Table 2 where the cosegregation of the hygromycin resistance character together with the ability for nopalin production is shown for each single plant in the progeny of three homozygous individuals (sublines originating from plants 9.1.19.4.1, 9.1.19.4.2 and 18.1.29) and two heterozygous plants (sublines from 18.1.25.20 and 18.1.30.13), respectively. These results together with the unexpected segregation patterns for the nos production strengthen the hypothesis, that in contrast to the homozygous offspring in hemizygous populations a more frequent loss of gene function may occur.

From the two hemizygous plants 18.1.25.20 (Fig. 6, lanes 1–4) and 18.1.30.19 (Fig. 6, lanes 5–8) four descendants at a time were randomly selected and the genomic DNA was subjected to Southern blot analysis. In Fig. 6a the blot was probed with the right border fragment, resulting in a hybridization signal of approximately 6.5 kbp that is specific for the junction region between the right part of the transgene and the plant DNA from clone 18 (see Figs. 2, 5). With the

Table 2. Cosegregation of the genes coding for hygromycin resistance and nopalin production in different sublines of clone 9 and clone 18 (phenotypical analysis)

Seed no.	9.1.19.4.1		9.1.19.4.2		18.1.29		18.1.25.20		18.1.30.13	
	hyg	nop	hyg	nop	hyg	nop	hyg	nop	hyg	nop
1	+	+	+	+	+	+	+	+	−	−
2	+	+	+	+	+	+	+	+	−	−
3	+	+	+	+	+	+	−	−	+	+
4	+	+	+	+	+	+	+	+	+	+
5	+	+	+	+	+	+	+	+	−	−
6	+	+	+	+	+	+	+	+	+	+
7	+	+	+	+	+	+	+	+	+	+
8	+	+	+	+	+	+	+	+	+	+
9	+	+	+	+	+	+	−	−	+	+
10	+	+	+	+	+	+	+	+	−	−
11					+	+	+	−	+	+
12					+	+	+	+	−	−
13					+	+	+	+	+	+
14					+	+	−	+	−	−
15					+	+	+	−	+	+
16					+	+	+	+	+	+
17					+	+	+	+	+	+
18					+	+	+	+	+	−
19							+	+		
20							+	+		

The plants 9.1.19.4.1, 9.1.19.4.2 and 18.1.29 and their progenies are homozygous with respect to the transgene. Plants 18.1.25.20 and 18.1.30.13 are hemizygous individuals. Frames indicate the loss of cosegregation.

exception of one plant, that was recessive with regard to the transmission of the transgene (lane 8 in Fig. 6) the bands were identical in size for all other plants. But they significantly varied in quantity, although exactly 10 μg plant DNA was applied to each slot. The two strong hybridization signals in lanes 1 and 4 could be explained by homozygosity of the appropriate plants, whereas the other quantitative differences could not be due to copy number effects. The quantitative differences might be due to modifications of the foreign DNA. More surprising results were obtained when the same filter shown in Fig. 6a was reprobed with both probes simultaneously, the right border fragment and the coding region of the hph gene.

Identical results were obtained with regard to the presence and the quantitative aspects of the junction fragment. But two candidates (lanes 1 and 5 in Fig. 6b) showing a 6.5-kbp hybridization signal for the junction fragment missed the signal at 7.5 kbp (which is specific for the internal *Hin*dIII-fragment of the transgene with the hph gene). In contrast, all seven plants that showed the 6.5-kbp hybridization signal turned out to be hygromycin resistant as well as positive in nopalin production. One explanation for this virtual loss of a single gene or part of a transgene found in Southern experiments may be the partial or variable modification of a transgene or parts of it. These partial/variable epigenetic

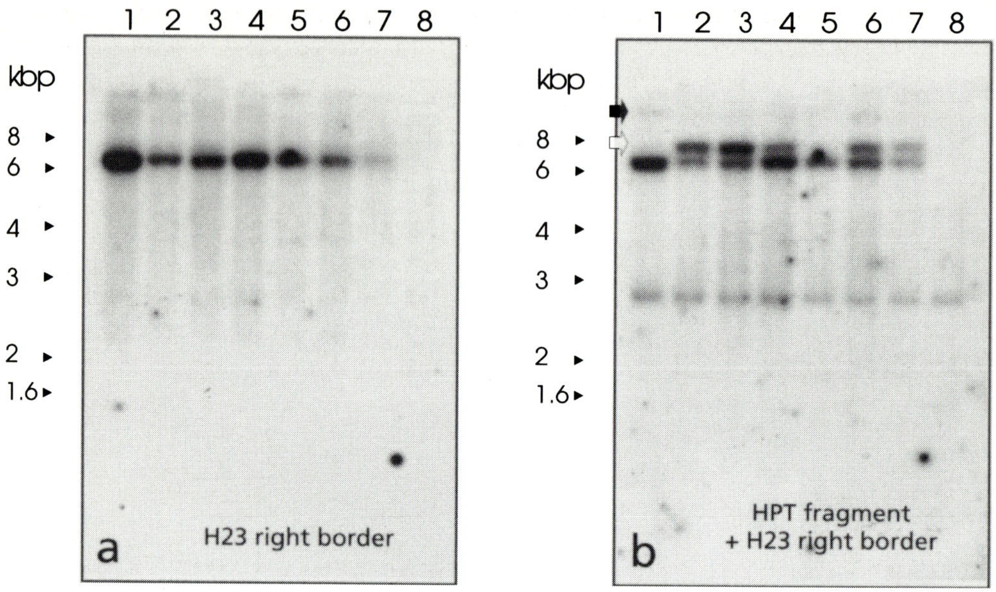

Fig. 6a,b. Southern blot analysis of parts of the R3 generation from two hemizygous sublines originating from the clone 18.1 progeny. Each 10 μg of total plant DNA was digested with the restriction enzyme *Hin*dIII and probed with the H23 right border fragment (**a**) or hybridized with the H23 right border fragment and the HPT fragment simultaneously (**b**, reprobed filter from **a**). *Lane 1,* plant no. 18.1.25.20.1; *lane 2,* plant no. 18.1.25.20.4; *lane 3,* plant no. 18.1.25.20.12; *lane 4,* plant no. 18.1.25.20.20; *lane 5,* plant no. 18.1.30.19.2; *lane 6,* plant no. 18.1.30.19.4; *lane 7,* plant no. 18.1.30.19.5; *lane 8,* plant no. 18.1.30.19.11. The *white arrow* indicates the position of loss of the hybridization signal in *lanes 1* and *5* (in **b**). The *black arrow* indicates a possible hybridization signal for the modified DNA, obtained in the high molecular weight fraction. The weak signal at 2.8 kbp is a unspecific cross-hybridization of the H23 right border fragment, often observed in Southern hybridization experiments with *Vicia narbonensis* (see also Pickardt et al. 1991)

changes lead to different modification states in the same locus in different cells or plant organs and to a bulk of various possible fragments as a consequence. Therefore, it may sometimes occur that specific fragments are not detectable in hybridization experiments, especially when large genomes are investigated.

The complex evaluation of the results obtained up to now showed:

- The two traits were stably expressed in the offspring from three transgenic lines up to the fifth generation.
- The transgene was stably inherited in the homozygous individuals of the progeny in all three lines investigated.
- The progeny analysis revealed a segregation that was 3:1 (phenotypically) or 1:2:1 (genotypically) as defined by statistical methods with some exceptions in sublines that remained hemizygous for a few generations.
- In the hemizygous individuals a more frequent loss of gene expression may have occurred when compared to the homozygous sublines, resulting in unexpected segregation ratios.

- When investigating the cosegregation of the two traits, perfect matching was observed in homozygous sublines,whereas the cosegregation in hemizygous plant material seems to be partly imperfect.
- Partial/variable modifications of transgenes possibly occurred especially in hemizygous sublines, which lead to a loss of specific hybridization signals in Southern experiments or to a loss/inactivation of gene functions.

Acknowledgements. The authors wish to thank Verena Schade for excellent technical assistance, as well as Thilo Schmidt-Rogge and Simone Planitzer for helpful discussions. We would like to thank Lothar Willmitzer for providing us with the *Agrobacterium*-vector pGV3850Hygro. The work was supported by the Bundesministerium für Forschung und Technologie/Germany to O.S.

References

Bianchi MW, Viotti A (1988) DNA methylation and tissue specific transcription of the storage protein genes of maize. Plant Mol Biol 11: 203–214

Carpenter R, Martin C, Coen ES (1987) Comparison of genetic behavior of the transposable element Tam3 at 2 unlinked pigment loci in *Antirrhinum majus*. Mol Gen Genet 207: 82–89

Deroles SC, Gardner RC (1988) Expression and inheritance of kanamycin resistance in a large number of transgenic petunias generated by *Agrobacterium*-mediated transformation. Plant Mol Biol 11: 355–364

Finnegan J, McElroy D (1994) Transgene inactivation: Plants fight back! Bio/Technology 12: 883–888

Hobbs SLA, Warkentin TD, De Long CMO (1993) Transgene copy number can be positively or negatively associated with transgene expression. Plant Mol Biol 21: 17–26

Linn F, Heidmann I, Saedler H, Meyer P (1990) Epigenetic changes in the expression of the maize A1 gene in *Petunia hybrida*: Role of numbers of integrated copies and state of methylation. Mol Gen Genet 222: 329–336

Martin, C, Mac Kay S, Lister C, Prescott A (1989) Activity of the transposon Tam3 in *Antirrhinum* and tobacco – possible role of DNA methylation. EMBO J 8: 997–1004

Matzke MA, Primig M, Trnovsky J, Matzke AJM (1989) Reversible methylation and inactivation of marker genes in sequentially transformed tobacco plants. EMBO J 8: 643–649

Matzke MA, Matzke AJM (1991) Differential inactivation and methylation of a transgene in plants by two suppressor loci containing homologous sequences. Plant Mol Biol 18: 821–830

Matzke AJM, Matzke MA (1993) Epigenetic variation as a consequence of homology dependent gene interactions in transgenic plants. Seminars in Developmental Biology 4: 83–89

Matzke MA, Matzke AJM, Mittelsten-Scheid O (1994) Inactivation of repeated genes – DNA-DNA interaction? In: Homologous recombination and gene silencing in plants Paszkowski JC (ed), pp 271–307, Kluwer Academic Publishers/Netherlands

Meixner M, Schneider U, Schieder O, Pickardt T (1995) Studies on the stability of foreign genes in the progeny of transgenic lines of *Vicia narbonensis*. In: Terzi M, Cella R, Falavigna A (eds) Current Issues in Plant Molecular and Cellular Biology. Kluwer Academic Publishers, pp 285–290

Meyer P, Linn F, Heidmann I, Meyer H, Niedenhof I, Saedler H (1992) Endogenous and environmental factors influence 35S-promoter methylation of a maize A1 gene construct in transgenic petunia and its color phenotype. Mol Gen Genet 231: 345–352

Mittelsten-Scheid O, Paszkowski J, Potrykus I (1991) Reversible inactivation of a transgene in *Arabidopsis thaliana*. Mol Gen Genet 228: 104–112

Ottaviani MP, Smits T, Häenisch ten Cate CH (1993) Differential methylation and expression of the β-glucuronidase and neomycin phosphotransferase genes in transgenic potato cv. Bintje. Plant Sci 88: 73–81

Otten L, Degreve H, Hernalsteens JP, van Montagu M, Schieder O, Straub J, Schell J (1981) Mendelian transmission of genes introduced into plants by Ti plasmid of *Agrobacterium tumefaciens*. Mol Gen Genet 183: 209–213

Pickardt T, Meixner M, Schade V, Schieder O (1991) Transformation of *Vicia narbonensis* via *Agrobacterium*-mediated gene-transfer. Plant Cell Rep 9: 535–538

Potrykus I, Paszkowski J, Saul MW, Petruska J, Shillito RD (1985) Molecular and general genetics of a hybrid foreign gene introduced into tobacco by direct gene transfer. Mol Gen Genet 199: 169–177

Pröls F, Meyer P (1992) The methylation patterns of chromosomal integration regions influence gene activity of transfered DNA in *Petunia hybrida*. Plant J 2: 465–475

Strickberger MW (1985) Genetics, 3rd edn. MacMillan, New York

Schmidt-Rogge T, Meixner M, Srivastava V, Guha-Mukherjee S, Schieder O (1993) Transformation of haploid *Datura innoxia* protoplasts and analysis of the plasmid integration pattern in regenerated transgenic plants. Plant Cell Rep 12: 390–394

van der Krol AR, Mur LA, Beld M, Mol JNM, Stuitje AR (1990) Flavonoid genes in *Petunia*: addition of a limited number of gene copies may lead to a suppression of gene expression. Plant Cell 2: 291–299

Walter C, Broer I, Hillemann D, Pühler A (1992) High frequency, heat treatment-induced inactivation of the phosphinotricin resistance gene in transgenic single cell suspension cultures of *Medicago sativa*. Mol Gen Genet 235: 189–196

25 Characterization of a Plant Scaffold-Attached Region from a T-DNA Integration Site

A. Dietz-Pfeilstetter[1], V. Kay[2], J. Landsmann[1], and J. Bode[2]

[1]Institut für Biochemie und Pflanzenvirologie, Biologische Bundesanstalt, Messeweg 11/12, 38104 Braunschweig, Germany
[2]Gesellschaft für Biotechnologische Forschung, Genetik von Eukaryonten, Mascheroder Weg 1, 38124 Braunschweig-Stöckheim, Germany

Introduction

Integration of foreign genes into plant genomes by the *Agrobacterium* T-DNA transfer system has been considered to occur at random. Also, the expression level of genes introduced into the plant genome is not predictable and can vary substantially between individual plants transformed with the same gene construct (Odell et al. 1987; Rogers et al. 1985), in some cases to the extent of loss of organ specificity (van der Hoeven et al. 1994). This variability of expression has been referred to as "position effect". It may partly be due to endogenous regulatory elements like promoters or enhancers close to the integration site and to the chromosomal structure/transcriptional state of the integration site.

The higher-order chromatin structure of the eukaryotic genome is thought to play an important role in replication and transcription processes. According to current models, chromatin is organized in chromosomal loop domains, which by means of their basal parts are attached to a proteinaceous scaffold called the nuclear matrix (Paulson and Laemmli 1977). Accordingly, the responsible DNA sequences have been called "scaffold-attached regions" (SARs) or matrix-attached regions (MARs). These S/MAR elements insulate chromatin domains from neighbouring loops with respect to their chromatin structure and torsional state and are thought to be responsible for a differential regulation of transcriptional processes (Bonifer et al. 1991; Goldman 1988). Chromatin loops encompass between five and more than 100 kilobases.

There are several points of evidence that T-DNA transferred into plant cells via *Agrobacterium tumefaciens* is preferentially integrated into transcriptionally active regions of the plant genome. The preferential T-DNA insertion into potentially transcribed genome regions may be mediated by the T-DNA integration mechanism which seems to require DNA replication and repair (Mayerhofer et al. 1991). Alternatively, the "open" and thus potentially active chromatin conformation typically found at the integration site may be a consequence of the in

E.R. Schmidt/T. Hankeln (Eds.)
Transgenic Organisms and Biosafety

vitro selection process, i.e. only transformants with T-DNA being integrated into permissive chromatin domains might be recovered during selection for the expression of the transferred antibiotic resistance markers.

To gain more insight into the molecular structure of T-DNA integration sites and its possible impact on gene expression, we have examined plant DNA sequences in the vicinity of T-DNA borders. Analysis of a transgenic petunia plant displaying leakiness in root specificity of the introduced hemoglobin PAR promoter from Parasponia andersonii (Landsmann et al. 1988) revealed a SAR close to one T-DNA end.

Sequence Analysis of a T-DNA Border Region

Petunia plants were transformed with the binary vector pPAR4925 (Landsmann et al. 1988). Some of the kanamycin resistant plants obtained not only displayed strong chloramphenicol acetyltransferase (CAT) enzyme activity in roots, but at times expressed the CAT reporter gene at a similar level in leaves as well. One of these transformants was chosen for further analysis and was found to contain three tightly coupled T-DNA copies (Dietz et al. 1994). Upon construction and screening of a genomic library recombinant clones containing right border (RB) or left border (LB) fragments, respectively, were obtained. The right border clone Nos1.1 was found to hybridize to a unique (single copy) petunia DNA region. Sequence analysis revealed that the plant DNA immediately adjacent to the right border is extremely A+T-rich (80% within the first 21 nucleotides), which is

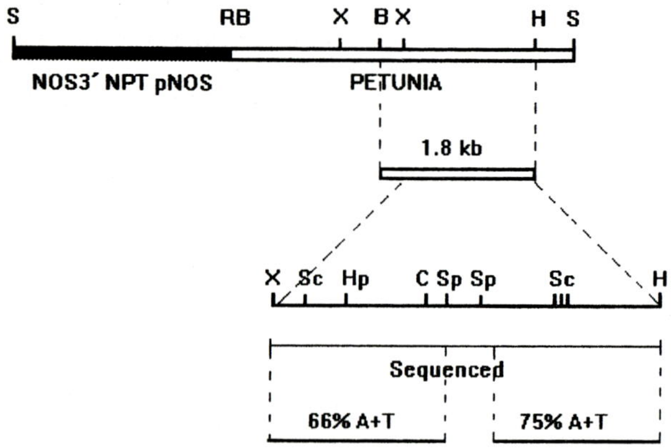

Fig. 1. The 1.8-kb subfragment of Nos1.1, identified by its hybridization characteristics. The *upper map* represents the insert of Nos1.1, while the *lower map* shows the 1.65-kb *Xba*I-*Hin*dIII petunia DNA fragment, which was subcloned and sequenced. *B*, *Bst*EII; *C*, *Cla*I; *H*, *Hin*dIII; *Hp*, *Hpa*I; *S*, *Sac*I; *Sc*, *Sca*I; *Sp*, *Spe*I; *X*, *Xba*I; *RB*, right border; *NPT*, neomycin phosphotransferase; *pNOS*, *NOS3′*, regulatory sequences from the *A. tumefaciens* nopalin synthase gene

consistent with previous reports on integration sites (Gheysen et al. 1987; Slightom et al. 1985). Distant from the right border integration site a 1.8-kb fragment showing base pairing to plant genomic DNA and to itself only at nonstringent conditions was identified by hybridization of border fragments with untransformed genomic DNA (Fig. 1). Scanning of the 1664-bp sequence obtained from the weakly hybridizing region revealed several motifs common to SARs.

In Vitro Binding to Nuclear Scaffolds

The assumption that the sequenced plant DNA close to the right T-DNA border represents a SAR element was verified by subjecting the 1.65-kb fragment and several of its subfragments to a standard in vitro binding assay, based on a scaffold preparation from murine cells (Mielke et al. 1990). For the equally sized *Xba*I-*Spe*I and *Spe*I-*Hin*dIII fragments a high (more than 80%) binding potential was found (Fig. 2), which can be ascribed to the presence of adequately spaced ATATTT motifs (Mielke et al. 1990). Subfragments of the *Xba*I-*Spe*I fragment displayed a lower binding potential, while the 180-bp *Spe*I fragment was released from the scaffold due to its small size.

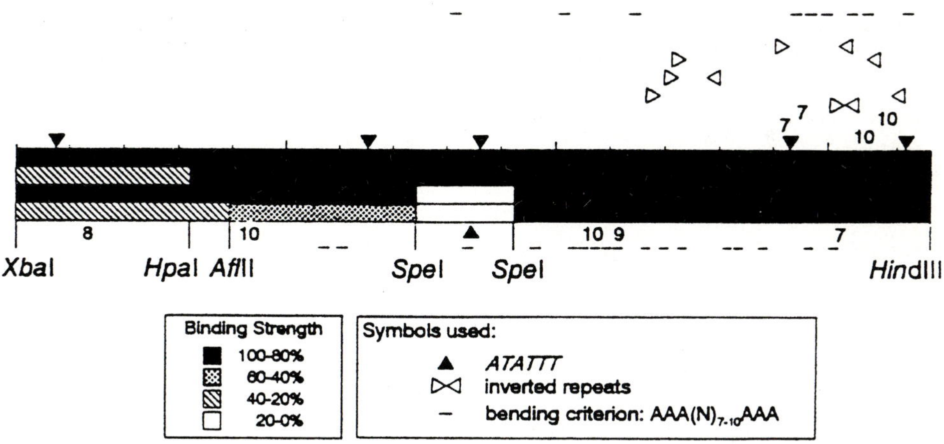

Fig. 2. Scaffold binding strength of petunia scaffold-attached region (SAR) and subfragments thereof for Ltk^- nuclear matrices. An autoradiograph from a standard reassociation using equimolar amounts of mouse Ltk^--derived scaffolds and different radiolabeled subfragments of the petunia SAR was scanned. Binding of the respective band was expressed as percentage binding, i.e. $cpm_p \times 100/(cpm_p + cpm_s)$ using SAR_{800} from the human interferon-β domain (Mielke et al. 1990) as a standard. Symbols adjacent to the bars mark ATATTT motifs (*vertical arrowheads*), inverted repeats (*horizontal arrowheads facing each other*), $(dA)_n$ tracts (*numbers* mark *n* values), segments conforming to the $AAA(N)_{7-10}AAA$ bending criterion (*dashes*)

To demonstrate scaffold binding of the petunia SAR when using a plant nuclear scaffold, scaffolds from pea leaves were prepared. Comparable binding affinities for the petunia SAR were found with pea and with mouse-L cell derived scaffolds (Dietz et al. 1994).

Petunia Scaffold-Attached Region Reduces the Variability of Expression Between Transformed Plants

SAR elements can confer a position independent gene expression to stably integrated genes in transgenic plants (Breyne et al. 1992). To test the petunia SAR (P-SAR) element for this feature, transformation vectors derived from PARGUSH (van der Hoeven et al. 1994) were constructed in which the β-glucuronidase (GUS) reporter gene was flanked by the petunia SAR (Fig. 3). Tobacco plants were transformed by the leaf disk method using *Agrobacterium tumefaciens* LBA4404. Between 19 and 45 kanamycin resistant plants were obtained from single transformations with the three different P-SAR gene constructs and were assayed for GUS enzyme activity at two to three time points. Tranformed plants containing a single P-SAR either 5′ or 3′ of the GUS gene showed variability of GUS expression in the roots similar to that observed with the "no SAR" vector PARGUSH (van der Hoeven et al. 1994). Variation was observed between different transformants as well as between different assay dates. Transformation with the double SAR vector PGSAR21, however, led to a narrow range of GUS expression in the roots between 20 and 70 nmol *p*-nitrophenol/mg protein/min for over 60% of the transformants. Figure 4 shows the distribution of GUS activities for the first assay time point in the different populations of transgenic plants.

P-SAR plants without GUS activity either did not contain the PARGUS gene construct or had rearrangements as was demonstrated by Southern blot analysis with the PAR promoter fragment as hybridization probe.

Discussion and Conclusions

Upon analysis of a transgenic petunia plant containing the CAT gene driven by the plant hemoglobin PAR promoter but not showing the expected root specific expression pattern, a SAR element was detected close to the T-DNA integration site. The 1.65-kb *Xba*I-*Hin*dIII fragment initially detected by its unusual behaviour in Southern blot hybridizations harbours many of the common SAR motifs, mostly concentrated in the part most distal to the integration site (Fig. 2). Its subfragments bind to a heterologous scaffold derived from cultured mouse cells in a predictable manner. Moreover, using a plant nuclear scaffold isolated from

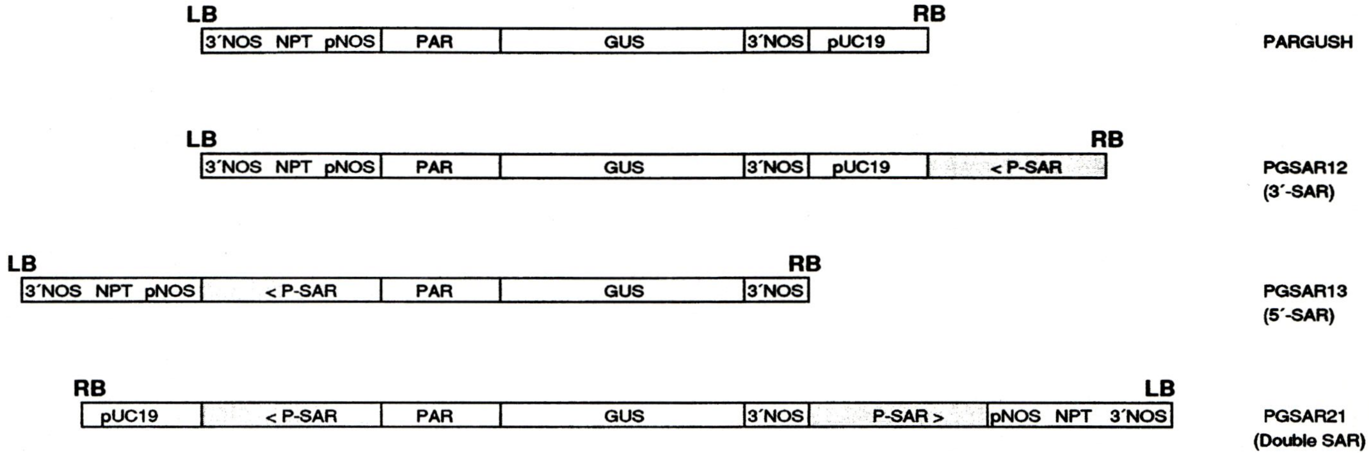

Fig. 3. T-DNA region of PARGUSH and of the derived scaffold-attached region (SAR) containing constructs. *PAR*, hemoglobin promoter from *Parasponia andersonii*; *GUS*, coding region of the β-glucuronidase gene; *RB*, right border; *LB*, left border

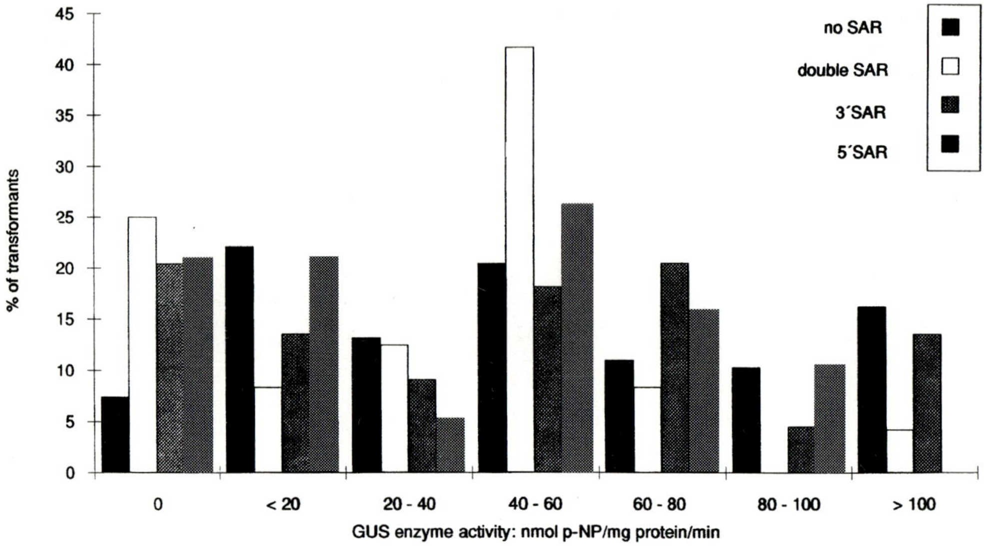

Fig. 4. Frequency distribution of GUS activities for the different T-DNA constructs. Percentages of transformants are plotted against classes of GUS activity (given in amounts of the product *p-NP* (*p*-nitrophenol) formed per milligram of total protein and per minute)

pea leaves comparable binding was observed for the 1.65-kb fragment and for an animal SAR element of similar length (Dietz et al. 1994). Therefore, one of the criteria for SAR elements (conservation across species boundaries; reviewed in Roberge and Gasser 1992) holds for the here identified petunia sequence.

Another feature of SAR elements is the shielding of an integrated transgene from influences of surrounding chromatin by flanking the gene at both ends (Klehr et al. 1991; Stief et al. 1989). This was confirmed for the petunia sequence by comparing the variability of expression in a number of transformants obtained with a "no SAR" and with different SAR transformation vectors (Figs. 3, 4). However, no elevation of transcription levels was observed after coupling the petunia SAR to a reporter gene in transgenic plants. This is in contrast to the enhancement effect which the element has in stably transfected mouse cells (Dietz et al. 1994). One possible explanation for this discrepancy might be the mechanism of T-DNA integration, which directs transgenes predominantly to active gene loci. SAR elements provided as part of the transgene construct should then cause no additional activation but mainly reduce the "position effect". Position independent but non-elevated gene expression has also been reported by Breyne et al. (1992) for tobacco calli transformed via *Agrobacterium tumefaciens* with a reporter gene flanked on both sides by a plant SAR.

Since the petunia SAR was detected at the single T-DNA integration site of a transformant in which the root specific PAR promoter was deregulated, it is conceivable that it bears a leaf specific enhancer. In that case all transgenic plants with this element flanking the GUS reporter gene should display root as well as leaf expression. However, a physiological and/or developmental influence on organ specificity similar to that reported by van der Hoeven et al. (1994) was observed.

It remains to be analyzed - by characterization of more transformants - whether the occurrence of SAR elements close to an integration site is as common for T-DNAs as it is for animal retroviruses (Mielke 1993).

Acknowledgments. We thank Antje Baumgartner for excellent technical assistance. This research was supported by a grant from the German Ministry of Research and Technology (BMFT, 321-7291-BCT396).

References

Bonifer C, Hecht A, Saueressig H, Winter DM, Sippel AE (1991) Dynamic chromatin: the regulatory domain organization of eukaryotic gene loci. J Cell Biochem 47: 99–108

Breyne P, Van Montagu M, Depicker A, Gheysen G (1992) Characterization of a plant scaffold attachment region in a DNA fragment that normalizes transgene expression in tobacco. Plant Cell 4: 463–471

Dietz A, Kay V, Schlake T, Landsmann J, Bode J (1994) A plant scaffold attached region detected close to a T-DNA integration site is active in mammalian cells. Nucleic Acids Res 22: 2744–2751

Gheysen G, Van Montagu M, Zambryski P (1987) Integration of *Agrobacterium tumefaciens* transfer DNA (T-DNA) involves rearrangements of target plant DNA sequences. Proc Natl Acad Sci USA 84: 6169–6173

Goldman MA (1988) The chromatin domain as a unit of gene regulation. BioEssays 9: 50–55

Klehr D, Maass K, Bode J (1991) Scaffold-attached regions from the human interferon beta domain can be used to enhance the stable expression of genes under the control of various promoters. Biochemistry 30: 1264–1270

Landsmann J, Llewellyn D, Dennis ES, Peacock WJ (1988) Organ regulated expression of the *Parasponia andersonii* haemoglobin gene in transgenic tobacco plants. Mol Gen Genet 214: 68–73

Mayerhofer R, Koncz-Kalman Z, Nawrath C, Bakkeren G, Crameri A, Angelis K, Redei G.P, Schell J, Hohn B, Koncz C (1991) T-DNA integration: a mode of illegitimate recombination in plants. EMBO J 10: 697–704

Mielke C, Kohwi Y, Kohwi-Shigematsu T, Bode J (1990) Hierarchical binding of DNA fragments derived from scaffold-attached regions: correlation of properties in vitro and function in vivo. Biochemistry 29: 7475–7485

Mielke C (1993) Dissertation, Technische Universität Braunschweig

Odell JT, Nagy F, Chua N-H (1987) Variability in 35S promoter expression between independent transformants. In: Key J, McIntosh L (eds) Plant gene systems and their biology, vol 62. Liss, New York, pp 321–329

Paulson JR, Laemmli UK (1977) The structure of histone-depleted metaphase chromosomes. Cell 12: 817–828

Roberge M, Gasser SM (1992) DNA loops: structural and functional properties of scaffold-attached regions. Mol Microbiol 6: 419–423

Rogers SG, OConnell K, Horsch RB, Fraley RT (1985) Investigation of factors involved in foreign protein expression in transformed plants. In: Zaitlin M, Day P, Hollaender A (eds) Biotechnology in plant science. Academic, London, pp 219–226

Slightom JL, Jouanin L, Leach F, Drong RF, Tepfer D (1985) Isolation and identification of TL-DNA/plant junctions in *Convolvulus arvensis* transformed by *Agrobacterium rhizogenes* strain A4. EMBO J 4: 3069–3077

Stief A, Winter DM, Straetling WH, Sippel AE (1989) A nuclear DNA attachment element mediates elevated and position-independent gene activity. Nature 341: 343–345

van der Hoeven C, Dietz A, Landsmann J (1994) Variability of organ-specific gene expression in transgenic tobacco plants. Transgenic Res 3: 159–165

26 Expression of Antibody Single-Chain Fragments with Specificities to Structural and Nonstructural Proteins of Beet Necrotic Yellow Vein Virus in Plants: An Alternative Approach To Produce Virus Resistant Plants

L.F. Fecker[1], U. Commandeur[1], A. Kaufmann[1], G. Himmler[2], R. Koenig[1], and W. Burgermeister[1]

[1]Biologische Bundesanstalt für Land- und Forstwirtschaft, Institut für Biochemie und Pflanzenvirologie, Messeweg 11, 38104 Braunschweig, Germany
[2]Institut für Angewandte Mikrobiologie, Universität für Bodenkultur, Nussdorfer Lände 11, 1190 Wien, Austria

Introduction

A novel approach to establish virus resistance in plants is based on recent progress to clone regions of immunoglobulin genes which encode the antigen binding domains of the variable regions of the heavy (V_H) and the light chains (V_L) of antiviral monoclonal antibodies (mAb) and to express these polypeptide chains connected by a linker peptide as single-chain fragments (scFv).

So far almost all genetically engineered virus resistances in plants are pathogen-derived (Sanford and Johnston 1985; Wilson 1993). The expression of a part of a viral genome, most frequently the viral coat protein gene, in transgenic plants may interfere with the normal replication cycle of a virus to such an extent that the multiplication of the virus becomes nearly impossible and the plants are thus rendered resistant. It has been argued, however, that pathogen-derived resistance is a risky approach because of the possibility of transcapsidation or, even worse, of recombination of the expressed viral genes with genes of other viruses.

Recently it has been reported that the intracellular expression of scFv specific for artichoke mottled crinkle virus (AMCV) coat protein in *Nicotiana benthamiana* or the extracellular expression of whole antibodies specific for tobacco mosaic virus (TMV) coat protein in *Nicotiana tabacum* greatly reduces the damage caused by these viruses (Tavladoraki et al. 1993; Voss et al. 1994). In both cases the transgenic plants showed a reduced level of infection with the respective virus and a delayed symptom development. In the case of the expression of antibodies directed against TMV the degree of resistance was related to the number of antibodies secreted to the apoplast. This antibody-based re-

E.R. Schmidt/T. Hankeln (Eds.)
Transgenic Organisms and Biosafety

sistance lacks the above-mentioned disadvantages of the pathogen-derived resistances.

We are now attempting to develop a scFv-dependent protection of plants against beet necrotic yellow vein virus (BNYVV), a fungus-transmitted virus causing rizomania, the presently most damaging virus disease of sugarbeet. The scFv we are using are derived from mAbs to structural and non-structural proteins of BNYVV.

mAb 14CG6 is specific for an epitope which has been localized by means of overlapping peptides near the carboxy-terminal end of the coat protein and which is accessible along the entire length of the viral particle (Commandeur et al. 1994). mAb 6G8 was produced against the viral 25 K protein encoded on BNYVV RNA 3 (Kaufmann 1994). RNA 3 is responsible for rizomania symptom development in sugarbeet (Tamada et al. 1990; Koenig et al. 1991). The 25 K protein is detectable in the cytoplasm and the nucleus of infected cells (Haeberlé and Stussi-Garaud 1994).

Expression of Single-Chain Fragments in Bacteria

For the production of scFv, cDNAs for the heavy and light chains of the two mAbs were synthesized using polyA(+)-RNA from hybridoma cell lines as template. The cDNA for V_H and V_L was amplified by polymerase chain reaction (PCR) with specific primers binding to conserved sequences at the ends of the V_H- and V_L-encoding regions (Orlandi et al. 1989). After treatment with appropriate restriction enzymes, the amplified V_H- and V_L-encoding sequences were integrated sequentially into the corresponding restriction sites of the expression vector pCOCK (Engelhardt et al. 1994). By this means, the V_H- and V_L-encoding sequences become connected by a DNA sequence which encodes the linker peptide. In addition, the lac promoter and coding sequences for a bacterial signal peptide (PelB) and a marker peptide (myc-tag) are fused to the 3′- or 5′-ends of the scFv-encoding sequences, respectively (Fig. 1A). The bacterial signal peptide is required for the secretion of the scFv to the periplasmic space, whereas the marker peptide is necessary for the immunological detection of the scFv. Thus, the recombinant plasmids pCOCKscFv14C and pCOCKscFv6G were obtained. By means of Western blotting we could demonstrate that *Escherichia coli* HB2151 transformed with these plasmids produced functional (antigen binding) scFv after induction with IPTG.

A detailed analysis of different fractions of *E. coli* cultures producing BNYVV coat protein-specific scFv revealed that variable amounts of scFv could be detected in the soluble periplasmic cell fraction and in the culture medium. However, most of the scFv were present in the insoluble cell fraction.

In order to compare the efficiencies of antigen detection by the coat protein-specific mAb 14CG6 and by the scFv derived from this mAb, the antigen titres were determined in an immuno-dot blot assay. Serial dilutions of the antigen

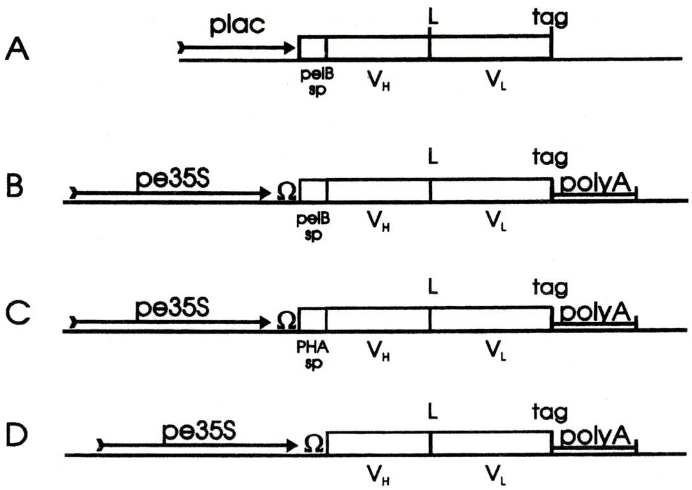

Fig. 1A–D. Constructs for the expression of single-chain fragments (scFv) in bacteria and plants in the plasmids pCOCK (**A**) and pRT213 (**B-D**). *plac*, promoter of *Escherichia coli* lac operon; pelB sp, PHA sp., V_H, L, V_L; tag, DNA fragments encoding signal peptide of pectate lyase, signal peptide of phytohemagglutinin, V_H, linker peptide, V_L and myc-tag peptide, respectively; pe35S, enhanced 35S promoter of CaMV; Ω, translational leader region of tobacco mosaic virus (TMV); polyA, polyadenylation signal of CaMV

were dotted onto a nylon membrane and after blocking, these membranes were incubated with mAb 14CG6 or a concentrated sample of the respective scFv-containing culture supernatant. At optimal scFv and mAb concentrations the sensitivity of antigen detection was only slightly lower with the scFv preparation than with the original mAb.

Expression of Single-Chain Fragments in *Nicotiana sp.*

As transformation of sugarbeet via *Agrobacterium tumefaciens* is very difficult to achieve, we have started to transform *Nicotiana benthamiana* and *N. tabacum* with different plant expression cassettes containing the coding sequences for the scFv.

For the construction of the expression cassettes, DNA fragments with the sequences for the 25 K and the coat protein specific scFv (V_H – linker peptide – V_L – marker peptide) were integrated between the cauliflower mosaic virus 35S promoter and polyadenylation signal into the pRT103 or pRT213 vectors (Toepfer et al. 1987), respectively. In the constructs prepared with the pRT213 vector (Fig. 1B–D), the translation enhancing omega leader region of TMV is present directly upstream of the open reading frame of the scFv and transcription is controlled by the enhanced 35S promoter. For each of the two vectors three types of contructs were made. For cytoplasmic expression the scFv encoding DNA was directly cloned behind the promoter region of pRT103 or the omega leader region of pRT213, respectively. In the constructs for an extracellular expression the scFv encoding sequence is preceded either by a sequence encoding the plant signal peptide of *Phaseolus vulgaris* phytohemagglutinin (PHA) (Dorel et al. 1989) or the bacterial signal peptide of pectate lyase B (Lei et al. 1987). The expression cassettes were subcloned into the binary vector pLX222 (Landsmann et al. 1988). *N. benthamiana* and *N. tabacum* were transformed via *A. tumefaciens* by the leaf disk method (Horsch et al. 1985).

Transformed plants were selected on the basis of their tolerance towards the antibiotic kanamycin and were analysed for the production of functionally active scFv. For this analysis a modified immunoblot assay was used in which a partially purified BNYVV preparation was dotted onto a small membrane disc. The discs were then transferred into holes of a microtiter plate and after treatment with blocking solution they were incubated with extracts of the kanamycin tolerant plants. Binding of the scFv produced from plants to immobilized viral coat protein was detected by mAb 9E10 specific for the myc-tag label. Many of the plants transformed with constructs for extracellular expression produced readily detectable amounts of functionally active, i.e. antigen binding scFv (Table 1). However, so far only a few of the plants transformed with constructs for cytoplasmic scFv expression were found to produce functionally active scFv in low concentrations. More sensitive techniques may be necessary to detect scFv in the cytoplasm.

Table 1. Detection of neomycin phosphotransferase (NPT II) levels by means of enzyme-linked immunosorbent assay (ELISA) and of single-chain fragments (scFv) by a modified dot blot immuno assay (see text) in extracts of kanamycin-tolerant regenerated *Nicotiana tabacum* and *N. benthamiana* plants

Species	Line	signal peptide	NPTII (ng/mg protein)	scFv*
N. tabacum	W 38–11	–	2.6	(+)
N. tabacum	W 38–12	–	0.9	n.d.
N. tabacum	W 38–4	PelB	46.2	+
N. tabacum	W 38–5	PelB	26.2	n.d.
N. tabacum	W 38–6	PelB	32.5	++
N. tabacum	W 38–36	PHA	3.4	++++
N. benthamiana	24	-	12.5	n.d.
N. benthamiana	27	-	18	(+)
N. benthamiana	28	-	0.8	n.d.
N. benthamiana	33	PelB	10.6	+
N. benthamiana	35	PHA	2.8	++
N. benthamiana	23	PHA	1.5	++++

The plants were grown in sterile culture medium (MSO) with 50 mg/l kanamycin and 500 mg/l claforan.
n.d., none, (+), questionable; +, very low; ++, low; ++++, high.

Some selected scFv-expressing *N. benthamiana* plants are now grown for seed production. Plants from the F1 generation will be analyzed for their ability to produce scFv and for their possible resistance to BNYVV infection. As scFv are foreign proteins which are not naturally produced by plants it will also be interesting to assess their location and stability in plant tissues.

Acknowledgements We are greatly indebted to the Deutsche Forschungsgemeinschaft (Project Ko 518/11 and 12) and the Gemeinschaft zur Förderung der Privaten Deutschen Pflanzenzüchtung e.V. (Project BR 29/91 FWW AIF) for financial support.

References

Commandeur U, Koenig R, Manteuffel R, Torrance L, Lüddecke P, Frank R (1994) Location, size and complexity of epitopes on the coat protein of beet necrotic yellow vein virus studied by means of synthetic overlapping peptides. Virology 198: 282–287

Dorel C, Voelker TA, Herman EM, Chrispeels MJ (1989) Transport of proteins to the plant vacuole is not by bulk flow through the secretory system, and requires positive sorting information. J Cell Biol 108: 327–337

Engelhardt O, Grabherr R, Himmler G, Rüker F (1994) Two-step cloning of antibody variable domains in a phage display vector. BioTechniques 17: 44–46

Haeberlé AM, Stussi-Garaud C (1994) In situ localization of the non-structural protein P25 encoded by beet necrotic yellow vein virus RNA3. J Gen Virol 76: 643–650

Horsch RB, Fry J, Hoffmann NL, Wallroth M, Eichholtz D, Rogers SG, Fraley RT (1985) A simple and general method for transferring genes into plants. Science 227: 1229–1231

Kaufmann A (1994) Molekularbiologische und serologische Untersuchungen über das beet necrotic yellow vein virus (BNYVV) und das beet soil borne virus (BSBV) und Erarbeitung der Grundlagen für eine gentechnologisch zu erzeugende immunologische Resistenz gegen das BNYVV. Ph.D. thesis, Technical University, Braunschweig ISBN3-930697-58-0

Koenig R, Jarausch W, Li Y, Commandeur U, Burgermeister W, Gehrke M, Lüddecke P (1991) Effect of recombinant beet necrotic yellow vein virus with different RNA compositions on mechanically inoculated sugarbeets. J Gen Virology 72: 2243–2246

Lei S-P, Lin H-C, Wang S-S, Callaway J, Wilcock G (1987) Characterization of the Erwinia carotovora pelB gene and its product pectate lyase. J Bacteriol 169: 4379–4383

Landsmann J, Llewellyn D, Dennis ES, Peacock WJ (1988) Organ regulated expression of the Parasponia andersonii haemoglobin gene in transgenic tobacco plants. Mol Gen Genet 214: 68–73

Orlandi R, Güssow DH, Jones PT, Winter G (1989) Cloning immunoglobulin variable domains for expression by the polymerase chain reaction. Proc Natl Acad Sci 86: 3833–3837

Sanford JC, Johnston SA (1985) The concept of parasite-derived resistance – deriving resistance genes from the parasite's own genome. J theor Biol 113: 395–405

Tamada T, Saito M, Kiguchi T, Kusume T (1990) Effect of isolates of beet necrotic yellow vein virus with different RNA components on the development of rizomania symptoms. Proceedings of the First Symposium of the International Working Group Plant Viruses with Fungal Vectors, Braunschweig, Germany, August 1990. Ulmer, pp 41–44

Tavladoraki P, Benvenuto E, Trinca S, De Martinis D, Cattaneo A, Galeffi P (1993) Transgenic plants expressing a functional single-chain Fv antibody are specifically protected from virus attack. Nature 366: 469–472

Töpfer R, Matzeit V, Gronenborn B, Schell J, Steinbiss H-H (1987) A set of plant expression vectors for transcriptional and translational fusions. Nucleic Acids Res 15: 5890

Voss A, Niersbach M, Hain R, Hirsch HJ, Liao YC, Kreuzaler F, Fischer R (1995) Reduced virus infectivity in N. tabacum secreting a TMV-specific full size antibody. Mol Breeding 1: Biol 39–50

Wilson TMA (1993) Strategies to protect crop plants against viruses: pathogen derived resistance blossoms. Proc Natl Acad Sci USA 90: 3134–3141

27 Gene Transfer in Aspen

M. Fladung and M.R. Ahuja

Federal Research Centre for Forestry and Forest Products, Institute of Forest Genetics, Sieker Landstrasse 2, 22927 Grosshansdorf, Germany

Introduction

Populus has been developed as a model system for biotechnological research in tree species. It has a relatively small genome, can be easily regenerated *in-vitro*, and genetically transformed with *Agrobacterium* strains. Certain genes that cause activation or inactivation of specific hormones could be of interest in terms of understanding plant growth and development. It is also interesting to determine if such genes, when introduced into long-lived tree species, would be stably expressed over a longer period of time.

Several genes of bacterial origin have been identified which inactivate hormones or liberate them from precursors (*rolC*, Estruch et al. 1991a; *rolB*, Estruch et al. 1991b; *isopentenyltransferase* (*ipt*), Barry et al. 1984; *indoleacetic acid lysine synthetase, iaaL*, Glass and Kosugue 1986). In transgenic tobacco and potato plants carrying one of the genes mentioned above, certain morphological and physiological responses have been identified (Spena et al. 1987; Schmülling et al. 1988; Spena et al. 1991; Fladung and Ballvora 1992; Fladung 1990; Fladung et al. 1993).

In this study, aspen (*Populus tremula*) and hybrid aspen (*P. tremula* × *tremuloides*) were transformed with genes under control of the 35S promoter of the cauliflower mosaic virus or the light-inducible promoter of the small subunit of ribulose-bisphosphate carboxylase gene of potato in presence or absence of the maize transposable element *Ac*. Data are presented on efficiency of transformation, phenotypic evaluation, and molecular analysis with polymerase chain reaction (PCR) using specific primers and Southern blot hybridisation.

E.R. Schmidt/T. Hankeln (Eds.)
Transgenic Organisms and Biosafety

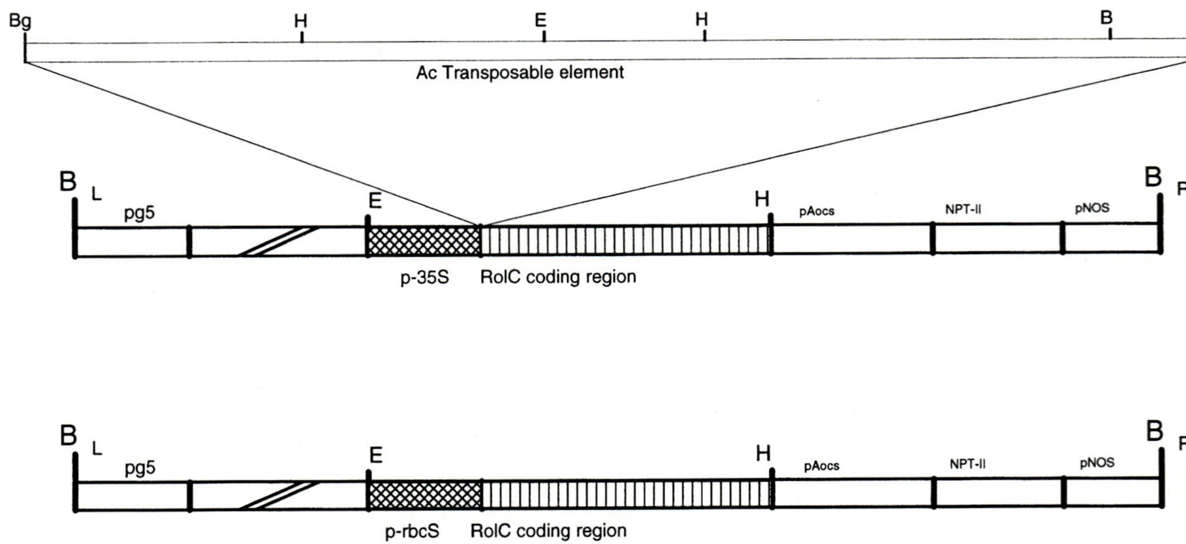

Fig. 1. The p35s-*Ac-RolC* (*top*) and the prbcS-*RolC* (*bottom*) constructions. The entire constructions are located between the *Eco*RI and *Hin*dIII sites of the binary vector pPCV002. p35S-*Ac-RolC:* The Ac transposable element is inserted between the 35S-RNA cauliflower mosaic virus promoter and the *RolC* coding region. p*rbcS-RolC*: The *rbcS*1 promoter of potato (Wolters et al. 1988) is inserted upstream of the *RolC* coding region. B_L, B_R, left and right border sequences of vector T-DNAs; *pg5*, truncated promoter of T_L-DNA gene 5; *pNOS*, promoter of nopaline synthase gene; *pAocs*, polyadenylation sequence of octopine synthase gene; *NPT-II*, neomycin phosphotransferase gene of transposon Tn5; *E*, *Eco*RI; *H*, *Hin*dIII; *B*, *Bam*HI; *K*, *Kpn*I. (Spena et al. 1989, Schmülling et al. 1993)

Results and Discussion

Based on a well-established tissue culture of aspen (*Populus tremula*) and hybrid aspen (*P. tremula*×*tremuloides*) and regeneration of mature plants (Ahuja 1986, 1987), a transformation method was developed, by which in high efficiencies (up to 60%) putative transgenic plantlets were regenerated on kanamycin containing regeneration medium. Transgenic plants were regenerated by the leaf disc transformation method. Two different aspen clones (W52, Brauna 11) and one hybrid clone (Esch5) were used for transformation experiments with eight different constructs (Table 1). For example, the plasmid pPCV002-rbcSC (Schmülling et al. 1993) carries the *rolC* gene from *Agrobacterium rhizogenes* under control of the light-inducible promoter of the small subunit of ribulose-bisphosphate carboxylase (*rbcS*) from potato (Wolter et al. 1988). The plasmid pPCV002-*Ac*-CaMVC carries the *rolC* gene under control of the 35S promotor of the cauliflower mosaic virus, with insertion of a transposable element *Ac* from *Zea mays* between the 35S promotor and the *rolC* coding region (Fig. 1).

The number of leaf disks infected with *Agrobacterium tumefaciens* strains containing different constructs and the number of independent regenerants obtained is shown in Table 2. A total of 1658 leaf disks were infected with *Agrobacterium* so far, of which 194 (12%) were independent regenerants obtained. Out of these, 184 remained which turned out to be kanamycin resistant.

Transgenic aspen carrying the *rolC* gene from *Agrobacterium rhizogenes* under control of the cauliflower-35S-promoter are reduced in size with smaller leaves, whereas aspen transgenic for the same *rolC* gene but under the control of light inducible *rbcS* promoter from potato are more or less similar to untransformed controls. An evaluation of the cellular components of the leaf, in terms of the length of upper and lower epidermis cells, stomata length and density revealed no difference between *35S-rolC* transgenic aspen and un-

Table 1. Plasmids used for transformation of two aspen (*Populus tremula*) clones and one hybrid aspen clone (*P.tremula* × *P.tremuloides*)

Plasmid	Transferred gene and promoter	P.tremula (W52, Brauna11)	P.tremula × P.tremuloides (Esch5)
pPCV002	—		Yes
(in GV3101)	35S-rolC	Yes	Yes
	35S-Ac-rolC	Yes	Yes
	rbcS-rolC	Yes	Yes
	35S-iaaL		Yes
	35S-Ac-iaaL		Yes
	35S-GAO		Yes
SLJ2962			
(in LBA4404)	rbcS-Ac-rolC		Yes

The plant selection marker in each case was kanamycin.

Table 2. Number of leaf disks infected with *Agrobacterium tumefaciens* strains containing different constructs, and number of independent regenerants and kanamycin-resistant and-sensitive clones, and plant phenotype or leaf color

Chimeric gene	Leaf disks (n)	Independent regenerants		Km-resistant clones		Phenotype or leaf color (no. of clones)
		(n)	(%)	+	–	
Esch5 clone						
—	181	10	6	10	0	Wild type
35S-rolC	163	16	10	15	1	15 light green, rolC$^+$
35S-Ac-rolC	197	28	15	28	0	16 variable/12 green
rbcS-rolC	108	19	18	17	2	15 light green/ 2 green
rbcS-Ac-rolC	256	23	9	21	2	2 variable/19green
35S-iaal	114	8	7	8	0	Drooping leaves
35S-ac-iaaL	35	22	63[a]	22	0	As wild type
35S-GAO	65	5	8	5	0	As wild type
W52 clone						
35S-rolC	39	12	31[a]	12	0	11 light green, rolC$^+$ 1 green, rolC$^-$
35S-Ac-rolC	85	29	34[a]	27	2	17 variable/10 green
rbcS-rolC	49	2	5	1	1	1 light green
Brauna 11 clone						
35S-rolC	109	7	6	6	1	4 light green, rolC$^+$ 2 green, rolC$^-$
35S-Ac-rolC	133	6	4	5	1	2 variable
rbcS-rolC	124	7	6	7	0	7 light green

Km, kanamycin.
[a]Improved transformation method.

transformed controls (Table 3). This indicates that in this case reduction in leaf size is not related to reduction in cell size. However, all clones carrying *35S-rolC* and *rbcS-rolC* genes revealed light green coloration of leaves when compared to untransformed aspen. Due to this special feature constructs were used in which expression of the *rolC* gene is inhibited by insertion of the transposable element *Ac* from maize (Spena et al. 1989). Transgenic aspen transformed with the *35S-Ac-rolC* and *rbcS-Ac-rolC* genes are morphologically somewhat similar to untransformed aspen. However, out of 51 independently regenerated *35S-Ac-rolC* transgenic aspen clones, 30 clones showed light green/dark green variegated leaves. This indicates the function of the transposable element *Ac* in transgenic aspen.

Molecular analysis done so far using the PCR method revealed the presence of the *rolC* and *Ac* genes in the genome of the transgenic aspen, and the first

Table 3. Measurements of cell length and width, stomata length and density of upper and lower epidermis of control (Esch5) and different *35S-rolC* transgenic clones of independent origin

	Upper epidermis				Lower epidermis			
	Cell length (μm)	Cell width (μm)	Stomata length (μm)	Stomata per 0.1 mm^2 (*n*)	Cell length (μm)	Cell width (μm)	Stomata length (μm)	Stomata per 0.1 mm^2 (*n*)
Esch5	59.4± 8.8	35.9± 4.1	31.3±2.9	1.4±1.3	64.4± 7.4	36.7± 5.4	32.2±3.1	11.5±1.9
Esch5	62.5± 7.1	38.4± 3.7	30.8±4.0	-	55.2± 9.2	31.6± 3.4	29.1±2.1	12.2±1.8
Esch5	65.3± 8.7	37.5± 4.0	29.7±3.2	0.9±1.1	60.8± 8.2	34.4± 6.8	30.7±3.3	10.9±1.4
E2-1-1	53.2±11.9	33.2± 6.1	32.5±3.4	-	53.9± 6.6	33.2± 5.2	36.0±4.0	-
E2-1-5	44.6± 7.8	27.4± 4.8	-	-	41.6± 5.5	26.1± 9.1	32.5±4.3	11.1±1.1
E2-2-2	72.5± 6.0	47.5± 5.5	36.7±2.2	2.4±1-3	70.4± 6.5	50.6± 6.1	36.8±3.8	12.2±1.4
E2-9-1	73.8± 7.1	45.5± 8.5	32.9±2.9	1.8±1.4	78.5±11.6	53.3± 6.5	32.3±3.1	9.2±1.5
E2-13-1	77.7±12.4	44.6±10.0	-	-	73.6±17.1	40.8±12.8	35.5±3.3	10.8±1.1
E2-16-2	66.4± 6.8	36.9± 5.2	32.3±2.5	3.4±0.92	71.1±10.0	40.0± 5.3	30.1±2.5	11.2±1.1

results in Southern hybridization experiments confirmed the results obtained by PCR and indicated the presence from zero to three or four copies of the transferred gene in the genome of the transgenic aspen.

The phenotype of the transgenic aspen carrying the *rolC* gene under control of the 35S promoter of the cauliflower-mosaic-virus different from the one observed in tobacco (Schmülling et al. 1988) and potato (Fladung 1990; Fladung and Ballvora 1992). In *35S-rolC* transgenic tobacco, potato and aspen, dwarf plants were obtained, but in tobacco lanceolate leaves were induced and in potato bushy plants containing many tillers with very small leaves were described. Transgenic aspen carrying the *35S-rolC* construct revealed a less capacity to increased tillering during vegetative plant growth compared to *35S-rolC* transgenic tobacco or potato plants, but show well-developed axillary buds.

In conclusion, it is suggested that in the case of the *35S-rolC*, one single gene under control of a specific promoter can have a dramatic effect on plant growth and development, but the same gene under control of a different promoter revealed a less pleiotropic effect in transgenic plants.

Acknowledgements. This project was partly supported by a project AIR2-CT94-1571 from the European Union, Brussels. We are grateful to Drs. A. Spena, T. Schmülling (Max-Planck-Institut für Züchtungsforschung, Köln, Germany) for providing *Agrobacterium* strains carrying the *rolC* and *iaa-L* constructs, Dr. J.D.G. Jones (Jones et al. 1992; John Innes Centre, Norwich, U.K.) for providing the *Agrobacterium* strain LBA4404, Mr. Olaf Nowitzki for excellent technical assistance, and Mr. H. Seuthe and W. Holzwart for greenhouse work.

References

Ahuja MR (1986) Aspen. In: Evans DA, Sharp WR, Ammirato PJ (eds) Handbook of Plant Cell Culture 4, Techniques and Applications, Macmillan Publishing Company, New York, 626–651

Ahuja MR (1987) In-vitro propagation of poplar and aspen. In: Bonga JM, Durzan DJ (eds) Cell and Tissue Culture in Forestry, Vol. 3, Martinus Nijhoff, Dordrecht, 621–651

Barry GF, Rogers SG, Fraley RT, Brand L (1984) Identification of a cloned cytokinin biosynthesis gene. Proc Natl Acad Sci USA 81: 4776–4780

Estruch JJ, Chriqui D, Großmann K, Schell J, Spena A (1991a) The plant oncogene *rolC* is responsible for the release of cytokinins from glucoside conjugates. EMBO J 10: 2889–2895

Estruch JJ, Schell J, Spena A (1991b) The protein encoded by the *rolB* plant oncogene hydrolyses indole glucosides. EMBO J10: 3125–3128

Fladung M (1990) Transformation of diploid and tetraploid potato clones with the *rolC* gene of *Agrobacterium rhizogenes* and characterization of transgenic plants. Plant Breeding 104: 295–304

Fladung M, Ballvora A (1992) Further characterization of *rolC* transgenic tetraploid potato clones, and influence of daylength and level of *rolC* expression on yield parameters. Plant Breeding 109: 18–27

Fladung M, Ballvora A, Schmülling T (1993) Constitutive or light regulated expression of the *rolC* gene in transgenic potato plants has different effects on yield attributes and tuber carbohydrate composition. Plant Mol Biol 23: 749–757

Jones JDG, Bishop G, Carroll, B, Dickinson, M, English J, Harrison K, Jones D, Scofield S, Thomas CM, (1992) Prospects for establishing a tomato gene tagging system using the maize transposon Activator (Ac). Proc Royal Soc Edinburgh 99B(3/4): 107–119

Glass NL, Kosuge T (1986) Cloning of the gene for indoleacetic acid-lysine synthetase from *Pseudomonas syringae* subsp. *savastanoi*. J. Bacteriol. 166: 598–603

Schmülling T, Schell J, Spena A (1988) Single genes from *Agrobacterium rhizogenes* influence plant development. EMBO J 9: 2621–2639

Schmülling T, Röhrig H, Pilz S, Walden R, Schell J (1993) Restoration of fertility by antisense RNA in genetically engineered male sterile tobacco plants. Mol Gen Genet 237: 385–394

Spena A, Schmülling T, Koncz C, Schell J (1987) Independent and synergistic activity of *rol A*, *B* and *C* loci in stimulating abnormal growth in plants. EMBO J 6: 3891–3899.

Spena A, Aalen RB, Schulze SC (1989) Cell-autonomous behavior of the *rolC* gene of *Agrobacterium rhizogenes* during leaf development: a visual assay for transposon excision in transgenic plants. Plant Cell 1: 1157–1164

Spena A, Prinsen E, Fladung M, Schulze SC, van Onckelen H (1991) The indoleacetic acid-lysine synthetase gene of *Pseudomonas syringae* subsp. savastanoi induces developmental alterations in transgenic tobacco and potato plants. Mol Gen Genet 227: 205–212

Wolter FP, Fritz CC, Willmitzer L, Schell J, Schreier P (1988) *RbcS* genes in *Solanum tuberosum*: Conservation of transit peptide and exon shuffling during evolution. Proc Natl Acad Sci USA 85: 846–850

28 Manufacture of Proteins Based on Recombinant Chinese Hamster Ovary Cells: Assessment of Genetic Issues and Assurance of Consistency and Quality

F.M. Wurm[1], C.J. Petropoulos[2], and J.V. O'Connor[2]

[1]Ecole Polytechnique Fédérale de Lausanne EPFL, Département de Chimie, Biotech Chem, IV, 1015 Lausanne, Switzerland
[2]Genentech Inc., Department of Process Sciences, South San Francisco, CA 94080, USA

Introduction

Chinese Hamster Ovary (CHO) cells are a popular mammalian host for the production of complex proteins for clinical applications. Today, more than 60% of candidate proteins for therapeutic and prophylactic purposes are being produced in mammalian cells, and CHO cells have served as pioneering hosts for recombinant cell culture technology on a large scale. Although other systems (prokaryotes, yeast, insects) are capable of producing large amounts of protein, the decision to employ mammalian cells has been driven by the need to obtain material with multiplicities in biological activity. These complex characteristics often depend on a number of post-translational events, i.e. correct folding and disulfide bridge formation, oligomerization, proteolytic processing, the addition of carbohydrate groups, secretion into the culture medium and more. Such activities are mediated, generally without major problems, by the protein producing machinery of mammalian cells. Reliable, high level, high quality and consistent productivity of cells in culture has been achieved through the development and use of novel techniques derived from many fields over the last 10 years. They range, to mention a few here, from the specifics of expression vectors to the complex biochemical composition of optimized cell culture media for large scale production processes and the engineering and design of adequate cell culture reactor equipment.

This article will discuss the status of recombinant DNA that has been introduced into CHO cells as well as the products derived from such transgenic DNA. Many readers will realize from this article a puzzling enigma: While the methods and technologies for gene transfer and expression in mammalian cells in large scale manufacturing processes have become efficient and reliable, little is

E.R. Schmidt/T. Hankeln (Eds.)
Transgenic Organisms and Biosafety

known about the "molecular" fate of DNA sequences introduced into the genome of mammalian cells. One of the major reasons for the lack of detailed knowledge in this area is the extreme complexity of the mammalian genome and scarcity and limitations of methods for the study of the mechanisms of transfer and the production of transgenic DNA in mammalian hosts in general.

On the other hand, a battery of methods is available for the analysis of purified proteins, and most of these methods have been optimized and tremendously fine-tuned over the last decade. When employed as routine test procedures during the manufacturing process, these methods assure the high quality and consistency of products derived from manufacturing processes based on recombinant mammalian cells (O'Connor et al. 1994; Garnick 1992; Lu et al. 1988; Chloupek et al. 1989; Ling et al. 1991; Stephenson and Clarke 1989).

Part 1 of this article represents a somewhat detailed summary of the most frequently applied methods used to establish high yielding cell culture production processes. It also contains presently available fragments of information about the molecular and cytogenetic characteristics of recombinant DNA within the host cells. In part 2 of this article we will discuss, in a more compressed format, the usual process strategies, regulations and quality control methods used in the manufacture of proteins whose synthesis is based on transgenic DNA in CHO cells. In particular, we will discuss our experience with an observed irregularity which was detected by the application of one of the most powerful methods in protein chemistry and which was traced back to a genetic heterogeneity in the cell line used for production of the protein in question. This example shows that presently available technology is able to detect, with a high degree of confidence, even very subtle heterogeneities in products derived from recombinant DNA.

Results and Discussion

Part 1

Introduction to the Basic Features of the Chinese Hamster Ovary Expression System

For those readers unfamiliar with this system, here is a short description: A mutant cell line, lacking enzyme activity for Dihydrofolate Reductase (Urlaub and Chasin 1980) can be cultivated in the presence of glycine, hypoxanthine and thymidine (GHT). When these cells are transfected with a functional expression vector for dihydrofolate reductase (DHFR), cell lines that have acquired a functional DHFR gene can be selected and expanded in media lacking GHT. A second expression cassette for a product of interest (for example, a protein with therapeutic value) can be included in the DHFR plasmid or co-transfected as part of a second vector (Ringold et al. 1981; Kaufman and Sharp 1982; Scahill et al.

1983). Screening a number of clones leads to the identification of cells which co-express the product of interest. This is thought to be the result of co-integration of transfected plasmid sequences into chromosomal DNA (Ringold et al. 1981; De Saint Vincent et al. 1981; Kaufman et al. 1985). Expression in CHO cells can be augmented by exposing recombinant cells to the drug methotrexate (MTX), a folate derivative which blocks DHFR activity completely and irreversibly. Such an exposure to MTX selects cells which express DHFR at an elevated level. The elevated DHFR expression is the result of increased copy numbers of the recombinant sequences (Nunberg et al. 1978; Kaufman et al. 1985; Schimke 1984). Treatments with stepwise elevated concentrations of MTX can be repeated several times, and may result in the isolation of cells which contain dramatically increased copy numbers of the transferred genes. As found by in situ hybridization and DNA blot hybridizations most of these sublines have amplified plasmid DNA integrated into one or more chromosomes to very high copy numbers (Pallavicini et al. 1990). Cell lines with 2000 or more copies have been found (Wurm et al. 1986). These cells often reveal extended and/or abnormal chromosome(s) and, more importantly, the elevated expression level for the DHFR genes is frequently accompanied by increased production of the gene product of interest. The same principles of DNA integration and amplification as those discussed above and in the following for the CHO system apply to other mammalian systems which are based on stable integration of exogenous DNA as well.

Integration of Exogenous DNA into the Chromosomes of Chinese Hamster Ovary Cells

Knowledge about the events leading to the integration of non-homologous plasmid DNA into chromosomal DNA of mammalian hosts is very limited. However, it has been known for some time now that transfected DNA molecules become physically and genetically linked (Wilson et al. 1982; Perucho et al. 1980). Analysis of transgene structures within the host genome indicates that plasmid molecules are subjected to endonuclease, exonuclease and ligase activity. Homologous recombination may occur between individual molecules before integration. Of course, integration itself requires an additional recombinase or integrase activity. Some of these issues are discussed below in more detail.

Usually supercoiled (circular) plasmid DNA is used for transfections. Such DNA must be converted into a linear structure prior to integration into the host cell DNA. Endonuclease activity is abundant in the nucleus and it generates, from circular molecules, molecules with free 5'- and 3' ends linearized at random locations (Finn et al. 1989). Such molecules are susceptible to further attack by exonuclease activity. As a result of these combined activities, a substantial fraction of plasmid molecules that enter the nucleus may be degraded before they have the opportunity to integrate into the genome (Loyter et al. 1982).

Another very important activity that affects the overall organization of plasmid-derived DNA within the receiving genome is mediated by nuclear ligases. Since mammalian cells have abundant end-joining capabilities nuclear ligases will act on all molecules with "free ends" including the linear plasmid sequences that result from endonuclease attack. By linking linear molecules together, ligases protect plasmid sequences from further degradation by exonuclease activity. This activity is also the molecular basis for the high success of co-integration upon co-transfection of two (or more) independent vectors, one carrying the selector gene (DHFR) and the other the expression cassette for the product of interest.

Results from our routine co-transfections performed over the last 5 years support the statements made above. We have found that a variable, but usually large fraction (20%–100%) of CHO clones established using calcium phosphate co-express co-transfected expression vectors. We assume that occasional transfections resulting in less than 50% of clones showing co-expression of the desired gene are due at least in part to insufficient sensitivity of the assays used for detection and are rarely due to actual non-expression of the product. Recently, we have performed a series of co-transfections (13 in total) and analyzed more than 2500 clones for the production of the production of TNFR-IgG, a chimeric protein. We found that 96% of these clones co-expressed the desired transgenes. Co-expression has been observed by other authors as well (Ringold et al. 1981; Kaufman and Sharp 1982; Kaufman et al. 1985; Wurm et al. 1986, 1992), however, variability in expression levels of the transgene of interest and lack of sensitive assays may have prevented a statistical analysis and may be the reason why many prefer vectors which incorporate both a selective gene and a gene of interest in one plasmid molecule.

The combined activity of endo- and exonucleases can result in the generation and integration of small plasmid fragments, within the host chromosome. Restriction enzyme analysis of genomic DNA prepared from recombinant cell lines support the presence of co-integrated plasmid fragments. Figure 1 shows the Southern analysis of several methotrexate selected subclones of CHO cells expressing the mouse c-myc protein (Wurm et al. 1986). The figure shows the hybridization banding patterns created by 4 independently established high copy subclones of those CHO cells. The main band represents the full length restriction fragment containing the c-*myc* gene (within the vector). One of the lanes (lane 4) shows two additional bands of smaller size. Since these smaller bands are equivalent in intensity to the main band, they are very likely to have arisen prior to amplification and are probably the result of fragmentation and/or rearrangement during or prior to the integration event. Folger and coworkers (1982) and Finn et al. (1989) identified and analyzed another mechanism – homologous recombination between plasmid molecules for fragmentation and rearrangement of transfected DNA in cells. Such recombinations can occur between individual identical plasmid molecules as well as between different vectors, since most of them share large regions (10%–50% of their size) of homologous sequences.

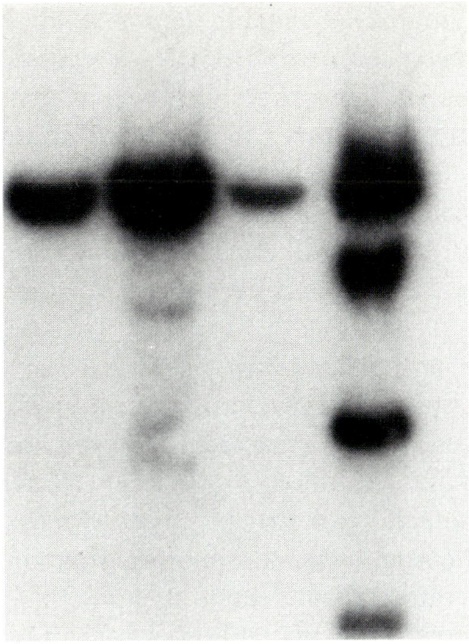

Fig. 1. Southern blot hybridization of four highly amplified subclones (selected at 320 μM methotrexate) of cell lines expressing the mouse c-*myc* gene. The major band in *lanes 1–3* represent the expected restriction fragment for the inserted DNA. The intensities of *bands 2* and *3* (from *top*) in *lane 4* are roughly equivalent to the one of *band 1* (by densitometric scan)

In summary, the structure of the DNA immediately prior to the primary integration event is affected by cellular nucleases, ligases and recombinases. These three activities contribute to the generation of DNA structures much larger and more complex than the original transfected plasmid(s). Obviously such structures may exist in an infinite number of combinations some fraction of which will include functional full-length expression cassettes. Obviously, the applied selection procedure, ensures the survival of only those cells that have incorporated functional structures for the selection gene. Additional screening can be used then to identify cell clones that express high levels of the desired protein product.

Structural Organization of Integrated DNA and Primary Integration Sites

It becomes obvious from the above discussion why there is very little known about the precise structure of integrated plasmid sequences. A complete analysis of an integrated DNA structure and its insertion site has not been reported because the information cannot be obtained by conventional means, for example restriction analysis. A complete answer to this question can only be derived by sequencing integrated plasmid structure along with the adjacent flanking cellular regions. Such data, which would give us more insights into the structure and chromosomal environments of integrated plasmid DNA in cell clones, have not yet been published. However, even an analysis of a single integration structure and its flanking host cell DNA in a cell line would be very laborious. For example, if the integrated sequences consisted of an undefined structure containing the equivalent of ten plasmid molecules and the plasmid was about 5 kb in size, a complete characterization of the locus would require sequencing across 50 kb, a daunting exercise even for laboratories that are geared toward DNA sequence analysis. As indicated by the findings of Robins and coworkers (1981) as well as our own limited studies (see below), integration structures may often contain many more than ten plasmids molecules which would make the sequence analysis even more laborious. Establishing sufficient data for conclusions with respect to organizational patterns of integrated structures will require sequencing a large number of independent insertion events, further multiplying the work involved.

In view of such a challenging situation we (and others) have used restriction enzyme digestion and Southern transfer analysis as a preliminary and very limited technique to acquire at least some insights into structural features and copy numbers of integrated DNA molecules in a number of primary clones (unamplified) that were generated by co-transfection of two plasmids. An example of such an analysis is shown here. It demonstrates that clones from the same (calcium phosphate) transfection may exhibit a wide range in number of integrated plasmids. Based on the densitometric scan of a photographic film exposure, we were able to estimate the range of plasmid copy number of six randomly chosen clones to be between 20 and 400 (Fig. 2). We caution that this quantitative determination was limited to the hybridization signal of a restriction fragment that contains a full-length gene carried in one of the plasmid vectors. The hybridization signals generated from other fragments that did not migrate at this position were not included. Because these additional fragments sometimes contained a significant fraction of the hybridization signal in the lane, we consider our 20–400 copy range estimate to be conservative.

Theoretically, integration may occur either randomly or at preferred sites in the genome. In the latter case, these sites may be defined by either structural or

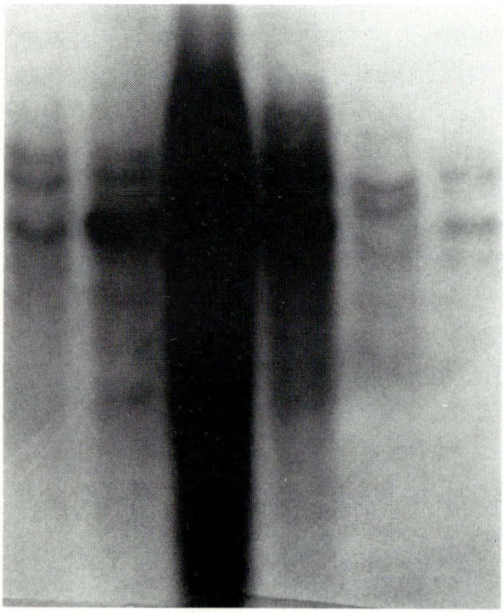

Fig. 2. Southern blot hybridization of six "best" producer cell clones (of 24 randomly chosen ones) with respect to the integrated plasmid sequences. The bands of highest intensity in *lanes 1, 2* and *6* (from *left* to *right*) indicate the position of a restriction fragment corresponding to the expected size

functional features of the chromosomal DNA. At present there is no experimental evidence to suggest that the integration of plasmid sequences delivered by standard transfection techniques occurs either at random or preferred sites. The observed expression level variations among randomly chosen clones (one to three orders of magnitude) may be based, at least in part, on influences exerted by the genomic "environment" or "neighborhood" of the integrated plasmid fragments. However, this is clearly not the only explanation for varying expression levels since, as pointed out above, the copy number of expression cassettes can also vary dramatically between clones.

We are presently trying to address the question of preferential integration (active versus non-active chromosome region) in an indirect way: We propose that the transcriptional activity of presumptive integration sites may influence the plasmid copy numbers observed in recombinant cell lines. If integration occurs in a purely random fashion, a majority of the integration events should take place in transcriptionally inactive regions (roughly 50% of mammalian DNA is single copy and only 0.1% of human or mouse genomic DNA contains coding sequences; Britten and Kohne 1968; Doolittle and Sapienza 1980; Little 1993). Integration within regions of low transcriptional activity will necessitate high plasmid copy numbers of a selectable marker, such as DHFR, in order to mediate sufficient expression for the cello's survival. Cells that have integrated a low number of plasmids in transcriptionally inactive regions will not survive selection. In contrast, cells that have acquired plasmid sequences within transcriptionally active regions will be more likely to survive even if the plasmid copy number is relatively low (even one). Thus, the frequency with which low copy number integrations are found in a large population of individual clonal cell lines may indicate indirectly whether integration occurs predominantly in transcriptionally active or inactive regions.

Recent studies indicate that defining the "integration structure" may be even more difficult than the direct cloning and sequencing of large stretches of DNA. Compelling evidence has been presented by the group of Wahl and co-workers (Carrol et al. 1987, 1988), that the "initial" integration locus might be quite unstable. It seems that this locus is frequently excised from the chromosome, taking with it large stretches of neighboring genomic DNA. According to this model, recombinant sequences become embedded in self replicating, episomal but submicroscopic structures which, at least in CHO cells and some other cell lines, appears to be "short lived", persisting only a few days. The episome can increase in size over time and, in some cell lines may be able to form "double minute" chromosomes. This extension in unit length per episome will result in an increase in copy number of the embedded exogenous DNA per cell. In CHO cells, these episomes will reintegrate into the chromosomal backbone, and remain there, possibly resulting from a selection mechanism which removes cells from the population which do not integrate the episome into a chromosome (Windle et al. 1991).

It becomes clear then from the discussion presented above that questions regarding primary integration loci are extremely difficult to address and, at this point in time, given the available technology, almost impossible to answer comprehensively. Up to now, no data have been presented which would indicate that structural or functional regions or epitopes of chromosomes may act as preferred targets for transfected exogenous DNA. We reason that it is more likely that most integrations occur at random sites, thus introducing DNA into regions of poor transcription rates. However, work is ongoing by a number of groups, including our own, to develop techniques and vectors which will target integration of transgenic DNA to transcriptionally active regions of the genome (Wurm et al. 1992, 1994; Reff 1993).

Number of Integration Events Per Cell

Depending on the cell line and the techniques used, in vitro a high percentage of cells are capable of taking up DNA and transporting it to the nucleus where it is actively transcribed (Jordan and Wurm 1994). Yet, only a very small percentage of these cells actually integrate functional DNA into their chromosomes. Felgner and Ringold (1989) reported on the fate of liposome and calcium phosphate transfected DNA in mouse L-cells. They found that an estimated 5×10^5 plasmid molecules were initially taken up by each cell, but that, on average, only 300 plasmid molecules, or ten plasmid molecules, could be detected in the nuclear fraction two days after liposome delivery or, calcium phosphate transfection, respectively. Thus, it would appear that only a small fraction of the molecules that enter the cell will become integrated into chromosomal DNA (Loyter et al. 1982). Perucho et al. (1980) studied the size of the chromosomally integrated concatemeric "package" and found it to be quite large, up to 1000 kilo base pairs. This apparent contradiction can be explained by the possibility that those few cells which emerge under selective growth conditions are exceptions. They might belong to a small fraction of cells that received and integrated a much larger quantity of DNA into their nucleus. However, the number of molecules or the size of the concatemeric structure within the nucleus prior to or after the integration does not address the question whether single, or multiple events occur predominantly.

Limited data from fluorescence in situ hybridization (FISH) studies suggest that the number of integration sites for any given cell is, in fact, small. When FISH was used to study integration in a set of 24 ("clonal") cell lines by a group collaborating with one of us (Glaab et al. 1995) single integration sites, each at unique chromosomal locations were found most. Robins et al. (1981) used standard in situ hybridization techniques to demonstrate that plasmids had integrated into one single chromosomal region, even in cells containing up to 100 integrated copies. Toneguzzo et al. (1988) reported on the integration pattern of exogenous DNA in lymphoid cells after electroporation. Using Southern blot hybridizations they found that 45 out of 50 clones generated a single restriction band containing the exogenous DNA when an enzyme was used which did not cut within the plasmid DNA. The remaining five clones had more than one band (up to four). They concluded that multiple integrations must have occurred in these clones. We are skeptical about this conclusion because there are several other explanations for the generation of more than one restriction fragment in cells which have been in culture for some periods after the integration event (for example chromosomal rearrangements in subpopulations of cells post transfection). We cannot totally rule out the possibility for multiple (primary) integration events in a single cell. However, we believe it occurs rarely, because of the many barriers exogenous DNA has to overcome before it can become stably integrated and expressed from a chromosomal location.

Genetic Stability and the Amplification of Stably Integrated DNA Using Methotrexate

The stepwise treatment of cells with increasing concentrations of MTX often results in the emergence of resistant CHO cell populations. These cells frequently contain elevated copy numbers for both the vector sequences expressing the gene of interest and fragments of the DHFR expression plasmid. It is assumed that an elevated number of DHFR gene copies produce higher amounts of DHFR, mediating resistance to MTX toxicity. Higher concentrations of MTX in the culture medium often correlate with higher production of the protein of interest (Ringold et al. 1981; Kaufman et al. 1982, 1985, 1986; Weidle et al. 1988). We have verified this observation in many cell lines producing various recombinant proteins. Here we show two examples. Table 1 shows expression levels of two clones (expressing a TNF-receptor-immunoadhesin) that were established by selection in 30 n*M* MTX. Subsequently, these clones were exposed to elevated concentrations of methotrexate. Expression of TNF-receptor immunoadhesin was improved (2.5- to sixfold) in subclones isolated at 1 μM and 5 μM MTX, albeit not to the same degree as MTX resistance (i.e. 50- to 150-fold).

The events taking place in cell populations which are exposed to methotrexate can be thought of as "Darwinian" in nature: infrequent genetic events (for example, amplification of DHFR genes) are selected for by cell survival in increasing concentrations of MTX. One should not expect that cells carrying am-

Table 1. Improvement of relative productivity of two clonal cell lines expressing TNFR-IgF following selection in elevated concentration of methotrexate (MTX)

Clone	Relative productivity		
	MTX concentration of first cloning	Recloning at 1000 m*M* MTX	Recloning at 5000 n*M* MTX
I	1 (30 n*M*)	2.5	6.1
II	0.9 (100 n*M*)	5.0	n.d

n.d, not determined
Productivity of clond I (cloned at 30 n*M* MTX) is given as 1.

plified DNA segments should be genetically stable strictu sensu immediately following the emergence of an amplification event. Cytogenetic studies by a number of laboratories (Stark et al. 1989; Milbrandt et al. 1983; Carrol et al. 1988) support the assumption that the majority of "species" in selected cell populations exhibit unusual chromosome structures. Based on their features they are thought to be genetically unstable. Following amplification and in the continual presence of MTX, cell (sub-)populations with stable chromosomal structures could emerge (again as a result of "Darwinian" selection).

The increased sensitivity and resolution of fluorescence in situ hybridization (FISH; Lichter et al. 1988; Pinkel et al. 1986; Lichter et al. 1990) has provided a new and direct approach to answer questions regarding the genetic effects of MTX selection. Briefly, we will summarize findings obtained by one of us (as part of a another team of collaborators) using FISH analysis of recombinant DNA in CHO cells (Pallavicini et al. 1990). For these studies cell lines were used that were initially established for the overproduction of the mouse c-*myc* gene, induced by heat shock (Wurm et al. 1986). These lines are ideal substrates for such studies since no c-myc protein is produced from this plasmid construction under "normal" circumstances, i.e. when the cells are kept at standard temperature (37 °C). This excludes the reasonable possibility of deleterious (i.e. selective) effects of overproduction of the non-selected gene in the system. Subsequently, these studies have been extended to cell lines with constitutive co-expression of DHFR and a gene of interest and three general observations were made (see also Table 2):

1. In the presence of MTX, a large subpopulation of cells grown contain multiple chromosome rearrangements, including extended regions of amplified sequences on unusual chromosomes. Often, these amplified sequences span very large regions of chromosomes (30%–70% of the total length). In many cases we observed hybridization signals extending to and over the far ends of chromosomes (telomeric regions). Such chromosomal abnormalities have been reported earlier and various models for amplification mechanisms have been suggested (Schimke 1984; Stark et al. 1989; Windle et al. 1991; Ma et al. 1993). The unusual rearrangements and structures observed, most likely result from the strong (chromosomal) mutagenic activity reported for MTX, since it

Table 2. Observed frequency of master integrations (MI) or alternate integrations (AI) in metaphase spreads prepared from cell populations of four Chinese hamster ovary (CHO) lines grown in the presence or absence of methotrexate (MTX)

Cell	Plux MTX		Minux MTX		Days
	MI (%)	AI (%)	MI (%)	AI (%)	
I	52	48	98	2	40
II	49	51	69	31	50
III	39	61	78	22	15
IV	43	57	67	33	12

is known to induce disturbances of chromatin, such as dicentric chromosomes, chromatid gaps, breakages, triradial structures and fragmentation in the DNA of human, mouse and hamster cells (Goulian et al. 1980; Barbi et al. 1984; Mondello et al. 1984; Raffetto et al. 1979).

2. In clonal cell lines, another subpopulation of cells was identified which contains one and sometimes two (occasionally quasi-tetraploid lines were found) unique, identifiable chromosome(s) exhibiting amplified sequences as stably defined patterns of modest size ("the master integration").
3. In the absence of MTX (or upon removal of MTX selection) the majority of cells showing variant, extended and unusual integration patterns disappeared from the culture within 15–30 days. In contrast, the unique chromosome carrying the "master integration" was stable and became the dominant representative of amplified sequences of the cell population.

The existence of two populations of cells with distinct types of amplified sequences and differential genetic stability may explain some of the contradicting results reported in the literature (Weidle et al. 1988; Kaufman et al. 1986). Individual cell lines may vary with respect to the fraction of amplified sequences represented by the unstable, variant chromosomal structures. Thus, loss of protein productivity may be more pronounced in some cell lines than others when methotrexate selection is removed for an extended period of time. Therefore it is advisable for the establishment of cell lines for long term production, to assess their productivity after at least 6 weeks of culture in the absence of methotrexate. Sometimes, but not necessarily regularly, an initial loss of productivity will be observed, followed by stabilization at a lower level, which may be sufficient for the projected needs.

Part 2

Assurance of Consistency and Reliability of Recombinant Product Formation from Processes Based on Chinese Hamster Ovary Cells

The manufacturer of a pharmaceutical protein has one major concern: Will a product that is tested in extensive and expensive clinical trials remain the same through many years of a defined manufacturing process? Even minor changes may result in loss or modification of activity and could result in an increased risk to the patient. In order to reduce this possibility, batteries of in-process controls and tests are an inherent part of the production of clinical proteins. In the following section, sensitive and robust techniques and methods are outlined and discussed. The objective is (a) to prevent even small changes in the procedures employed to produce the product of interest and (b) to enable the detection and exclusion of protein products differing from the one tested in clinical trials.

Adherence to the Good Manufacturing Practice Protocol

The first level of protection against inadvertent changes of the product rests in effective management and control of the production process over time. The challenge is that of any mature industry: to produce large amounts of material at competitive cost while ensuring that product consistency and quality are maintained. Obviously, manufacturing teams and their supervisors undertake every effort to reproduce the manufacturing process to the utmost detail in every production "run". This is achieved by defining and describing each of the various steps, in the form of detailed protocols. Almost every aspect of the procedure is documented and establishes, in the form of GMP (good manufacturing practice) protocols, the basis for the overall procedure. A "sign off" procedure by supervisors represents an integral part, assuring that the operating personnel for the manufacturing process are in fact controlling, assessing and executing it according to the established protocols. In many instances, the signatures are prerequisites to allow the progression of the process to the next steps.

Knowledge about the particular parameters and steps that can result in product changes is usually acquired during the period of cell line and process development leading to the establishment of the manufacturing process. Once certain limits of process conditions have been identified (within which no change was observed) the GMP protocol is designed and finalized. Certain criteria, which have been defined as part of the manufacturing protocols, when observed by the operating personnel can trigger a more elaborate investigation by senior personnel and can at times result in an interruption of the manufacturing process.

Quality Control for the Detection of Variant Product Forms

Quality control (QC) is an integral part of the manufacturing process for recombinant products. A comprehensive approach, utilizing independent validated techniques, is applied to assess the quality and identity of the product from various angles (for a review, see Anicetti and Hancock 1994). It is the goal of QC efforts to assure the manufacturer (and the patient) that products produced over years of manufacturing will meet the stated specifications in terms of identity, quantity, activity and purity.

In principle, it is no longer so difficult to produce large quantities of highly purified recombinant proteins. However, the method to produce these proteins is still considered a young technology, since it is dependent on the manipulation of genetic material in the laboratory. Those manipulations involve the creation of plasmid vectors, based on fragments of genetic information derived from various sources, and the transfer of such DNA to or into the genetic material of a complex organism such as *Escherichia coli*, yeast or, as discussed here, mammalian cells cultivated in vitro. A major concern has been the fidelity (amino acid sequence identity) of the final product, particularly in view of the high degree of ignorance with respect to gene transfer mechanisms and its consequences, as summarized in the first part of this article. Based on our experiences with this technology (more than 10 years now), we argue that opinionated statements suggesting the need for nucleic acid based tests and assessments (Galibert 1990) appear to be far overstated and overrated: Paralleling the rapid development of techniques allowing manipulations and transfers of DNA molecules to various hosts, powerful new methods have emerged in analytical protein chemistry which have increased the capacity to characterize purified protein preparations to an extremely high degree of sensitivity and resolution. These methods have provided manufacturers of recombinant proteins efficient tools to assess identity, quantity, potency and purity of the product at the level where it is most needed: just prior to its administration (delivery) to the patient. The results of quality control assessments from thousands of lots of our approved products are very encouraging with respect to the (amino acid sequence) identity: No lot has been called into question based on a QC finding that has raised a suspicion that a mutant gene has emerged within the host cells, resulting in the appearance of a variant product version as the major form of molecule in the preparation.

The following, however, will discuss an interesting finding with one of our candidate therapeutic products during the early phases of development of the manufacturing process. In this instance we detected a "mutated" version that "contaminated" the gene pool for the product of interest. This mutation most likely arose in the early stages of the generation of the producer cell line. In this context we will discuss peptide mapping in combination with mass spectrometry for the (amino acid sequence) analysis of purified proteins. This technique, while being powerful and most convincing in linking the projected amino acid sequence of a known protein with data generated from purified populations of protein molecules, is not the only method of establishing quality and consistency of products. Standard methods like polyacrylamide gel electrophoresis for the determination and verification of the electrophoretic characteristic of the protein in question or isoelectrofocusing for the analysis of charge distribution heterogeneities (glycosylation) will continue to be useful to a certain degree. However, we will not discuss these methods here since they are well known, both with respect to their potential and their limitations as techniques for the characterization and identification of recombinant proteins.

Also, we will not elaborate on more specialized assays which have to be developed in a customized way to assess the biological activity of the protein in question. Such assays, when applied on a routine basis for lot release are frequently based on in vitro systems which "look" for the particular activity of the protein which is thought to be the principle activity mediating its function in the patient (i.e. enzymatic, blocking, binding, etc.). We exclude these assays from our discussion here for the following reason: Changes which are so dramatic that they will abolish or reduce to a significant degree (>20%) biological (physicochemical) activity in a test sample most likely occur rarely. If and when they occur, they should be easily detectable, since activity based assays are very simply validated and performed to a high degree of precision. Much more challenging in this respect are subtle changes, for example resulting in the exchange of a single amino acid which does not affect activity, conformation, electrophoretic behavior etc., yet might contribute, for example to an altered antigenicity of the product in the patient.

Peptide mapping is extremely powerful for product characterization purposes for the following four reasons: (1) peptide mapping analyzes the protein therapeutic, not the DNA that encodes for the protein, (2) post-translational modifications such as glycosylation, phosphorylation and hydroxylation cannot be predicted from the DNA sequence, (3) deamidation, oxidation of methionine and other natural modes of protein degradation are easily detected, and (4) fermentation conditions can cause amino acid substitutions that are independent of the DNA sequence (e.g., norleucine for methionine in *E. coli*-derived interleukin-2; Lu et al. 1988).

The first step in peptide mapping is the digestion of the protein of interest with an appropriate enzyme (peptidase) which results in a defined set of peptides. The choice of the enzyme(s) is dependent on the size and amino acid sequence of the protein: it is of course desirable to create a mixture of molecules that can be separated and fractionated easily on an appropriate chromatographic resin (usually a reversed phase column). High-pressure/performance liquid chromatography (HPLC) systems are used to generate a gradient for the separation of the peptides.

As an example for peptide mapping we will discuss the analysis of rhuMAB HER2, a recombinant humanized monoclonal antibody directed against the surface antigen HER2 (which is overexpressed in certain breast cancer patients). A version of this protein, which is presently (1995) being evaluated in phase III clinical trials, is produced in Chinese hamster ovary cells in a large scale manufacturing process in reactors (for the cell culture aspect) of up to a volume of 12 000 l. We chose to discuss this protein here because it served as a first example in our hands where we detected, with an early version of the protein during the initial phases of process development an irregularity with respect to the expected amino acid sequence.

For the purpose of peptide mapping of rhuMAB HER 2 the enzyme trypsin was chosen for digestion. Injection of the mixture into a Nucleosil C-18 (HPLC) column (Fig. 3) resulted in 77 collectable fractions. Each of these fractions were collected and subjected to amino acid (composition) analysis. Due to the small size of the peptides (resulting in unambiguous mixtures of quantifiable amounts of individual amino acids), most of the 77 fractions could be assigned to short stretches of amino acid sequence within the antibody molecule, using just the data from composition analysis. Some co-eluting or ambiguous peaks were further analyzed by Edman degradation and/or electrospray mass spectrometry. Overall it was possible to confirm 98% of the primary structure of the protein. However one peptide consisting of 22 amino acids (fraction 58), represented by a small peak in the HPLC profile, partially matched the expected amino acid sequence of another peptide, also consisting of 22 amino acids (fraction 61). The amino acid composition of peptide 61 was in agreement with the correct se-

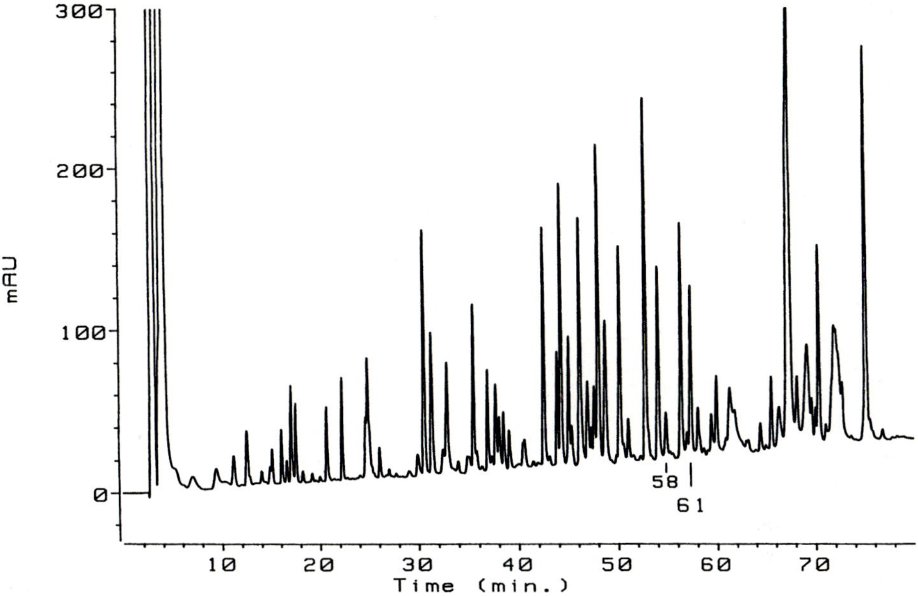

Fig. 3. Reverse-phase high-performance liquid chromatography (HPLC) profile of a tryptic peptide digestion of a recombinant humanized monoclonal antibody (mAb) against the antigen HER-2 (rhuMAb-HER2). The numbers *58* and *61* indicate the elution position of those peptides discussed in the text. *Fraction 61* represents the expected peptide, and *fraction 58* represents the "variant" peptide

quence of a fragment from rhuMAB HER2, whereas the one of fraction 58 suggested a "mutation" from a tyrosine to a glutamine (or glutamic acid) at position 376 of the protein (Table 3).

The smaller size of fraction 58 suggests that protein which was the originator of this fragment was present in the total population of molecules at a frequency of about 15%. We tested many lots from a manufacturing campaign and found the ratio of the "correct" fragment to the (variant) "false" fragment to be variable. We verified, through additional tests that in fact a tyrosine to glutamine mutant form of human recombinant anti-HER2 was present in the population of molecules and that the origin of this mutant could be traced back to one particular transfection resulting in the establishment of the cell line used for the manufacturing of the material studied during the early phase of process development (Harris et al. 1993). We concluded, and later verified through PCR assisted DNA analysis, that the cell line chosen for the production of this material contained and expressed, in somewhat variable but significant quantities, two sets of expression vectors. One set represents the original DNA construct whereas the other contained a mutant form in which a single nucleotide has been exchanged, resulting in a code alteration from tyrosine to glutamine. The origin of this phenomenon is still under investigation.

Table 3. Amino acid analysis of peptides containing heavy chain residues 374–395

	Amino acid residues/mol		
	Expected value	Fraction 61 observed	Fraction 58 observed
Cys	0	0.0	0.0
Asx	4	3.7	3.7
Thr	0	0.1	0.1
Ser	2	1.8	1.8
Glx	4	4.0	5.2
Pro	2	1.9	2.0
Gly	2	2.0	2.0
Ala	1	1.2	1.2
Val	1	1.2	1.2
Meth	0	0.0	0.0
Ile	1	1.1	1.0
Leu	0	0.1	0.2
Tyr	2	2.0	1.0
Phe	1	1.1	1.0
His	0	0.0	0.1
Lys	1	1.0	1.0
Arg	0	0.0	0.0

More frequently than the example discussed here, where the product heterogeneity was based on the genetic exchange of one amino acid by another, are heterogeneities in protein preparations resulting from influences (physical, chemical, enzymatic) excerpted during the manufacturing process. The exposure of the recombinant protein to these conditions can result, for example, in partial deamidation or oxidation of amino acids. It has been shown by one of us that peptide mapping is the method of choice for the detection of such heterogeneities as well (O'Connor 1993).

The example discussed above shows, we think convincingly, that peptide mapping is in fact a very robust and sensitive technique allowing the detection of single amino acid mutants in preparations of purified proteins. In combination with batteries of tests assessing the quality, identity and purity of protein preparations from angles different from the one discussed here it is possible, and has been shown already for thousands of lots of recombinant protein pharmaceuticals, to provide these products to the patient (and physician) with a very high degree of confidence.

References

Anicetti V, Hancock WS (1994) Analytical considerations in the development of protein purification processes. In: Harrison R (ed) Protein purification process engineering. Dekker, New York

Barbi G, Steinbach P, Vogel W (1984) Nonrandom distribution of methotrexate-induced aberrations on human chromosomes. Detection of further folic acid sensitive fragile sites. Hum Genet 68: 290–294

Britten RJ, Kohne D (1968) Repeated sequences in DNA. Science 161: 529–540

Carrol SM, DeRose ML, Gaudray P, Moore CM, Neeha-Vandevanter DR, Von Hoff DD, Wahl GM (1988) Double minute chromosomes can be produced from precursors derived from a chromosomal deletion. Mol Cell Biol 8: 1525–1533

Carrol SM, Gaudray P, DeRose ML, Emery JF, Meinkoth JL, Nakkim E, Subler M, Von Hoff DD, Wahl GM (1987) Characterization of an episome produced in hamster cells that amplify a transfected CAD gene at high frequency: functional evidence for a mammalian replication of origin. Mol Cell Biol 7: 1750–1759

Chloupek RC, Harris RJ, Leonard CK, Keck RG, Keyt BA, Spellman MW, Jones AJS, Hancock WS (1989) Study of the primary structure of recombinant tissue plasminogen activator by reversed-phase high-performance liquid chromatographic tryptic mapping. J Chromatog 463: 375–396

De Saint Vincent BR, Delbrück S, Eckhart W, Meinkoth J, Vitto L, Wahl G (1981) The cloning and reintroduction into animal cells of a functional CAD gene, a dominant amplifiable genetic marker. Cell 27: 267–277

Doolittle WF, Sapienza C (1980) Selfish genes, the phenotype paradigm and genome evolution. Nature 284: 601–603

Felgner PL, Ringold GM (1989) Cationic liposome-mediated transfection. Nature 337: 387–388

Finn GK, Kurz BW, Cheng RZ, Shmookler Reis RJ (1989) Homologous plasmid recombination is elevated in immortally transformed cells. Mol Cell Biol 9: 4009–4017

Folger KR, Wong EA, Wahl G, Capecchi MR (1982) Patterns of integration of DNA microinjected into cultured mammalian cells: evidence for homologous recombination between injected plasmid molecules. Mol Cell Biol 2: 1372–1387

Galibert F (1990) Stability of a gene recombinant: what does it mean and how to check for it. Biologicals 18: 221–224

Garnick RL (1992) Peptide mapping for detecting variants in protein products. Dev Biol Standard 76: 117–130

Glaab F, Scherthan H, Wurm FM (1995) Paper in preparation

Goulian M, Bleile B, Tseng BY (1980) Methotrexate-induced misincorporation of uracil into DNA. Proc Natl Acad SciUSA 77: 1956–1960

Harris RJ, Murnane AA, Utter SL, Wagner KL, Cox ET, Polastri GD, Helder JC, Sliwkowski MB (1993) Assessing genetic hetereogeneity in production cell lines: detection by peptide mapping of a low level Tyr to Gln sequence variant in a recombinant antibody. Bio/Technology 11: 1293–1297

Jordan M, Wurm FM (1994) High level transient expression in mammalian cells: Identification and optimization of physico-chemical parameters of the calcium phosphate transfection method. (to be published in the Proceedings of ESACT meeting 1994, Velthoven, Holland)

Kaufman RJ, Sharp P (1982) Amplification and expression of sequences cotransfected with a modular dihydrofolate reductase complementary DNA gene. J Mol Biol 159: 601–621

Kaufman RJ, Wasley LC, Furie BC, Shoemaker CB (1986) Expression, purification and characterization of recombinant-carboxilated factor IX synthesized in Chinese hamster ovary cells. J Biol Chem 261: 9622–9628

Kaufman RJ, Wasley LC, Spiliotes AJ, Gossels SD, Latt SA, Larsen GR, Kay RM (1985) Coamplification and coexpression of human tissue-type plasminogen activator and murine dihydrofolate reductase in Chinese hamster ovary cells. Mol Cell Biol 5: 1750–1759

Lichter P, Chieh-Ju CT, Call K, Hermanson G, Evans GA, Housman D, Ward DC (1990) High-resolution mapping of human chromosome 11 by in situ hybridization with cosmid clones. Science 247: 64–69

Lichter P, Cremer T, Chieh-Ju CT, Watkins PC, Manuelidis L, Ward DC (1988) Rapid detection of human chromosome 21 aberrations by in situ hybridization. Proc Natl Acad Sci USA 85: 9664–9668

Ling V, Guzzetta AW, Canova-Davis E, Stults JT, Hancock WS, Covey TR, Shushan BI (1991) Characterization of the tryptic map of recombinant DNA derived tissue plasminogen activator by high-performance liquid chromatography-electrospray ionization mass spectrometry. Anal Chem 63: 2909–2915

Little P (1993) Small and perfectly formed. Nature 366: 204–205

Loyter A, Scangos GA, Ruddle FH (1982) Mechanisms of DNA uptake by mammalian cells: Fate of exogenously added DNA monitored by the use of fluorescent dyes. Proc Natl Acad Sci USA 79: 422–426

Lu HS, Tsai LB, Kenney WC, Lai P-H (1988) Identification of unusual replacement of methionine by norleucine in recombinant interleukin-2 produced by E. coli. Biochem Biophys Res Commun 156: 807–813

Ma C, Martin S, Trask B, Hamlin JL (1993) Sister chromatid fusion initiates amplification of the dihydrofolate reductase gene in Chinese hamster cells. Genes Dev 7: 605–620

Milbrandt JD, Azizkhan JC, Hamlin JL (1983). Amplification of a cloned Chinese hamster dihydrofolate reductase gene after transfer into a dihydrofolate reductase-deficient cell line. Mol Cell Biol 3/7: 1274–1282

Mondello CR, Giorgi, Nuzzo F (1984) Chromosomal effects of methotrexate on cultured human lymphocytes. Mutat Res 139: 67–70

Nunberg JH, Kaufman RJ, Schimke RT, Urlaub G, Chasin LA (1978) Amplified dihydrofolate reductase genes are localized to a homogeneously staining region of a single chromosome in a methotrexate-resistant Chinese hamster ovary cell line. Proc Natl Acad Sci USA 75: 5553–5556

O'Connor JV (1993) The use of peptide mapping for the detection of heterogeneity in recombinant DNA-derived proteins. Biologicals 21: 111–117

O'Connor JV, Keck RG. , Harris RJ, Field MJ (1994) New techniques in protein chemistry. In: Brown F, Lubiniecki AS (eds) Genetic stability and recombinant product consistency. Dev Biol Stand 83: 165–173

Pallavicini MG, DeTeresa PS, Rosette C, Gray JW, Wurm FM (1990) Effects of methotrexate (MTX) on transfected DNA stability in mammalian cells. Mol Cell Biol 10: 401–404

Perucho M, Hanahan D, Wigler M (1980) Genetic and physical linkage of exogenous sequences in transformed cells. Cell 22: 309–317

Pinkel D, Straume T, Gray JW (1986) Cytogenetic analysis using quantitative, high sensitivity fluorescence hybridization. Proc Natl Acad Sci USA 83: 2934–2938

Raffetto, Parodi GS, Faggin P, Maconi A (1979) Relationship between cytotoxicity and induction of sister-chromatid exchanges in mouse foetal cells exposed to several doses of carcinogenic and non-carcinogenic chemicals. Mutat Res 63: 335–343

Reff ME (1993) High-level production of recombinant immunoglobulins in mammalian cells. Curr Opin Biotechnol 4: 573–576

Ringold G, Dieckmann B, Lee F (1981) Co-expression and amplification of Dihydrofolate Reductase cDNA and the Escherichia coli XGPRT gene in chinese hamster ovary cells. J Mol Appl Genet 1: 165–175

Robins DM, Ripley S, Henderson A, Axel R (1981) Transforming DNA integrates into the host chromosome. Cell 23: 29–39

Scahill SJ, Devos R, van der Heyden J, Fiers W (1983) Expression and characterization of the product of a human immune interferon cDNA gene in Chinese hamster ovary cells. Proc Natl Acad Sci USA 80: 4654–4658

Schimke RT (1984) Gene amplification in cultured animal cells. Cell 37: 705–713

Stark GR, Debatisse M, Giulotto E, Wahl GM (1989) Recent progress in understanding mechanisms of mammalian DNA amplification. Cell 57: 901–908

Stephenson RC, Clarke S (1989) Succinimide formation from aspartyl and asparaginyl peptides as a model for the spontaneous degradation of proteins. J Biol Chem 264: 6164–6170

Toneguzzo W, Keating A, Glynn S, McDonald K (1988) Electric field gene transfer: characterization of DNA transfer and patterns of integration in lymphoid cells. Nucleic Acids Res 16: 5515– 5532

Urlaub G, Chasin LA (1980) Isolation of Chinese hamster cell mutants deficient in dihydrofolate reductase activity. Proc Natl Acad SciUSA 77: 4216–4220

Weidle UH, Buckel P, Wienberg J (1988) Amplified expression constructs for human tissue-type plasminogen activator in Chinese hamster ovary cells: instability in the absence of selective pressure. Gene 66: 193–203

Wilson JH, Berget PB, Pipas JM (1982) Somatic cells efficiently join unrelated DNA segments end to end. Mol Cell Biol 2: 1258–1269

Windle B, Draper B.W, Yin Y, O'Gorman S, Wahl GM (1991) A central role for chromosome breakage in gene amplification, deletion formation, and amplicon integration. Genes Dev 5:160–174

Wurm FM, Gwinn KA, Kingston RE (1986) Inducible overexpression of the mouse c-myc protein in mammalian cells. Proc Natl Acad USA 83: 5414–5418

Wurm FM, Johnson A, Lie Y, Etcheverry TM, Anderson KP (1992) Host cell derived retroviral sequences enhance transfection and expression efficiency in CHO cells. In: Spier, Griffiths, MacDonald (eds) Animal cell technology: developments, processes and products. Butterworth-Heinemann, Oxford, pp 35–41

Wurm FM, Johnson A, Etcheverry T, Lie YS, Petropoulos CJ (1994) Retrotargeting: use of defective retroviral DNA fragments to improve recombinant protein production in mammalian cells. In: Spier, Griffiths, Berthold (eds) Animal cell technology: products for today and the future. Butterworth-Heinemann, Oxford, pp 24–29

IV Field Release of Transgenic Plants

29 Out in the Open: Field Trials with Genetically Modified Crop Plants in Four European Countries, Chile and the United States

T.A.W.M. Saat

VanderHave Research, P.O. Box 1, 4410 AA Rilland, The Netherlands

Introduction

Any European directive can be implemented in different ways in different countries. Directive EC/90/220, dealing with introducing genetically modified organisms (GMOs) into the environment, is no exception in this respect. The international consensus underlying legislation regarding GMOs is intended to result in world-wide harmonization. Consensus in this context of course only refers to common opinions among competent authorities and national experts. Public acceptance of and appreciation for scientific work within the broad field of biotechnology are far less unanimous (e.g. Zechendorf 1994). It is public opinion, or better still, public opinions, that will determine whether or not genetically modified novel foods, medicines and other products will reach the market successfully in the foreseeable future. Not less important, the public will have its say about continuing the funding of biotechnology research carried out at present at universities and public institutes. The words "out in the open" in the title therefore literally mean "for everyone to see and judge". Field trial activity reports, like the example presented in Table 1, should be open to public discussion. When comparing field trials in six different countries, it is difficult not to overemphasize differences. This is especially appealing, when in all of these countries but the United Kingdom the same crop and trait, maize harbouring glufosinate ammonium resistance, were field tested.

Intermezzo: Gene Transfer and Access to "New" Genetic Information

One of the biosafety principles that can be regarded as being generally recognized, is that the genetic trait determines a possible risk and not the method

E.R. Schmidt/T. Hankeln (Eds.)
Transgenic Organisms and Biosafety

Table 1. Overview of all genetically modified organism (GMO) field trials carried out, considered or being prepared by the Royal Vanderhave Group

Crop	Traits	Country									
		NL	BE	FR	UK	GE	SP	IT	CHI	ARG	USA
Maize	Gu	91+	93+	91+	–	95?	–	95?	93+	–	93+
Maize	Ms Gu	95?	–	95?	–	–	–	–	94	–	–
Beet	Vr	92+	95?	93+	93+	–	–	95?	95	–	–
Beet	Gy	92+	95?	93+	93+	–	–	95?	95	–	–
Beet	Gu	94+	–	–	–	–	–	–	–	–	–
Beet	Fp	94+	–	–	–	–	–	–	–	–	–
Beet	Tp	95?	–	–	–	–	–	–	–	–	–
Beet	Sp	95?	–	–	–	–	–	–	–	–	–
Oilseed rape	Ma	91+	–	–	–	–	–	–	–	–	–
Oilseed rape	Gu	93+	–	92+	–	–	–	–	–	–	–
Oilseed rape	Ms	Gu	–	–	92	–	–	–	–	–	–
Oilseed rape	Ir	–	–	–	93	–	–	–	–	–	–
Oilseed rape	Fp	95?	–	–	–	–	–	–	–	–	–
Swede	Gy	94+	–	–	–	–	–	–	–	–	–
Swede	Fp	95?	–	–	–	–	–	–	–	–	–
Fodder rape	Gy	94+	–	–	–	–	–	–	–	–	–
Fodder rape	Fp	95?	–	–	–	–	–	–	–	–	–
Sunflower	Ma	–	–	94	–	–	–	–	–	–	94
Sunflower	Gu	95?	–	–	–	–	–	–	–	–	–
Sunflower	Fr	–	–	95?	–	–	–	–	–	–	–
Sunflower	Pq	–	–	95?	–	–	–	–	–	–	–
Sunflower	Fp	–	–	–	–	–	95?	–	–	–	–
Sunflower	Tp	–	–	–	–	–	95?	–	–	–	–
Sunflower	Ge	–	–	–	–	–	95?	–	–	–	–
Sunflower	Ir	–	–	–	–	–	–	–	–	94	95?

L, Netherlands; BE, Belgium; FR, France; UK, United Kingdom; GE, Germany; SP, Spain; IT, Italy; CHI, Chile; ARG, Argentina; USA, United States.
In addition to marker genes, the following traits were introduced: fructan production (Fp), fungus resistance (Fr), vegetative growth enhancement (Ge), glufosinate ammonium resistance (Gu), glyphosate resistance (Gy), insect resistance (Ir), marker genes only (Ma), male sterility (Ms), protein quality enhancement (Pq), sugar production enhancement (Sp), trehalose production (Tp), virus resistance (Vr). 91+ indicates the year 1991 and subsequent year(s); 95? indicates those experiments that are presently under consideration or in preparation.

by which it has been acquired by an organism (e.g. Eucarpia 1989). Despite wide official recognition, application of this principle when asked about the risks of transgenic organisms or (horizontal) transgene transfer regularly turns out to have a provocative effect. Therefore this part of our contribution to an open and frank discussion hides behind the eminent authority of professor Arthur Rörsch, who heads the Dutch Board for Agronomical Research (NRLO). According to one of his lectures, if half of the earth's four billion known species became extinct, only few biochemical functions would disappear (Rörsch 1994). There are a great many good reasons to preserve as many of the earth's creatures as possible, but apparently saving biochemical functional diversity seems not to be one of them.

He illustrated his controversial statement by taking a wide class of enzymes as an example. There are about five thousand different natural enzymes that are -ases, ordered into sixty different sub-classes. All these sub-classes turn out to have already been present more than one billion years ago, when life on earth consisted almost exclusively of microorganisms. In other words: the genes for the different basic biochemical functions were shared before plants, fungi, animals and other phyla branched off. Subsequently, evolution of regulatory mechanisms of such basic functions resulted in ever new and refined creatures. This process bears some resemblance to continuously refining the regulated interaction of only a limited array of functional electronic parts (resistances, capacities, transistors etc.), resulting in a wide variety of products, ranging from door bells to computers. Some classes of organisms have retained many of the genes coding for basic biochemical building blocks though sometimes, as in the case of plants, expressing them at a low level. Other organisms seem to have lost the ability to perform many of these functions. Within this model, it's the regulation and fine-tuning of biochemical processes that determines any species' identity, not the basic functional capabilities themselves. Although resulting in a plea for more fundamental, regulation-oriented research, one could use Rörsch's argumentation to put discussions about (horizontal) (trans)gene transfer in a different light. Introducing a -glucuronidase (GUS) or a neomycin phospho-transferase (NPT-II) into a plant species, and eventually into its relatives, should perhaps be seen more as a reintroduction of a trait lost during evolution. Treating such an event as if a whole new, or even unnatural, situation were created might need some amendment. In fact, some plant species still exhibit local GUS activity at some stages in their development (Hu et al. 1990). Even if Rörsch's arguments presented above were valid, it would probably not be of much use in helping to overcome fears of gene transfer and genetic modification in general. According to another of Rörsch's lectures, aircrafts have enabled half of the fifty plant diseases not endemic to Australia before 1945 to reach this continent by now (van Damme 1993). Still, the general public regards biotechnology a greater threat to the integrity of nature than botanical gardens, hospitals or tourism, to name but a few sources of new genetic information. That is why it is in our own interest as biotechnologists to prevent, as much as is practically possible, negative or unforeseen side effects of genetic modification. This remains valid, even when we realize that the same amount of effort could, when applied elsewhere, result in a far greater overall risk reduction.

Role of Field Trials in Plant Breeding

The field trials of a plant breeding firm aim at testing potential new varieties under normal agricultural growing conditions. This has been the case for conventional varieties for a great many years, and this ought to become the case for crops that happen to be genetically modified. In theory, the statement mentioned above saying that it's the trait that counts, and not the method by which it has been acquired should also be applied for field trials. In reality, for instance, all of a plant's herbicide resistances are equal, but some turn out to be more equal than others. Breeding companies are used to performing field evaluations and seed productions wherever and whenever they are useful. Field evaluations usually take place in the heart of the various regions of cultivation. When a new disease or pathotype is spotted, special trial fields aimed at detecting a resistance to the disease early on, may be situated way off. For sowing seed production, companies have to spread risks by producing each variety in more than one region simultaneously. For breeding purposes the southern hemisphere provides an opportunity to alternate two growing seasons each year. In some regions in Asia and Africa climatic conditions are such that breeders are enabled to work all year round. When selecting locations for the various types of field work, infrastructure and the presence of a skilled staff have to be taken into account as well. For one of the ten or fifteen largest seed companies in the world, the considerations listed above add up to having numerous seed productions and tens of thousands of field trial plots in at least twenty-five countries on five continents. For genetically modified crops, it is still quite a different story altogether. An overview of all field trial activities in which VanderHave or its plant material are involved, realized as well as planned, can still be presented on a single sheet of paper, as is demonstrated in Table 1. For most of the company's major products, with the exception of some monocotyledonous crops like onions, cereals and grasses, at least some developments have reached a stage at which field trials are in progress. According to a recent, unpublished version of the GIBiP database counting all field trials with transgenic plants from 1986 until 1993 (Ahl-Goy et al. 1994; P. Ahl-Goy et al., unpublished), VanderHave was the third largest notifier in Europe. Active only since 1991, it had carried out 22 of the 391 field trials on this continent. All European public institutes and universities put together performed only 70 field trials with plants, during the entire seven year period. The figures presented above not only indicate that there is considerably less experience with field trials involving GMO's than with conventionally bred varieties. They also point out that, while most members of recombinant DNA commissions are scientists from universities and institutes, experience with and knowledge of field trials is mostly generated elsewhere. It is therefore crucial that there is a steady and open exchange of information.

Comparison Between Countries: Results

The reason that field trials are carried out in many different parts of the world has nothing to do with biotechnology. New varieties have to be adapted to local conditions. The only way to verify whether or not this requirement is met, is by testing material on the spot. At first, field trials are often exclusively aimed at the expression of the transgenic trait of interest. Such an initial round of trials often has to be performed under conditions that have too little in common with normal agriculture to be of much practical value to breeders. A small number of plants, delayed sowing, sowing by hand, removing inflorescences and other isolation and safety measures hardly interfere with the experimental goals at that early stage. Step by step and case by case, however, the subsequent trials may become more similar to ordinary breeding trials. At that point, the transgenic character has almost been reduced in interest to that of any other important trait. Depending on the crop, the focus usually lies on such characteristics as yield, lodging, earliness, conventional disease resistances and digestibility. The plots on which the level of an introduced resistance to a herbicide is still verified, in for instance a crop like maize, are an appendix to the main trial in which only conventional methods and herbicides can be used for weed control. As is the case with any herbicide resistance, the genetically modified crop should be able to withstand receiving a double dose in order to qualify as a commercial variety. During weed treatment by farmers, some strips of the field might accidentally be sprayed twice with herbicide due to overlapping beams. Even in this case no significant crop damage should occur. On all 19 locations in all five countries where glufosinate ammonium resistant maize has been tested so far, the level of glufosinate resistance proves to be sufficient for use in commercial varieties. Traits such as glufosinate resistance in maize and glyphosate resistance in sugar beets are passed to offspring as single dominant alleles in accordance with Mendel's laws in all six countries.

A summary of the quantitative data available from several rounds of trials with several types of material in four of these countries is presented in Table 2. The data in the first part of Table 3 refer to material derived from a single original transformant, called T14. The two plants of type A not scored as unharmed in Table 3 did not necessarily succumb to the herbicide treatment. They might just as well have been eaten by rodents or trampled during spraying. The three unharmed plants of type D most likely are the result of the conditions during seed production in Chile. The batches of seeds sown in 1994 were produced on a single field, using isolation measures intended to obtain a seed purity of at least 99%. Among offspring from such small scale experimental seed productions some aberrant phenotypes are the rule. Production of type E seeds did not involve transgenic plants and therefore could be carried out at a more distant location. Any aberrants in type E material therefore did not involve transgenic pollinator plants and died unnoticed after spraying glufosinate ammonium. The data given in the bottom row of Table 3 reflect results with maize plants derived from another single transformed plant, called T25. When comparing the data concerning T14- and T25-derived maize in Table 3, the inheritance pattern seems to be the same whether the locus consists of a single copy of the vector, hardly wider than the DNA of interest (material derived from T25, in some documents designated TR25), or contains multiple copies of the entire vector (material derived from T14/TR14). Data from glyphosate-resistant sugar beet plants field tested in the United Kingdom are presented in Table 4 for comparison's sake. Because sugar beet is generally a self-incompatible crop, only few viable selfing seeds were available for the trial. Although the low numbers do not

Table 2. Inheritance pattern of an introduced trait of maize in the Netherlands, Belgium and France from 1991 to 1994 (see Table 2a,b) and of sugar beet in the United Kingdom in 1993 (Table 2c)

Type	Material	Segregation expected (% resistant)
0	Original transformant	100
A	Offspring of homozygous GMO parent(s)	100
B	Offspring of hemizygous GMO parents	75
C	Offspring of hemizygous GMO parent	50
D	Offspring of homozygous recessive GMO parent(s)	0
E	Offspring of non-GMO parents	0

Table 3. Segregation pattern with regard to glufosinate resistance (see Table 2)

Type	Sprayed	Unharmed		Expected (%)
		(*n*)	(%)	
A	455	453	99.6	100
B	339	258	76.1	75
C	3633	1826	50.3	50
D	716	3	0.4	0
Wild-type maize				
E	1747	0	0.0	0
T25-derived maize				
C	383	202	52.7	50

allow any definite conclusions to be drawn, the transgenic character in sugar beets in Great Britain shows a segregation pattern that seems to be in accordance with Mendel's laws. Only for maize, yield data from hybrid material are available.

In Tables 5 and 6 a comparison is presented of results of several versions of two hybrid maize varieties, tested on three and five locations in the Netherlands respectively. Both sibling hybrids that had lost the introduced trait by previous crossing as well as the registered conventional version of these hybrids were used as reference. The presence or expression of the introduced gene did not result in yield reduction: Table 5 shows that the selection indexes (SI1) of the varieties with the extra resistance are most certainly not lower than those for the non-resistant segregants or the original conventional maize varieties. Other characteristics used by breeders to select new varieties, some of which are also listed in Table 5, show no measurable or significant change due to presence and expression of a transgene. From other locations and years no quantitative data are available or no hybrid material was tested, though all the general observations on performance and inheritance are in accordance with the results presented. With sugar beets, only transgenic parental lines of hybrid varieties-to-be have been

Table 4. Segregation pattern with regard to glyphosate resistance of clones and offspring of two independent transformants of sugar beet (see Table 2)

Type	Sprayed	Unharmed		Expected (%)
		(*n*)	(%)	
0	11	11	100.0	100
B	7	4	57.1	75
B	118	92	78.0	75

Table 5. Comparison of breeding characteristics of maize hybrids equivalent to two commercial hybrid varieties

Type	SI1	YLDS	DM A	DM R	L2	EH	PH
Scarlet variety							
A	101	102	100	29.4	6.1	144	226
D	98	98	101	29.8	6.0	149	226
E	100	101	100	29.5	6.0	145	226
Accent variety							
A	103	104	98	28.5	9.0	138	221
D	100	101	96	27.9	8.9	140	225
E	97	95	106	30.9	9.0	126	215

Types A and D are derived from a transgenic parental line homozygously containing the introduced gene or without it, respectively. Type E reflects the conventional variety produced under comparable seed production conditions. The following breeding characteristics are presented: overall selection index (SI1), fresh yield for use as silage (YLDS), dry matter silage absolute/relative (DM A/DM R), lodging resistance (L2), and ear/plant height (EH/PH).

tested so far, so no relevant yield data from hybrid material have become available yet.

Comparison Between Countries: Unexpected Results

From the summarized results of the Dutch 1993 maize trial presented in Table 3, the data obtained from the field in Wieringermeer in 1993 have been omitted because they do not seem to fit within the theory. In Table 7 the results of all ten locations of that year's trial are presented individually. The percentage of plants that remained unharmed at Wieringermeer lies well below the expected 50% for this type of plant material as well as below the 49% observed over the other nine trial sites. In Table 8 the segregation pattern of individual genotypes that the trial in Wieringermeer consisted of are listed along with the results for the same

Table 6. Segregation patterns of T14-derived maize hybrids isogenic to the Scarlet variety of maize (see Table 5) after spraying with glufosinate ammonium

Type	Sprayed	Unharmed		Expected (%)
		(*n*)	(%)	
A	106	104	98.1	100
C	117	52	48.4	50
D	124	0	0.0	0

batches of seed when sown on other locations. From Table 8 it becomes clear that the unexpected results have to do with the location only and have nothing to do with the transgenic nature of the plants. The same batches of seed that gave on average 2% unharmed plants at the Wieringermeer location, resulted in on average 48% unharmed plants on other locations. Untreated, non-modified plants too were severely stunted and showed a yellow and purple coloration (results not shown). Investigations revealed that the Wieringermeer field had been sown within a winter wheat plot partly cleared by using glyphosate. Before being cleared with roundup however, the wheat had already been treated with two other herbicides. The harm that was caused to the maize plants can most likely be attributed to one of these herbicides, called Ally. This compound, metsulfuron-methyl, is a sulfonylurea herbicide against various kinds of weeds, to be used with cereals only. Still, when a GMO field trial involving herbicide resistance results in severely damaging almost all so-called resistant plants, even unbiased visitors would probably have drawn obvious but incorrect conclusions, blaming the transgenic nature of the plants. If such misinterpretations had been presented to the public in an unfavourable light by a third party, it might have

Table 7. Overview of data from the 1993 round of experiment B/NL/92/08 with glufosinate ammonium-resistant maize in ten locations in the Netherlands

Site	Type	Sprayed	Unharmed		Expected (%)
			(*n*)	(%)	
Reimerswaal RE	C	170	85	50	50
Emmen EM	C	156	80	51	50
Wieringermeer WI	C	155	3	2	50
Dinteloord DI	C	172	87	51	50
Dongeradeel DO	C	77	85	48	50
Bellingwedde BE	C	82	88	48	50
Dronten FL	C	183	93	51	50
Susteren SU	C	129	63	49	50
Udenhout UD	C	127	62	49	50
Kloosterhaar KL	C	164	76	46	50
Total (without WI)	C	1460	719	49	50

Table 8. Specification of 1993 data from the location Wieringermeer (WI) compared to results with identical material sown at other locations during the same trial (see Table 7)

Site	Genotype	Sprayed	Unharmed	
			(*n*)	(%)
WI	10 219538-K	26	1	4
FL	10 219538-K	29	15	52
KL	10 219538-K	31	13	42
WI	16 219562-K	24	1	4
EM	16 219562-K	24	10	42
DI	16 219562-K	32	14	44
BE	16 219562-K	31	12	39
KL	16 219562-K	28	17	61
WI	19 219574-K	27	0	0
FL	19 219574-K	32	15	47
WI	27 219610-K	25	0	0
SU	27 219610-K	19	12	63
WI	32 219630-K	28	0	0
DI	32 219630-K	25	16	64
SU	32 219630-K	23	6	26
WI	34 219638-K	25	0	0
BE	34 219638-K	29	17	59
UD	34 219638-K	23	11	48
Total (WI)	–	155	3	2
Total (others)	–	326	158	48

proven very difficult, even with a full set of data at hand, to explain them away in a reaction. When reported by the notifier, at the Mainz' workshop as well as in our yearly progress report to the Dutch competent authority, this unexpected event hasn't drawn more attention than it deserves, being just an anecdote in the margin. In fact, it should be reassuring that, even though glufosinate ammonium resistant plants are often referred to as herbicide resistant plants, they have not become more resistant to herbicides in general. They have merely acquired one extra resistance besides the ones already present, as was intended.

Comparison Between Countries: Procedures

One can compare the hardware of the various countries: the laws and other parts of the legal framework. As a notifier, the software, or how the people involved perform their job within these legal boundaries, is more relevant. As an example, there exist striking differences between countries in the amount of time it takes the competent authorities to decide on similar applications (Table 9). Belgium and France manage to decide well within 90 days as is required by Directive EC/

Table 9. Overview of application periods in days for field trials with glufosinate-resistant, genetically modified maize

Country	Year			
	1991	1992	1993	1994
France	52	15	42	50
Belgium	n.a.	n.a.	42	75
Netherlands	111+30	186+30	145+30	133+42
U.K.[a]	n.a.	n.a.	149	96
U.S.A.	n.a.	n.a.	60	21
Chile	n.a.	60	30	30

n.a., not applicable.
[a]Glyphosate- and virus-resistant, genetically modified sugar beet.

90/220. If delay due to requests for additional information is taken into account, the United Kingdom too manages to decide within 3 months. Of the six countries tried so far, Chile and the United States are the most swift. By far the largest amount of patience from notifiers waiting for the authorities to come to a decision is asked in the Netherlands. Even after a positive answer, a waiting period of 30–42 days (see Table 9) has to be observed before activities may commence. That is, if the permit hasn't been suspended by the State Council in the meantime, resulting in a further delay of two to seven months. It is our experience that the differences in procedure length, such as occurred in 1994, are no reflection of differences in efficiency or efficacy, but are the consequence of distinctive legal frameworks. All six countries in this comparison provide notifiers with very good or even excellent help and advice. So far, the differences encountered haven't led to moving part of our research activities to other countries. Attention to field trial procedures is declining anyhow. Nowadays, many notifiers from industry are mainly focusing their efforts on the final step: market introduction of novel foods and other GMO products.

Comparison Between Countries: Simplification

All around the globe there are initiatives to simplify procedures, now that experience is teaching all sides involved what risks should and should not be taken into extra account. In the Netherlands one can apply for a trial on more than one site for more than 1 year. Notifiers that want to carry out trials in Belgium can apply for a continuation permit from the second year on. In such a fast track procedure little more than the new aspects of the prolongation have to be presented in full. Even though 5-year permits are issued by the Netherlands, in practice notifiers still have to file a new application every year, since the number of sites, constructs and genotypes increases with time. The Belgian fast track procedure therefore seems superior. In the Dutch situation however, many permits are temporarily suspended by the State Council. Multi-year permits from a year before can bridge gaps until the suspension of the most recent permits is lifted. In fact, both the Belgian and Dutch procedure are superior, under their respective circumstances. When it comes to simplification, notifiers are extremely difficult to please. The Dutch authorities regularly change the forms that have to be used when applying. Their latest improvement resulted in a form exclusively for use with plants. Instead of cheering, some grumpy scientists complained about having to enter data for all their crops on a new form once again.

Comparison Between Countries: Public Opinion

In Belgium, France, Chile, and the state of Wisconsin, the general public, apart from some officials, is not informed. The seemingly slower countries in Europe, the United Kingdom and the Netherlands, actively encourage or even demand that notifiers inform society. Reporting on an international seminar of competent authorities and non-governmental organizations that took place in July 1994, Schenkelaars noted: "...most European countries have their own legal provisions and mechanisms on public information and participation, often reflecting Europe's cultural and political diversity. This diversity ranges from governments that show little intention of making information available, to governments that take concrete steps to seek public opinion." From the viewpoint of a notifier any coin has two sides. And the other side to openness about field trial activities is vulnerability to sabotage. In 1991 and 1992 our Dutch field trials with GMO maize were destroyed by anonymous activists. It is not this risk of having one's trials destroyed that has kept VanderHave from carrying out GMO trials in Germany. Until recently, the manpower to provide the anticipated amount of information required by colleagues and the German general public was simply not available. The possibilities of having field trials in Kenya and India have been investigated briefly. At present it remains unclear whether governments and their officials in Africa and Asia would see such developments as opportunities or as threats. On the one hand developing countries voice their concern over being kept out of yet another promising modern technology. On the other hand they express their fear of possibly risky experiments being carried out in their backyards. Under such uncertain circumstances long-established companies will be reluctant to perform field trials in these parts of the world.

Summary and Discussion of the Comparison

As was to be expected, the plant's phenotype with regard to general glufosinate ammonium resistance and its pattern of inheritance do not vary from one country to another. Trials at different sites aim at identifying the best new agronomical varieties for each growing area. Over the years and in the various countries, the introduced agronomic traits as such are expressed at a sufficiently high level for commercial use. Application procedures for field trials with genetically modified organisms vary considerably in the time and effort required to obtain permission, but this turns out to be a reflection of the various types of national legislation. Many of the differences in legislation and obligations to inform of the general public coincide with distinctions in national character. This is one of the main reasons why not all of the industrial notifiers still have international harmonization high on their agendas (e.g. Dodet 1994).

Closing Remarks

In the author's personal opinion, support of notifiers by the competent authorities is good or excellent in all countries under consideration. Provided such support is continued, fully simplified and harmonized directives and laws that have to be rebuilt from scratch have low priority. At least for some notifiers there is no need for simplifications beyond developing "fast track" procedures for field trials and market introductions, and regularly updating "exemption lists". Harmonization of legislation is not expected to iron out submerged differences in public attitude and national character anyhow. Even under identical laws, German consumers, who reportedly show 80% disapproval of novel foods because of anticipated health effects (DPA 1994), will have to be approached differently from American consumers, of which only 25% do not accept the role of biotechnology in food production (Anonymous 1994). The time and costs to be saved by simplification and harmonization probably pale next to the worldwide effort needed to explain why certain applications of this new technology might be regarded as positive achievements.

References

Anonymous (1994) Amerikanen niet vies van biotechnologie. Agrarisch Dagblad, October 28

Ahl-Goy P, Chasseray E, Duesing J (1994) Field trials of transgenic plants: an overview. Agro-Food-Industry Hi-Tech March/April 1994, pp 10–15

Dodet B (1994) Industrial perception of EC biotechnology regulations. TibTech 12: 373–476

DPA (1994) Duitsers kopen geen genvoedsel. Agrarisch Dagblad, June 17

Eucarpia (1989) Statements of Eucarpia on risk assessment regarding the release of transgenic plants. Eucarpia Bulletin 18

Hu CY, Chee PP, Chesney RH, Zhou JH, Miller PD, O'Brien W (1990) Intrinsic GUS-like activities in seed plants. Plant Cell Reports 9: 1–5

Rörsch A (1994) Vooruitzichten in de biotechnologie anno 1994 "back to the future". Lecture at annual meeting of NIABA, 27 April 1994 (Dutch association for industrial and agronomical biotechnology, P.O. Box 443, Leidschendam, The Netherlands)

Schenkelaars P (1994) Letter to the editor. Bio/technology 12: 1048–1049

Van Damme R (1993) Toekomst ziet er uiterst somber uit zonder biotechnologie. De Stem, March 26

Zechendorf B (1994) What the public thinks about biotechnology. Better than synthetic food but worse than organ transplantation: a survey of opinion polls. Bio/technology 12: 870–875

30 How Will Transgenic Sugar Beets Behave in Natural Plant Communities?

D. Bartsch, A. Emonds, H. Gluth, C. Haag, C. Morak,
M. Pohl-Orf, and M. Schmidt

Department of Biology V (Ecology, Ecotoxicology and Ecochemistry),
Technical University of Aachen, Worringerweg 1, 52056 Aachen, Germany

Problems with Transgenic Plants: Are There Ecological Long-Term Effects?

According to Regal (1993), there are three arguments that genetic engineering differs from traditional breeding methods in its potential to alter plants:

1. Phylogenetic leapfrogging is a true ecological novelty. New attribute qualities can be introduced into organisms.
2. The genetic modification can be carried out without traditional debilitating trade-offs. Instead of substitutional changes of alleles in traditional techniques, gene technology is able to add radical adaptive improvements.
3. Gene technology offers access to the non-Mendelian portions of the genome. The basic genetic command system, which may be hidden for altering in conventional manipulations, is now open for reprogramming.

However, there are some important reasons why ecologically incapacitated organisms should not, in any event, become supercompetitive. But if an ecologically adaptive or superior trait is – directly or indirectly – transferred into an ecologically competitive host, environmental problems might soon occur.

The key question is: Will the transgenic attribute present an adaptive or superior advantage in ecosystems where the engineered organism or some of its genetically competent relatives can naturalize?

It has become apparent that naturalization only represents an initial step on the road to environmental problems, not damage per se. However, the velocity of the implication occurrence differs in biological time and space (Fig. 1). A look at the implications on the various biological organisation levels is helpful for locating long-term effects. The velocity of the ecological effects (if they occur!) decreases with the increasing complexity of the structure of the biological system and process.

E.R. Schmidt/T. Hankeln (Eds.)
Transgenic Organisms and Biosafety

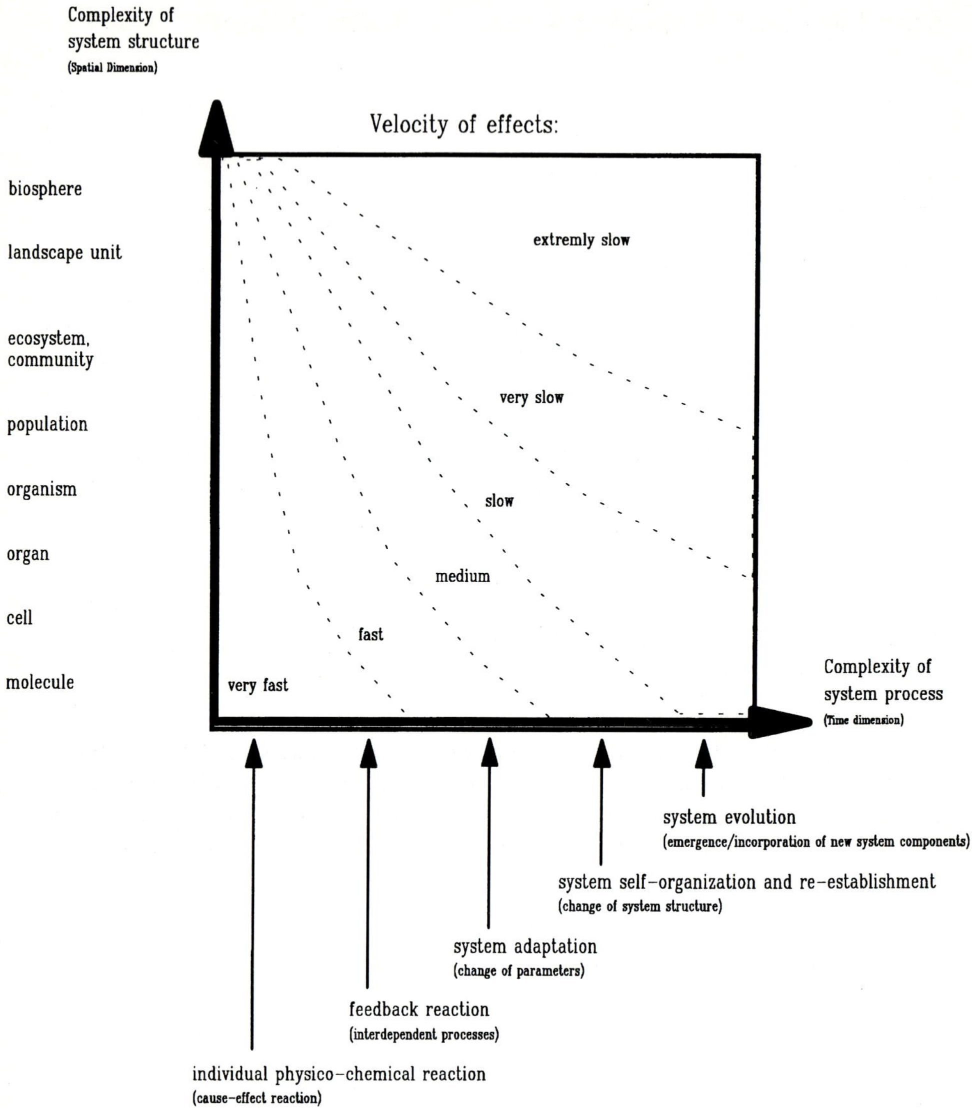

Fig. 1. Hierarchical levels in biological systems relating to structure and function (O'Neill et al. 1986). The chance of detecting biological alterations depends on the velocity of effects in biological systems

The expected effects should be seen separately from the hierarchical levels of biological organisation. This is very important for risk research experiments. Both the spatial and the temporal dimensions determine the possibility of getting data. Short-term experiments aiming to evaluate the influence of transgenic organisms on complex community structure alterations are impossible to carry out.

Step by Step and Case by Case: The PLANTA Sugar Beet

In the last 20 years biosafety paradigms have been developed: risk assessment has to depend both on the attributes of the used organism and of the engineered gene(s). Experimental research has to be carried out step by step and case by case.

The starting point for our risk research project in Aachen was a BNYV-virus resistance in sugar beets created by the company PLANTA in Einbeck (Germany) using an *Agrobacterium tumefaciens* mediated gene transfer. Besides containing the coat protein of the BNYVV, the transgenic sugar beets also contain two genetical markers: The Phosphinothricine-Acetyl-transferase (bar) gene determines resistance to the herbicide BASTA (from the company Hoechst, Frankfurt). The NPTII gene enables plant growth in the presence of the antibiotic Kanamycin. PLANTA has made this plant material (sugar beet, *Beta vulgaris* L. subsp. *rapacea* var. *altissima* (Koch) (Döll), Rothmaler 1990) available to interested scientists.

Case Studies: Potential Behaviour in Natural Plant Communities

The species *Beta vulgaris* L. includes cultivars (spinach-, sugar-, red- and fodder beet) and wild forms (sea- or wild-beet) which are freely compatible in terms of pollination. Therefore, in order to demonstrate the potential behaviour of transgenic sugar beets, case studies of three different plant communities are presented (Fig. 2).

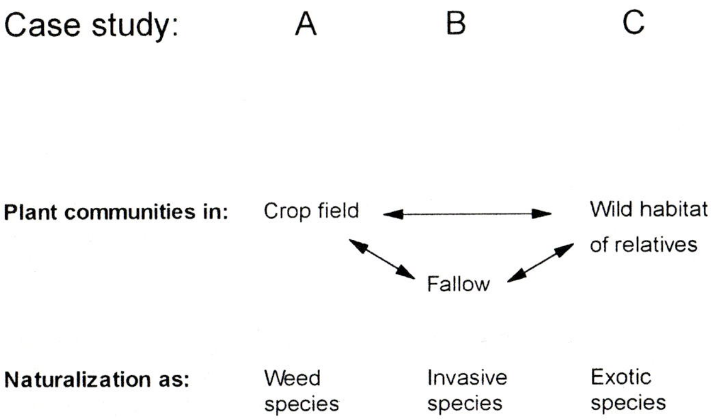

Fig. 2. Target habitats (*A–C*) according to ecological research of long-term effects of transgenic sugar beets in natural plant communities. Basic scientific questions should focus on naturalization in familiary habitats and invasiveness studies in potential habitats

Ecological impacts could arise through factors which improve the acclimatisation and dispersal of new genotypes. One important factor is, according to Crawley (1990), tolerance of close proximity to other plants (e.g. allelopathy, tolerance of competitors and abiotic environmental factors). A genetically engineered trait, which leads to resistance to a (virus) disease, offers an ecological advantage per se. The first and direct advantage will appear in the early life stage of beet. The virus naturally infects young seedlings by the vector funghi *Polymyxa betae*. However, the fitness of non-infected or resistant plants is increased in comparison with virus-infected plants. In order to evaluate the degree of this advantage, autoecological experiments were developed to compare resistant, susceptible and tolerant hybrids of sugar beet.

Case A: Agricultural Crop/Weed Communities

Several examples demonstrate that the exchange of genes between cultivated, wild and weed populations of beet (*Beta vulgaris* L.) can lead to undesirable effects. An overview is shown in Fig. 3. Weed beet populations in Northern France originated from the unintentional in-pollination of attributes of wild beet (*Beta vulgaris* L. subsp. *maritima*, Arcang; Rothmaler 1990). The genetically dominating occurrence of annuality in wild beet populations is one of the most important attributes for the weed infestation. Boudry et al. (1993) found a frequency of 90% annuality in weed beet populations and a similar frequency of cytoplasmatic maternity attributes originated from cultivar forms. Most plants of wild beet populations are annual, however the cultivar lines are always biennial. This phenomenon leads to the assumption that F1-Hybrids are formed in seed production areas by the wild beet pollination of a sugar beet breeding line. Locations where breeding districts overlap with wild beet habitats in Central Europe are presented in Fig. 3. The product, a wild beet/cultivar beet hybrid, is able to 'bolt' among the biennial sugar beet varieties in the cultivation period. Problems with annual weed beets are known for example in Northern France and Belgium (Fig. 3). If the bolting plants are not taken away before seeding, stable weed beet complexes are quickly formed. As an example, a weed beet population was built up in only 12 years on a field in Eschweiler near Aachen (Germany) in a threefold crop rotation (sugar beet, wheat, maize). Historically, we ascertained that the first cultivation of sugar beet in this location began in 1981 on a freshly recultivated brown coal surface mining area free of beets in the seed bank. In four periods of sugar beet cultivation without mechanical cutting of bolters, the weed beet population increased from just a few plants to over 80 000 plants per hectar. Such an increase is possible due to the enormous production ability of up to 20 000 seeds per flowering beet plant. Weed populations were not observed on neighbouring fields which had bolters mechanically removed.

The data show that transgenic beets will naturalize per se in sugar beet crop rotation weed communities. The establishment is independent of ecological advantage due to a genetically engineered trait in general. Of course, an increase

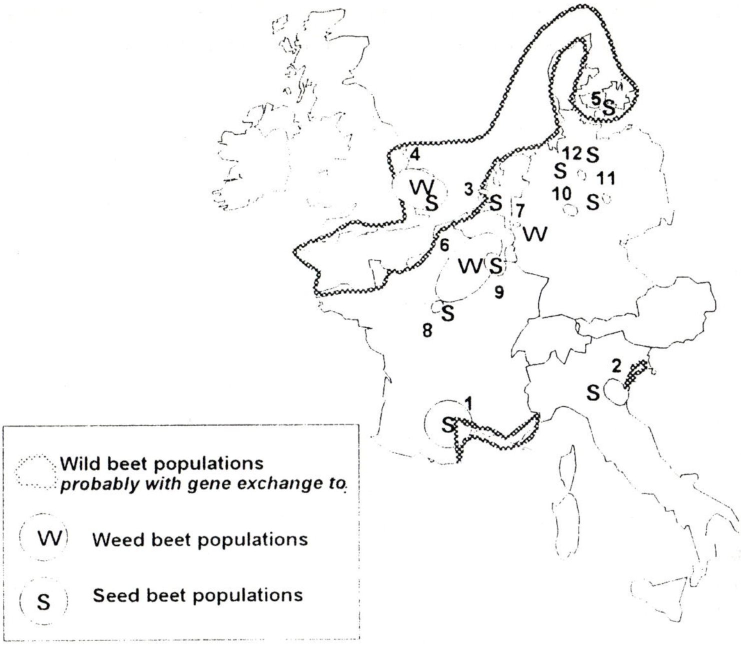

Fig. 3. Distribution of breeding districts (with seed beet populations), agricultural areas (with weed beet populations) and natural habitats (with wild beet populations) is shown for some examples in Central Europe. An overlap of seed beet/wild beet populations is given in numbers *1–5*, some weed beet populations are illustrated in numbers *4*, *6*, and *7*, and a few isolated locations of breeding companies (with seed beet populations) are found in numbers *8–12*

in the beets fitness will increase the amount of naturalization. Nevertheless the most important factor will be the periodic mechanical removel of bolters in young stage of weed development to avoid unintended long-term effects.

Case B: Successional Communities in Young Fallow

In 1993, a first field experiment was conducted in order to evaluate a test system describing the competitiveness phenomena of sugar beet in young fallow (Bartsch et al. 1994). The competitiveness was determined in degrees of biomass production in a planting system of sugar beet with different densities of *Chenopodium album*. The experimental programme focuses on the investigation of the ecological interaction of sugar beet in the target ecosystem. Young fallow is one of the most interesting area in the target agro-ecosystem where an unwanted survival of sugar beets could realistically occur. Sugar beet is a relatively weak competitor and, at the same time, is dependent on the high nutrient levels present in the test field. We assume that a small number of overwintering beets (e.g. dormant seed) will have to struggle with an upcoming competitor. The growth of the sugar beet was assessed in a situation where it was competing with *Chenopodium album*. The genus *Chenopodium* was chosen on the basis of the following criteria:

- Weed properties in pure sugar beet culture (common habitat exists)
- Occurrence in young fallow
- Similar ecological niches to the cultured plants (common adaptation strategies in disturbed habitats with possibly similar alternative habitats outside of agricultural cultures).

The methods used in experiments are described in detail in Bartsch et al. (1994). As seen in Fig. 4, the sugar beet genotypes differ significantly in their biomass production on a fresh weight basis at the virus-infected site (Oberviehhausen). The highest biomass production is exhibited by the normal virus-tolerant variety, the lowest by the susceptible unchanged hybrid. The biomass production of the transgenic virus-resistant hybrid is significantly higher in comparison to the susceptible control hybrid at Oberviehhausen, whereas no difference is seen at the virus-free site (Wetze).

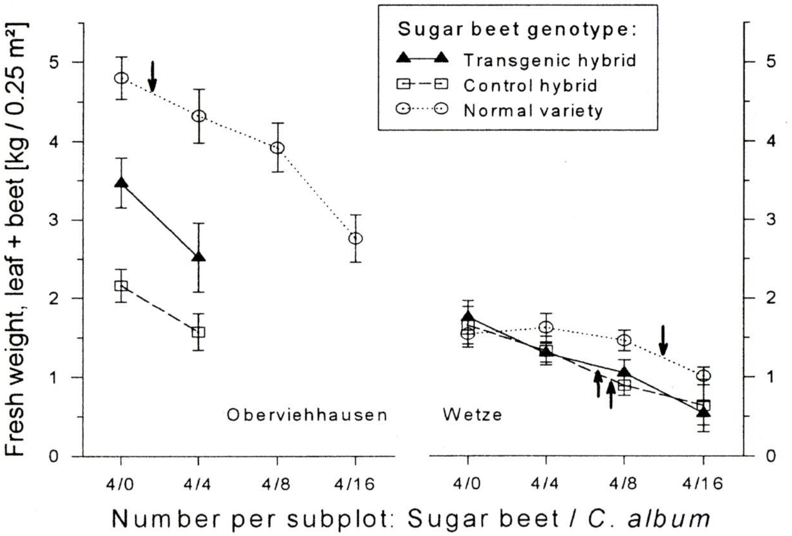

Fig. 4. Vegetative biomass production (kg fresh weight/subplot) of different sugar beet genotypes (transgenic virus-resistant hybrid, control susceptible hybrid, normal virus-tolerant variety) grown in four different competition densities with *Chenopodium album* at a site with (Oberviehhausen) and a site without beet necrotic yellow vein (BNYV) virus infection (Wetze). *Bar lines* represent the standard deviation from the mean (n = 20 parallels, four plants each). *Arrows* symbolize the biomass of the sugar beet genotypes grown in competition with the naturally occurring weeds (control variant with different densities and species). Some experimental variants of transgenic and control hybrids are missing due to a decline in seed emergence in spring

The biomass production decreases for all sugar beet genotypes at the test sites with an increase of the *C. album* competitor. The normal variety, however, had a significantly higher productivity at both test sites with the exception of the control plot without *C. album* competition in Wetze.

The sugar beet productivity results of the control variant (competition with the natural weed flora) lay inside the interval of *C. album* competition results (Fig. 4). The variance of data received from plants grown with natural weeds was up to 50% higher than those plants grown with *C. album*.

The relative growth (Fig. 5) revealed a decrease in the relative biomass production for all genotypes in Oberviehhausen. No difference is seen between the transgenic and the susceptible control hybrid at both test sites. The competitiveness of the normal variety is significantly higher.

A comparison between the transgenic and the susceptible control hybrid reveals no unexpected results. Due to the breeding effect, the cultivar variety has the highest absolute and relative biomass production.

As a result of genetic engineering, the virus resistant hybrid showed a higher absolute productivity than the unaltered genotype under conditions of virus-infestation. But the relative productivity of both is unaltered in general and when under competitive stress. Both test hybrids showed no difference under virus-free field conditions, where the potential ecological advantage of the transgenic genotype is absent.

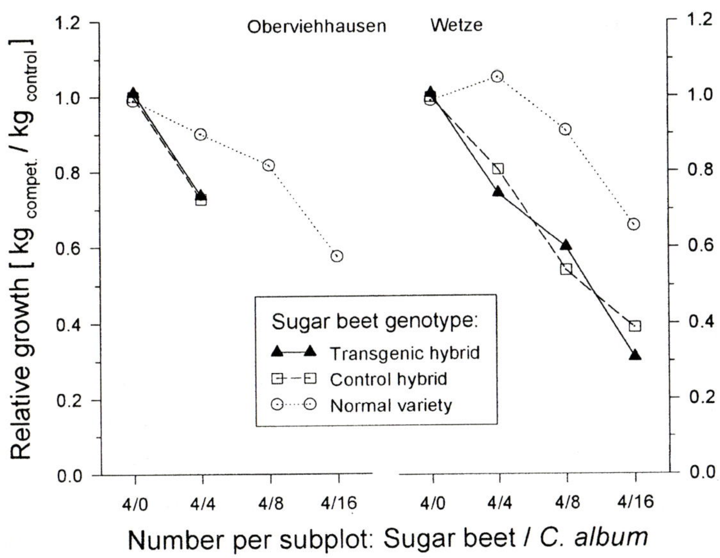

Fig. 5. Relative biomass production of sugar beet genotypes (transgenic virus-resistant hybrid, control susceptible hybrid, normal virus-tolerant variety) grown in four different competition densities with *Chenopodium album* at a site with (Oberviehhausen) and a site without beet necrotic yellow vein (BNYV) virus infection (Wetze). Some experimental variants of transgenic and control hybrids are missing due to a decline in seed emergence in spring

The phenomenon of low productivity at Wetze and the low competitive situation could be attributed to the long drought. The low seed emergence rate at Oberviehhausen was probably due to pre-cultivation of sugar beet a year before which increased soil-born seed/seedling diseases.

In comparison to the results in 1993 (Bartsch et al. 1994), the competitiveness of sugar beet hybrids is higher than sugar beet breeding lines. In this field study the influence of transgenic virus resistance on the ecologically relevant factors (according to Crawley 1990), the internal (vegetative) growth rate and the density tolerance have been clarified.

The experiments are conducted in a step-by-step fashion. Today there is still a lack of data on the whole life cycle fitness of the transgenic sugar beets. We aim to continue competition experiments to analyse data on the following parameters: frost tolerance, generative biomass development, seed dormancy, germinative ability of seed, and plant dispersal. However, the degree of naturalization in fallow plant communities is unknown. The establishment will depend on the ecological advantage of genetically engineered traits and on the natural ability to survive in fallow plant communites. However, the natural sugar beet survival rate is as far as we know very low (e.g. data from Waldhardt 1994).

Case C: Communities in Natural Habitats of Beets

A genetical introgression from seed beet to wild beet populations has been observed in Europe (Santoni and Berville 1993). Most of the sugar beet seed production areas are located in Eastern England, in Suffolk, Norfolk and Lincolnshire (Longden 1974; Hornsey and Arnold 1979; Longden 1989), in Southern France near Toulouse (Boudry et al. 1993) or in Northern Italy near Cesena (Fig. 3). The production areas contain wild and/or weed beet populations seperated by only a few kilometres. Due to the wind pollination and the alien pollination due to the common gynodiecy of wild beets (Boutin et al. 1987; Cuguen et al. 1994), a genetical exchange could appear.

An ecological inventory of 'undisturbed' wild beet habitats would help to characterize a future input of transgenic gene immigration. In 1994, a first examination was carried out to locate and sample wild beet populations near seed production areas in North-Eastern Italy. The results are seen in Fig. 6. This part of Italy, north of the river Po, was selected because it is the area where the BNYVV was first detected, in the early 1950s. We assumed that if there is a wild beet area infected with the virus, it would with most probability be in the Po estuary.

Root material sampled from coastal wild beet habitats (selected root hairs from six different locations, ten plants each, three parallels per plant) was tested with the ELISA method (coat protein test from Boehringer, Mannheim) for BNYVV infestation. Non-infected transgenic sugar beets with low prodution of virus coat protein were used as controls. According to our results, the coat

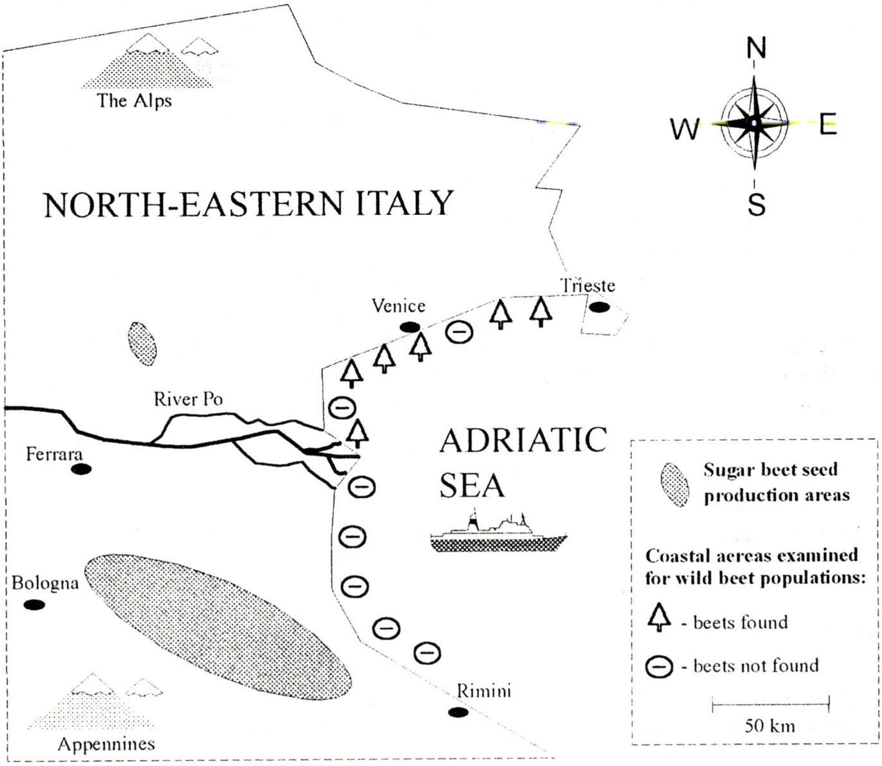

Fig. 6. Overview of breeding areas (with seed beet populations) more than 50 km apart from natural habitats (with wild beet populations) in North-Eastern Italy. Although the possibility of an unintended pollen exchange cannot be excluded, an ecological advantage of virus resistance was not apparent. There was no infestation of the wild beet populations in 1994

protein was detectable in the transgenics, whereas none of the 60 tested wild beet plants from 6 different locations showed a virus infestation.

Theoretically, two strongly diverse natural wild beet plant population developments will be observable. In the 'model of introgression', the genetically engineered genes could spread and integrate into the gene pool without altering the genetical constitution of the population. This model would demonstrate a long-term effect without ecological damage, if other plant species are not dramatically displaced due to superior competitiveness. On the other side, the 'model of displacement' would cause unwanted ecological effects: Superior competitiveness would – in this case – lead to the displacement and loss of genetical variability when displaced by the transgenics.

However, we need more data about the undisturbed situation in wild communities in order to detect future implications for transgenic sugar beets.

Conclusions

The following conclusions can be made:

- Transgenic crop species will naturalize in natural plant communities in the same way as conventional plants.
- Risk assessment should take into account that the new genes can potentially spread to all plant communities, where crossable relatives occur. Biosafety research has to focus on the additional effects of the transgenic attributes.
- It is more difficult to determine the naturalization ability of the transgenes in some neighbouring communities where relatives have been excluded. If the genetically engineered trait offers a dramatical new ecological advantage, problems might occur.
- However, an ecological inventory of the 'undisturbed' wild relative habitats would help to characterize a future input of transgenic gene immigration.

Acknowledgements. The authors wish to thank Dr. Martyn Sly-Jex for reviewing the English manuscript. The work was funded by the German Ministry of Science and Technology (BMFT-Grant No. 0310532A).

References

Bartsch D, Haag C, Morak C et al (1994) Autoecological studies of the competitiveness of transgenic sugar beets. Verhandlungen der Gesellschaft für ökologie, vol 23, pp 435–444

Boudry P, Mörchen M, Saumitou-Laprade P et al (1993) The origin and evolution of weed beets: consequences for the breeding and release of herbicide-resistant transgenic sugar-beets. Theor Appl Genet 87: 471–478

Boutin V, Pannenbecker G, Ecke W et al (1987) Cytoplasmic male sterility and nuclear restorer genes in a natural population of *Beta maritima*: genetical and molecular aspects. Theor Appl Genet 73: 625–629

Crawley MJ (1990) Die ökologie genetisch veränderter Organismen: eine Einschätzung der Umweltrisiken. Arbeitsmaterialien Nr. 2 der Arbeitsstelle für Technikfolgenabschätzung und -bewertung der Universität Hamburg 1991; originally printed in: Mooney HA, Bernardi GA (eds) Introduction of genetically modified organisms into the environment. Chichester, pp 133–150

Cuguen J, Wattier R, Saumitou-Laprade P et al (1994) Gynodiecy and mitochondrial DNA polymorphism in natural populations of *Beta vulgaris* ssp. *maritima*. Genet Sci Evol 26: 87–101

Hornsey KG, Arnold MH (1979) The origins of weed beet. Ann Appl Biol 92: 279–285

Longden PC (1974) Sugar beet as a weed. Proceedings 12th British Weed Control Conference, pp 301–308

Longden PC (1989) Effects of increasing weed-beet density on sugar-beet yield and quality. Ann Appl Biol 114: 527–532

O'Neill RV, De Angelis DL, Waide JB Allen TFH(1986) A hierarchical concept of ecosystems. Princeton: Princeton Univ. Press

Regal PJ (1993) Scientific principles for ecologically based risk assessment of transgenic organisms. Mol Ecol 3(1): 5–14
Rothmaler W (1990) Exkursionsflora von Deutschland. In: Schubert R, Vent W (eds) Volk und Wissen, Berlin
Santoni S, Berville A (1993) Characterization of the nuclear ribosomal DNA units and phylogeny of *Beta* L. wild forms and cultivated beets. Theor Appl Genet 83: 533–542
Waldhardt R (1994) Flächenstillegungen und Extensivierungsmaβnahmen im Ackerbau – Flora, Vegetation und Stickstoff-Haushalt. Dissertation, Göttingen University

Subject Index